Steel Pipe—A Guide for Design and Installation

AWWA MANUAL M11

Fourth Edition

Science and Technology

AWWA unites the drinking water community by developing and distributing authoritative scientific and technological knowledge. Through its members, AWWA develops industry standards for products and processes that advance public health and safety. AWWA also provides quality improvement programs for water and wastewater utilities.

MANUAL OF WATER SUPPLY PRACTICES—M11, Fourth Edition

Steel Pipe—A Guide for Design and Installation

Project Manager and Technical Editor: Melissa Christensen
Copy Editor: Mart Kelle
Production Editor: Carol Stearns

Library of Congress Cataloging-in-Publication Data

Steep pipe : a guide for design and installation.-- 4th ed.
p. cm. -- (AWWA manual ; M11)
Includes bibliographical references and index.
ISBN 1-58321-274-4
1. Water pipes--Design and construction--Handbooks, manuals, etc. 2. Pipe, Steel--Design and construction--Handbooks, manuals etc. I. American Water Works Association. II. Series.

TD491.A49 S74
628.1'5--dc22

2004043748

Printed in the United States of America

American Water Works Association
6666 West Quincy Avenue
Denver, CO 80235-3098

ISBN 1-58321-274-4

Printed on recycled paper

Contents

Figures

Tables

Foreword

This manual was first authorized in 1943. In 1949, committee 8310D appointed one of its members, Russel E. Barnard, to act as editor in chief in charge of collecting and compiling the available data on steel pipe. The first draft of the report was completed by January 1957; the draft was reviewed by the committee and other authorities on steel pipe. The first edition of this manual was issued in 1964 with the title *Steel Pipe-Design and Installation*.

The second edition of this manual was approved in June 1984 and published in 1985 with the title *Steel Pipe—A Guide for Design and Installation*.

The third edition of the manual was approved in June 1988 and published in 1989.

This fourth edition of the manual was approved March 2003. Major revisions to the third edition included in this edition are (1) the manual was metricized and edited throughout; (2) a discussion of Chemistry, Casting and Heat Treatment (Sec. 1.3) and a discussion of stress evaluation in spiral-welded pipe (Sec. 1.12) were added to chapter 1; (3) Table 4-1 was revised to reflect new steel grades and Charpy test requirements for pipe with wall thicknesses greater than ½ in. (12.7 mm); (4) calculations for external fluid pressure (Sec. 4.4) was revised to include consideration of pipe stiffness added by the cement–mortar coating and lining; (5) in Table 6-1, values of E' used for calculation of pipe deflection were revised to reflect increasing soil stiffness with increasing depth of cover; (6) in chapter 7, the discussion of ring girder design was revised, and a design example was added; (7) chapter 9, Fittings and Appurtenances, was revised to reflect the provisions of AWWA C208-96; (8) a new section on installation of flanged joints was added to chapter 12; and (9) thrust-restraint design calculations in chapter 13 were revised.

This manual provides a review of experience and design theory regarding steel pipe used for conveying water, with appropriate references cited. Application of the principles and procedures discussed in this manual must be based on responsible judgment.

This page intentionally blank.

Acknowledgments

This revision of Manual M11 was made by the following members of the Steel Water Pipe Manufacturers Technical Advisory Committee (SWPMTAC). The Steel Water Pipe Manufacturers Technical Advisory Committeee Task Group on updating the manual M11 had the following personnel at the time of revision:

Dennis Dechant, *Task Group Chairman*

H.H. Bardakjian, American International, Rancho Cucamonga, Calif.

R.J. Card, Victaulic Depend-O-Lok Inc., Atlanta, Ga.

R.R. Collins, JCM Industries Inc., Nash, Texas

D.H. Eaton, Romac Industries Inc., Bothell, Wash.

B. Kane, Cascade Waterworks Manufacturing Company, Yorkville, Ill.

B.D. Keil, Continental Pipe Manufacturing Company, Pleasant Grove, Utah

M. Mintz, M-Square Associates Inc., Elmont, N.Y.

R.N. Satyarthi, Baker Coupling Company, Inc., Los Angeles, Calif.

K.L. Shaddix, Smith-Blair Inc., Texarkana, Texas

B. Spotts, RTLC Piping Products Inc., Kosse, Texas

J.C. Taylor, Piping Systems Inc., Fort Worth, Texas

M. Topps, Glynwed Piping Systems, Hixson, Tenn.

R. Warner, National Welding Corporation, Midvale, Utah

This revision was reviewed and approved by the Standards Committee on Steel Pipe. The Standards Committee on Steel Pipe had the following personnel at the time of approval:

George J. Tupac, *Chairman*

John H. Bambei Jr., *Vice Chairman*

Dennis Dechant, *Secretary*

Consumer Members

G.A. Andersen, NYC Bureau of Water Supply, Little Neck, N.Y.

J.H. Bambei Jr., Denver Water Department, Denver, Colo.

D.W. Coppes, Massachusetts Water Resources Authority, Southborough, Mass.

R.V. Frisz, US Bureau of Reclamation, Denver, Colo.

T.R. Jervis, Greater Vancouver Regional District, Burnaby, B.C.

T.J. Jordan, Metropolitan Water District of Southern California, La Verne, Calif.

T.A. Larson, Tacoma Public Utilities, Tacoma, Wash.

G.P. Stine, San Diego County Water Authority, Escondido, Calif.

Milad Taghavi, Los Angeles Department of Water & Power, Los Angeles, Calif.

J.V. Young, City of Richmond, Richmond, B.C.

General Interest Members

W.R. Brunzell, Brunzell Associates Ltd, Skokie, Ill.
R.L. Coffey, Kirkham Michael & Associates, Omaha, Neb.
H.E. Dunham, MWH Americas Inc., Bellevue, Wash.
*K.G. Ferguson,** MWH Americas Inc., Parker, Ariz.
S.N. Foellmi, Black & Veatch Corporation, Irvine, Calif.
J.W. Green, Alvord Burdick & Howson, Lisle, Ill.
K.D. Henrichsen, HDR Engineering Inc., St. Cloud, Minn.
*M.B. Horsley,** Black & Veatch Corporation, Overland Park, Kan.
J.K. Jeyapalan, Pipeline Consultant, New Milford, Conn.
Rafael Ortega, Lockwood Andrews and Newnam, Houston, Texas
A.E. Romer, Boyle Engineering Corporation, Newport Beach, Calif
H.R. Stoner, Consultant, North Plainfield, N.J.
C.C. Sundberg, CH2M Hill Inc., Bellevue, Wash.
G.J. Tupac, G.J. Tupac & Associates, Pittsburgh, Pa.
J.S. Wailes,† Standards Engineer Liaison, AWWA, Denver, Colo.
L.W. Warren, Seattle, Wash.
W.R. Whidden, Post Buckley Schuh & Jernigan, Orlando, Fla.

Producer Members

H.H. Bardakjian, Ameron International, Rancho Cucamonga, Calif.
Mike Bauer, Tnemec Company, Inc., North Kansas City, Mo.
R.J. Card, Victaulic Depend-O-Lok Inc., Atlanta, Ga.
R.R. Carpenter, American Cast Iron Pipe Company, Birmingham, Ala.
Dennis Dechant, Northwest Pipe Company, Denver, Colo.
J.E. Hagelskamp,† American Cast Iron Pipe Company, Birmingham, Ala.
B.D. Keil, Continental Pipe Manufacturing Company, Pleasant Grove, Utah
*J.L. Luka,** American SpiralWeld Pipe Company, Columbia, S.C.
*B.F. Vanderploeg,** Northwest Pipe Company, Portland, Ore.
J.A. Wise, Canus International Sales Inc., Langley, B.C.

---

*Alternate
†Liaison

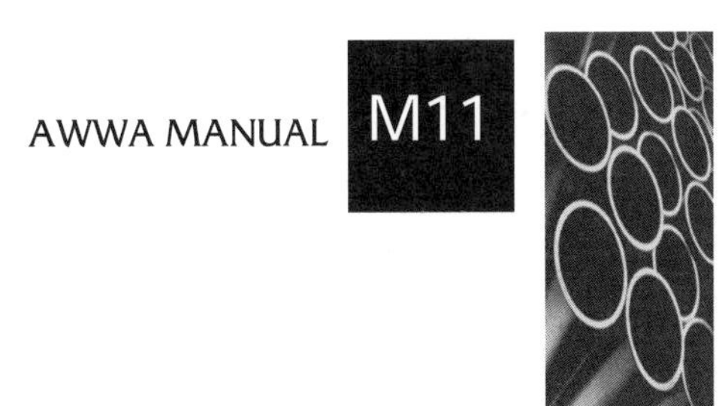

Chapter 1

History, Uses, and Physical Characteristics of Steel Pipe

HISTORY

Steel pipe has been used for water lines in the United States since the early 1850s (Elliot 1922). The pipe was first manufactured by rolling steel sheets or plates into shape and riveting the seams. This method of fabrication continued with improvements into the 1930s. Pipe wall thicknesses could be readily varied to fit the different pressure heads of a pipeline profile.

Because of the relatively low tensile strength of the early steels and the low efficiency of cold-riveted seams and riveted or drive stovepipe joints, engineers initially set a safe design stress at 10,000 psi (68.95 MPa). As riveted-pipe fabrication methods improved and higher strength steels were developed, design stresses progressed with a 4-to-l safety factor of tensile strength, increasing from 10,000 (68.95) to 12,500 (86.18), to 13,750 (94.8), and finally to 15,000 psi (103.42). Design stresses were adjusted as necessary to account for the efficiency of the riveted seam. The pipe was produced in diameters ranging from 4 in. (100 mm) through 144 in. (3,600 mm) and in thickness from 16 gauge to 1.5 in. (38 mm). Fabrication methods consisted of single-, double-, triple-, and quadruple-riveted seams, varying in efficiency from 45 percent to 90 percent, depending on the design.

Lock-Bar pipe, introduced in 1905, had nearly supplanted riveted pipe by 1930. Fabrication involved planing 30-ft (9.1-m) long plates to a width approximately equal to half the intended circumference, upsetting the longitudinal edges, and rolling the plates into 30-ft (9.1-m) long half-circle troughs. H-shaped bars of special configuration were applied to the mating edges of two 30-ft (9.1-m) troughs and clamped into position to form a full-circle pipe section.

According to the general procedure of the times, a 55,000-psi (379.2 MPa) tensile-strength steel was used. With a 4-to-1 safety factor, this resulted in a 13,750-psi (94.8 MPa) design stress. Lock-Bar pipe had notable advantages over riveted pipe: it had only one or two straight seams and no round seams. The straight seams were considered 100-percent efficient as compared to the 45-percent to 70-percent efficiency for riveted seams. Manufactured in sizes from 20 in. (500 mm) through 74 in. (1,850 mm), from plate ranging in thickness from 3⁄16 in. (4.8 mm) to 1⁄2 in. (12.7 mm), Lock-Bar played an increasingly greater role in the market until the advent of automatic electric welding in the mid 1920s.

By the early 1930s, both the riveting and Lock-Bar methods gradually were replaced by welding. Pipe produced using automatic electric-fusion welding was advantageous because fewer pieces were used, fewer operations were performed, and because of faster production, smaller seam protrusion, and 100-percent welded-seam efficiency. The fabricators of fusion-welded pipe followed similar initial production sequences as for Lock-Bar. Through the 1930s and into the 1940s, 30-ft (9.1-m) plates were used. By the 1950s, some firms had obtained 40-ft (12.2-m) rolls, and a few formed 40-ft (12.2-m) lengths in presses.

In the 1930s, a new approach was used to design stresses. Prior to that time, it had been common practice to work with a safety factor of 4-to-1 based on the tensile strength. As welded pipe became predominant, using 50 percent of the material yield stress became widely accepted.

Helically formed and welded pipe was developed in the early 1930s and was used extensively in diameters from 4 in. (100 mm) through 36 in. (875 mm). Welding was performed using the electric fusion method. After World War II, German machines were imported, and subsequently, domestic ones were developed that could spirally form and weld through diameters of 144 in. (3,600 mm).

USES

Steel water pipe meeting the requirements of appropriate AWWA standards has been found satisfactory for many applications, some of which follow:

- Aqueducts
- Supply lines
- Transmission mains
- Distribution mains
- Penstocks
- Treatment-plant piping (Figure 1-1)
- Self-supporting spans
- Force mains
- Circulating-water lines
- Underwater crossings, intakes, and outfalls

The installation of pipe in this plant was simplified using the specially designed fittings and lightweight pipe.

Figure 1-1 Steel pipe in filtration plant gallery

General data on some of the notable steel pipelines have been published (Cates 1971, Hinds 1954). Data on numerous others have appeared in the *Journal AWWA* and other periodicals, as well as in many textbooks and engineering handbooks.

CHEMISTRY, CASTING, AND HEAT TREATMENT

General

The properties of steels are governed by their chemical composition, by the processes used to transform the base metal into the shape, and by their heat treatment. The effects of these parameters on the properties of steels are discussed in the following sections.

Chemical Composition

Constructional steels are a mixture of iron and carbon with varying amounts of other elements—primarily manganese, phosphorus, sulfur, and silicon. These and other elements are unavoidably present or intentionally added in various combinations to achieve specific characteristics and properties of the finished steel products. The effects of the commonly used chemical elements on the properties of hot-rolled and heat-treated carbon and alloy steels are presented in Table 1-1. The effects of carbon, manganese, sulfur, silicon, and aluminum will be discussed.

Carbon is the principal hardening element in steel because each additional increment increases the hardness and tensile strength of the steel. Carbon has a moderate tendency to segregate, and increased amounts of carbon cause a decrease in durability, toughness, and weldability.

Manganese increases the hardness and strength of steels but to a lesser degree than carbon. Manganese combines with sulfur to form manganese sulfides, therefore decreasing the harmful effects of sulfur.

Table 1-1 Effects of alloying elements

Aluminum (Al)
- Used to deoxidize or "kill" molten steel

Boron (B)
- Small amounts (0.0005 %) increases hardenability in quenched and tempered steels
- Used only in aluminum killed steels
- Most effective at low carbon levels

Carbon (C)
- Principal hardening element in steel
- Increases strength and hardness
- Decreases ductility, toughness, and weldability
- Moderate tendency to segregate

Chromium (Cr)
- Increases strength
- Increases atmospheric corrosion resistance

Copper (Cu)
- Primary contributor to atmospheric corrosion resistance

Manganese (Mn)
- Increases strength
- Controls harmful effects of sulfur

Nickel (Ni)
- Increases strength and toughness

Nitrogen (N)
- Increases strength and hardness

Phosphorus (P)
- Increases strength and hardness
- Decreases ductility and toughness
- Considered an impurity, but sometimes added for atmospheric corrosion resistance

Silicon (Si)
- Used to deoxidize or "kill" molten steel
- Decreases durability and toughness

Sulfur (S)
- Considered undesirable except for machinability
- Decreases ductility, toughness, and weldability
- Adversely affects surface quality
- Strong tendency to segregate

Vanadium (V) and Columbium (Nb)
- Small additions increase strength

Sulfur is generally considered an undesirable element except when machinability is an important consideration. Sulfur adversely affects surface quality, has a strong tendency to segregate, and decreases ductility, toughness, and weldability.

Silicon and aluminum are the principal deoxidizers used in the manufacture of carbon and alloy steels. Aluminum is also used to control and refine grain size.

Casting

The traditional steel-making process is performed by pouring (teeming) molten steel into a series of molds to form castings known as ingots. The ingots are removed from the molds, reheated, then rolled into products with square or rectangular cross sections. This hot-rolling operation elongates the ingots and produces semifinished products known as blooms, slabs, or billets. All ingots exhibit some degree of nonuniformity of chemical composition known as segregation. This is an inherent characteristic of the cooling and solidification of the molten steel in the mold.

The initial liquid steel contacting the relatively cold walls and bottom of the mold solidifies very rapidly, having the same chemical composition as the liquid steel entering the mold. However, as the rate of solidification decreases away from the mold sides, crystals of relatively pure iron solidify first. The first crystals to form contain less carbon, manganese, phosphorus, sulfur, and other elements than the liquid steel from which they were formed. The remaining liquid is enriched by these elements that are continually being rejected by the advancing crystals. Consequently, the last liquid to solidify, which is located around the axis in the top half of the ingot, contains high levels of rejected elements and has a lower melting point than the poured liquid steel. This segregation of the chemical elements is frequently expressed as a local departure from the average chemical composition. In general, the content of an element that has a tendency to segregate is greater than average at the center of the top half of an ingot and less than average at the bottom half of an ingot.

Certain elements tend to segregate more than others. Sulfur segregates to the greatest extent. The following elements also segregate, but to a lesser degree, and in descending order are phosphorus, carbon, silicon, and manganese. The degree of segregation is influenced by the composition of the liquid steel, the liquid temperature, and the ingot size. The most severely segregated areas of the ingot are removed by cropping, which is the process of cutting and discarding sufficient material during rolling.

Continuous casting is the direct casting of steel from the ladle into slabs. This steel-making process bypasses the operations between molten steel and the semifinished product that are inherent in making steel products from ingots. During continuous casting, molten steel is poured at a regular rate into the top of an oscillating water-cooled mold with a cross-sectional size corresponding to the desired slab. As the molten metal begins to solidify along the mold walls, it forms a shell that permits the gradual withdrawal of the strand product from the bottom of the mold into a water-spray chamber where solidification is completed. The solidified strand is cut to length and then reheated and rolled into finished projects, as in the conventional ingot process. Continuous casting produces a smaller size and higher cooling rate for the strand, resulting in less segregation and greater uniformity in composition and properties for steel products than for ingot products.

Killed and Semikilled Steels

The primary reaction involved in most steel-making processes is the combination of carbon and oxygen to form carbon monoxide gas. The solubility of this and other gases dissolved in the steel decreases as the molten metal cools to the solidification temperature range. Excess gases are expelled from the metal and, unless controlled, continue to evolve during solidification. The oxygen available for the reaction can be eliminated and the gaseous evolution inhibited by deoxidizing the molten steel using additions of silicon or aluminum or both. Steels that are strongly deoxidized do not evolve any gases and are called killed steels because they lie quietly in the mold. Increasing gas evolution results in semikilled, capped, or rimmed steels.

In general, killed steels are less segregated and contain negligible porosity when compared to semikilled steels. Consequently, killed-steel products usually exhibit a higher degree of uniformity in composition and properties than semikilled steel products.

Heat Treatment for Steels

Steels respond to a variety of heat treatments that produce desirable characteristics. These heat treatments can be divided into slow cooling treatments and rapid cooling treatments. Slow cooling treatments, such as annealing, normalizing, and stress relieving, decrease hardness and promote uniformity of structure. Rapid cooling treatments, such as quenching and tempering, increase strength, hardness, and toughness. Heat treatments of base metal are generally mill options or ASTM requirements.

Annealing. Annealing consists of heating the steel to a given temperature followed by slow cooling. The temperature, the rate of heating and cooling, and the amount of time the metal is held at temperature depends on the composition, shape, and size of the steel product being treated and the desired properties. Usually steels are annealed to remove stresses, induce softness, increase ductility and toughness, produce a given microstructure, increase uniformity of microstructure, improve machinability, or to facilitate cold forming.

Normalizing. Normalizing consists of heating the steel to between 1,650°F and 1,700°F (899°C and 927°C) followed by slow cooling in air. This heat treatment is commonly used to refine the grain size, improve uniformity of microstructure, and improve ductility and fracture toughness.

Stress relieving. Stress relieving of carbon steels consists of heating the steel to between 1,000°F to 1,200°F (538°C to 649°C) and holding for the appropriate amount of time to equalize the temperature throughout the piece followed by slow cooling. The stress-relieving temperature for quenched and tempered steels must be maintained below the tempering temperature for the product. Stress relieving is used to relieve internal stresses induced by welding, normalizing cold working, cutting, quenching, and machining. It is not intended to alter the microstructure or the mechanical properties significantly.

Quenching and tempering. Quenching and tempering consists of heating and holding the steel at the appropriate austenizing temperature (about 1,650°F [899°C]) for a significant amount of time to produce a desired change in microstructure, then quenching by immersion in a suitable medium (water for bridge steels). After quenching, the steel is tempered by reheating to an appropriate temperature, usually between 800° F and 1,200° F (427°C and 649°C), holding for a specified time at that temperature, and cooling under suitable conditions to obtain the desired properties. Quenching and tempering increases the strength and improves the toughness of the steel.

Controlled rolling. Controlled rolling is a thermomechanical treatment performed at the rolling mill. It tailors the time-temperature-deformation process by controlling the rolling parameters. The parameters of primary importance are (1) the temperature at the start of controlled rolling in the finished strand; (2) the percentage reduction from the start of controlled rolling to the final plate thickness; and (3) the plate-finishing temperature.

Hot-rolled plates are deformed as quickly as possible at temperatures above about 1,900°F (1,038°C) to take advantage of the workability of the steel at high temperatures. In contrast, controlled rolling incorporates a hold or delay time to allow the partially rolled slab to reach the desired temperature before the start of final rolling. Controlled rolling involves deformation at temperatures ranging between 1,500°F and 1,800°F (816°C and 982°C). Because rolling deformation at these low temperatures increases the mill loads significantly, controlled rolling is usually restricted to less than 2-in. (51-mm) thick plates. Controlled rolling increases the strength, refines the grain size, improves the toughness, and may eliminate the need for normalizing.

Controlled finishing-temperature rolling. Controlled finishing-temperature rolling is a less severe practice than controlled rolling and is aimed primarily at improving notch toughness of plates up to 2½-in. (64-mm) thick. The finishing temperatures in this practice (about 1,600°F [871°C]) are higher than required for controlled rolling. However, because heavier plates are involved than in controlled rolling, mill delays are still required to reach the desired finishing temperatures. By controlling the finishing temperature, fine grain size and improved notch toughness can be obtained.

PHYSICAL CHARACTERISTICS

Steel is widely used because of the following properties: great strength; ability to yield or deflect under a load while resisting it; ability to bend without breaking; and resistance to shock. The design engineer should understand these properties, how they are measured, what they will do, and how reliable they are.

DUCTILITY AND YIELD STRENGTH

Solid materials can be divided into two classes: ductile and brittle. Engineering practice treats these two classes differently because they behave differently under load. A ductile material exhibits a marked plastic deformation or flow at a fairly definite stress level (yield point or yield strength) and shows a considerable total elongation, stretch or plastic deformation before final breakage. With a brittle material, the plastic deformation is not well defined, and the ultimate elongation before breakage is small. Steels, as listed in Table 4-1, are typical of the ductile materials used for steel water pipe.

It is ductility that allows comparatively thin-walled steel pipe, even though decreased in vertical diameter two to five percent by earth pressures, to perform satisfactorily when buried in deep trenches or under high fills, provided the true required strength has been incorporated in the design. Because of ductility, steel pipe with theoretically high, localized stresses at flanges, saddles, supports, and joint-harness lug connections has performed satisfactorily for many years.

Designers who determine stress using formulas based on Hooke's law find that the calculated results do not reflect the integrity exhibited by the structures discussed in this manual. These discrepancies occur because the conventional formulas apply only up to a certain stress level and not beyond. Many eminently safe structures and parts of structures contain calculated stresses above this level. A full understanding of the performance of such structures requires that the designer empirically examine the actual behavior of steel as it is loaded from zero to the breaking point.

The physical properties of steel (yield strength and ultimate tensile strength) used as the basis for design and purchase specifications are determined from tension tests made on standard specimens pulled in a tensile-testing machine. The strength of ductile materials, in terms of design, is defined by the yield strength as measured by the lower yield point, where one exists, or by the American Society for Testing and Materials (ASTM) offset yield stress, where a yield point does not exist. For steel typically used in water pipe, the yield strength is determined by the specification as the stress caused by a load creating a 0.5-percent extension of the gauge length. The point is shown in Figure 1-2. The yield strength of steel is considered the same for either tension or compression loads.

Ductility of steel is measured as an elongation, or stretch, under a tension load in a testing machine. Elongation is a measurement of change in length under the load and is expressed as a percentage of the original gauge length of the test specimen.

STRESS AND STRAIN

In engineering, stress is a value obtained by dividing a load by an area. Strain is a length change per unit of length. The relation between stress and strain, as shown on a stress–strain diagram, is of basic importance to the designer.

A stress–strain diagram for any given material is a graph showing the strain that occurs when the material is under a given load or stress. For example, a bar of steel is pulled in a testing machine with suitable instrumentation for measuring the load and indicating the dimensional changes. While the bar is under load, it stretches. The change in length under load per unit of length is called strain or unit strain; it is usually expressed as percentage elongation or, in stress analysis, microinches (min) per inch, where 1 min = 0.000,001 in. (25.4 nm). The values of strain are plotted along the horizontal axis of the stress–strain diagram. For purposes of plotting, the load is converted into units of stress (pounds per square inch) by dividing the load in pounds by the original cross-sectional area of the bar in square inches. The values of stress are plotted along the vertical axis of the diagram. The result is a conventional stress–strain diagram.

Because the stress plotted on the conventional stress–strain diagram is obtained by dividing the load by the original cross-sectional area of the bar, the stress appears to reach a peak and then diminish as the load increases. However, if the stress is calculated by dividing the load by the actual cross-sectional area of the bar as it decreases in cross section under increasing load, it is found that the true stress never decreases. Figure 1-3 is a stress–strain diagram on which both true stress and true strain have been plotted. Because conventional stress–strain diagrams are used commercially, only conventional diagrams are used for the remainder of this discussion.

Figure 1-2 shows various parts of a pure-tension stress–strain curve for steel such as that used in water utility pipe. The change in shape of the test piece during the test is indicated by the bars drawn under the curve. As the bar stretches, the cross section decreases in area up to the maximum tensile strength, at which point local reduction of area (necking in) takes place.

Many types of steel used in construction have stress–strain diagrams of the general form shown in Figure 1-2, whereas many other types used structurally and for machine parts have a much higher yield and ultimate strengths, with reduced ductility. Still other useful engineering steels are quite brittle. In general, low-ductility steels must be used at relatively low strains, even though they may have high strength.

The ascending line on the left side of the graph in Figure 1-2 is straight or nearly straight and has a recognizable slope with respect to the vertical axis. The break in the slope of the curve is rather sudden. For this type of curve, the point where the first deviation from a straight line occurs marks the proportional limit of the steel. The yield strength is at some higher stress level. Most engineering formulas involving

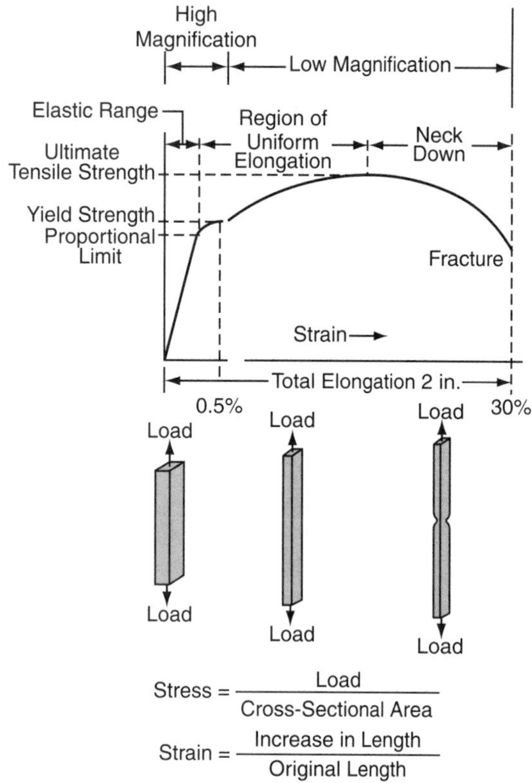

The change in shape of the test piece of steel, which occurred during the test, is shown by the bars drawn under the curve.

Figure 1-2 Stress–strain curve for steel

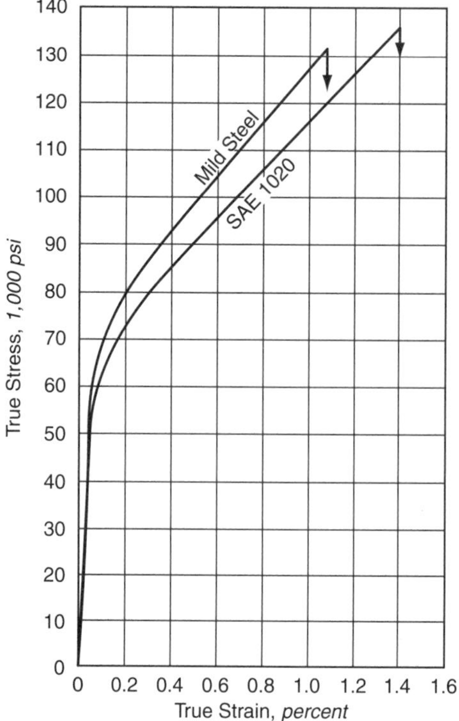

Unlike conventional stress–strain curves, both true stress and true strain have been calculated for the curves shown.

Figure 1-3 True stress–strain for steel

stress calculation presuppose a loading such that working stresses will be below the proportional limit.

Stresses and strains that fall below the proportional limit—such as, those that fall on the straight portion of the ascending line—are said to be in the elastic range. Steel structures loaded to create stresses or strains within the elastic range return precisely to their original length when the load is removed. Exceptions may occur with certain kinds and conditions of loading not usually encountered in water utility installations. Within this range, stress increases in direct proportion to strain.

The modulus of elasticity is defined as the slope of the ascending straight portion of the stress–strain diagram. The modulus of elasticity of steel is about 30,000,000 psi (762 TPa), which means that for each increment of load that creates a strain or stretch of 1 min per inch (1 mm/m) of length, a stress of 30 psi (762 kPa) is imposed on the steel cross section (30,000,000 × 0.000,001 = 30).

Immediately above the proportional limit, between it and the 0.5-percent extension-under-load yield strength of the material (Figure 1-2) lies a portion of the stress–strain diagram that is termed the elastic-plastic range of the material. Typical stress–strain curves with this portion magnified are shown in Figure 1-4 for two grades of carbon steel used for water pipe. Electric-resistance strain gauges provide a means of studying the elastic-plastic segment of the curve. These and associated instruments allow minute examination of the shape of the curve in a manner not possible before development of these instruments.

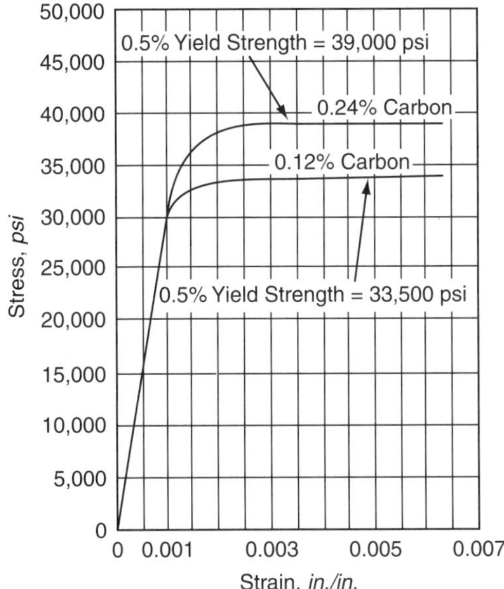

The curves show the elastic-plastic range for two grades of carbon steel.

Figure 1-4 Stress–strain curves for carbon steel

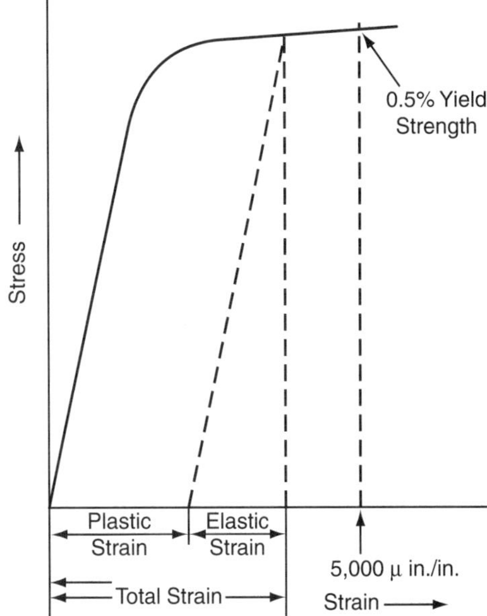

Shown are the elastic and plastic portions of a stress–strain curve for a steel stressed to a given level.

Figure 1-5 Plastic and elastic strains

The elastic-plastic range is becoming increasingly important to the designer. Analysis of this range was necessary, for example, to determine and explain the successful performance of thin steel flanges on thin steel pipe (Barnard 1950). Designs that load steel to within the elastic-plastic range are safe only for certain types of apparatus, structures, or parts of structures. For example, designing within this range is safe for the hinge points or yield hinges in steel ring flanges on steel pipe; for hinge points in structures where local yielding or relaxation of stress must occur; and for bending in the wall of pipe under earth load in trenches or under high fills. It is not safe to rely on performance within this range to handle principal tension stress in the walls of pipe or pressure vessels or to rely on such performance in other situations where the accompanying deformation is uncontrolled or cannot be tolerated.

Figure 1-5 shows the elastic and plastic portions of a stress-strain curve for a steel stressed to a given level. Figure 1-6 shows graphically how a completely fictitious stress is determined by a formula based on Hooke's law, if the total strain is multiplied by the modulus of elasticity. The actual stress is determined using only the elastic strain with the modulus of elasticity, but there is at present no way to separate theoretically the elastic and plastic strains in a structure. The only alternative is to take the total measured strain as indicated by strain gauges and then determine the actual stress from the stress–strain curve, as shown in Figure 1-7.

STRAIN IN DESIGN

Analysis of a structure becomes more complete when considering strain as well as stress. For example, it is known that apparent stresses calculated using classic formulas based on the theory of elasticity are erroneous at hinge-point stress levels. The magnitude of this error near the yield-strength stress is demonstrated in the next paragraph, where the classically calculated result is compared with the measured performance.

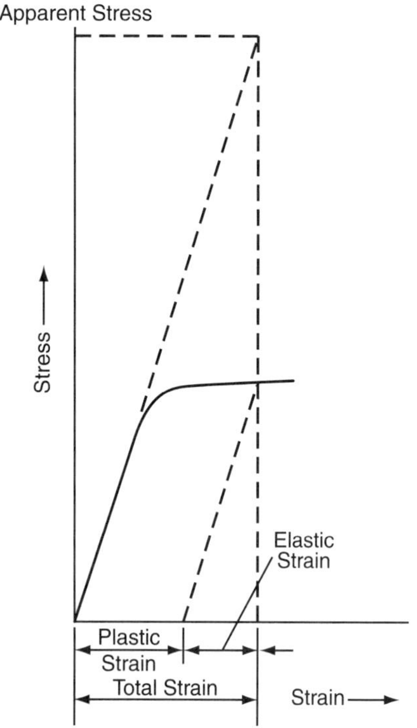

If the total strain is multiplied by the modulus of elasticity, the stress determined by use of a formula based on Hooke's law is fictitious

Figure 1-6 Actual and apparent stresses

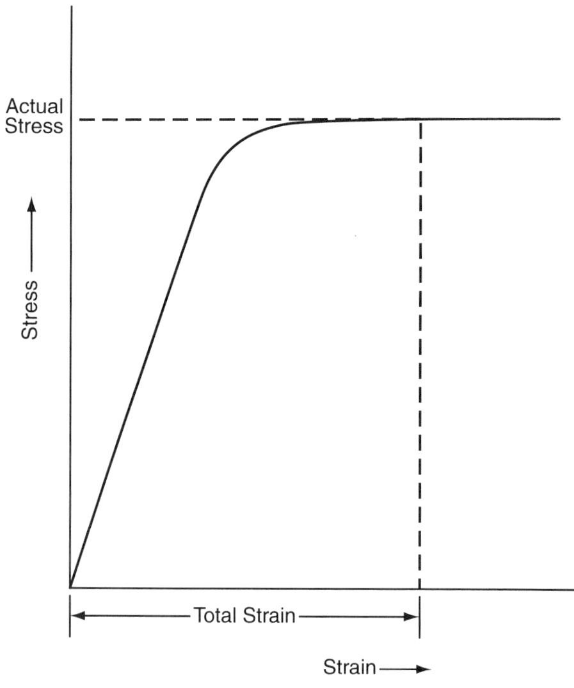

When the total measured strain is known, the actual stress can be determined by use of the stress–strain curve.

Figure 1-7 Determination of actual stress

By definition, the yield-strength load of a steel specimen is that load which causes a 0.5 percent extension of the gauge length. In the elastic range, a stress of 30 psi (762 kPa) is imposed on the cross-sectional area for each microinch-per-inch increase in length under load. Because an extension of 0.5 percent corresponds to 5,000 min/in. (5 nm/m), the calculated yield-strength stress is 5,000 × 30 = 150,000 psi (1,034 MPa). The measured yield-strength stress, however, is approximately 30,000–35,000 psi (207–241 MPa) or about one fourth of the calculated stress.

Similarly varied results between strain and stress analyses occur when the performance of steel at its yield strength is compared to the performance of its ultimate strength. There is a great difference in strain between the yield strength of low- or medium-carbon steel at 0.5-percent extension under load and the specified ultimate strength at 30-percent elongation. This difference has a crucial bearing on design safety. The specified yield strength corresponds to a strain of 5,000 min/in. (5 nm/m). To pass the specification requirement of 30 percent elongation, the strain at ultimate strength must be not less than 0.3 in./in. or 300,000 min/in. (300 nm/m). The ratio of strain at ultimate strength to strain at yield strength, therefore, is 300,000:5,000 or 60:1. On a stress basis, from the stress–strain diagram, the ratio of ultimate strength to yield strength is 50,000:30,000 or only 1.67:1.

Steels, such as those used in waterworks pipe, show nearly linear stress–strain diagrams up to the yield level, after which strains of 10 to 20 times the elastic-yield strain occur with no increase in actual load. Tests on bolt behavior under tension substantiate this effect (Bethlehem Steel Co. 1946). The ability of bolts to hold securely and safely when they are drawn into the region of the yield, especially under vibration conditions, is easily explained by the strain concept but not by the stress concept. The bolts act somewhat like extremely stiff springs at the yield-strength level.

ANALYSIS BASED ON STRAIN

In some structures and in many welded assemblies, conditions permit the initial adjustment of strain to working load but limit the action automatically, either because of the nature of the loading or because of the mechanics of the assembly. Examples are, respectively, pipe under earth load and steel flanges on steel pipe. In these instances, bending stresses may be in the region of yield, but deformation is limited.

In bending, there are three distinguishable phases that a structure passes through when being loaded from zero to failure. In the first phase, all fibers undergo strain less than the proportional limit in a uniaxial stress field. In this phase, a structure will act in a completely elastic fashion, to which the classic laws of stress and strain are applicable.

In the second phase, some of the fibers undergo strain greater than the proportional or elastic limit of the material in a uniaxial stress field; however, a more predominant portion of the fibers undergo strain less than the proportional limit, so that the structure still acts in an essentially elastic manner. The classic formulas for stress do not apply, but the strains can be adequately defined in this phase.

In the third phase, the fiber strains are predominantly greater than the elastic limit of the material in a uniaxial stress field. Under these conditions, the structure as a whole no longer acts in an elastic manner. The theory and formulas applicable in this phase are being developed but have not yet reached a stage where they can be generally used.

An experimental determination of strain characteristics in bending and tension was made on medium-carbon steel (<0.25 percent carbon) similar to that required by AWWA C200, Standard for Steel Water Pipe 6 Inches and Larger (latest edition). Results are shown in Figure 1-8. Note that the proportional-limit strains in bending are 1.52 times those in tension for the same material. Moreover, the specimen in bending showed fully elastic behavior at a strain of 1,750 min/in., which corresponds to a calculated stress of 52,500 psi (361.98 MPa) ($1{,}750 \times 30 = 52{,}500$) using the modulus of elasticity. The specimens were taken from material having an actual yield of 39,000 psi (268.9 MPa). Therefore, this steel could be loaded in bending to produce strains up to 1,750 min/in. (1.75 nm/m) and still possess full elastic behavior.

Steel ring flanges made of plate and fillet welded to pipe with a comparatively thin wall have been used successfully for many years in water service. The flanges ranged from 4 in. through 96 in. (100 mm through 2,400 mm) in diameter. Calculations were made to determine the strain that would occur in the pipe wall adjacent to the flanges. Table 1-2 shows the results.

Note that from the table, in practice, the limiting strain was always below the commonly recognized yield-strength strain of 5,000 min/in. (5.0 nm/m) but did approach it closely in at least one instance. All of these flanges are sufficiently satisfactory, however, to warrant their continued use by designers.

Designing a structure on the basis of ultimate load capacity from test data rather than entirely on allowable stress is a return to an empirical point of view, a point of view that early engineers accepted in the absence of knowledge of the mathematics and statistics necessary to calculate stresses. The recent development of mathematical processes for stress analysis has, in some instances, overemphasized the importance of stress and underemphasized the importance of the overall strength of a structure.

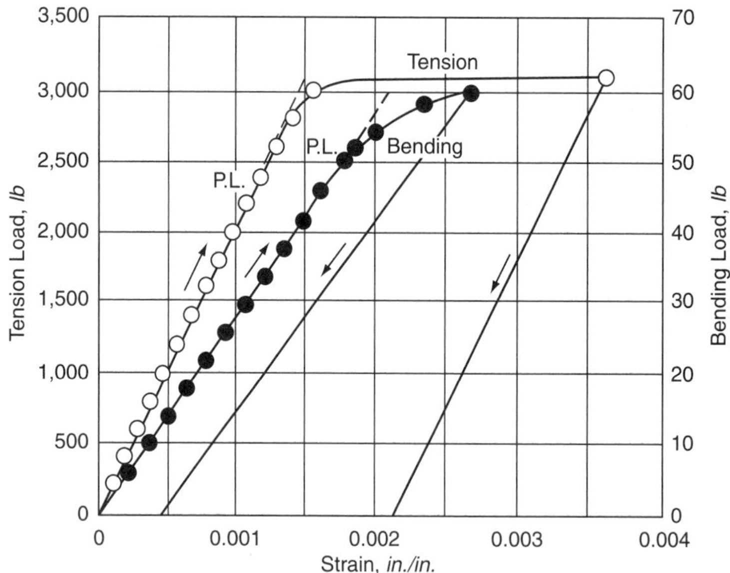

The proportional limit (P.L.) strains in bending are 1.52 times those in tension for the same material.

Figure 1-8 Experimental determination of strain characteristics

Table 1-2 Maximum strain in pipe wall developed in practice

	Operating Pressure		Max. Strain	
Standard Flange	*psi*	*kPa*	*μin./in.*	*(mm/m)*
A	75	(517.1)	1,550–3,900	(1.55–3.90)
	150	(1,034.2)	2,200–4,650	(2.20–4.65)
B	150	(1,034.2)	1,100–3,850	(1.10–3.85)

Source: Barnard, R.E., Design of Steel Ring Flanges for Water Works Service—A Progress Report. Jour. AWWA, 42:10:931 *(Oct. 1950).*

DUCTILITY IN DESIGN

The plastic, or ductile, behavior of steel in welded assemblies may be especially important. Allowing the stress at certain points in a steel structure to go beyond the elastic range is successful current design practice. For many years, in buildings and in bridges, specifications have allowed the designer to use average or nominal stresses because of bending, shear, and bearing, resulting in local yielding around pins and rivets and at other points. This local yield, which redistributes both load and stress, is caused by stress concentrations that are neglected in the simple design formulas. Plastic action is and has been depended on to ensure the safety of steel structures. Experience has shown that these average or nominal maximum stresses form a satisfactory basis for design. During the manufacturing process, the steel in steel pipe has been forced beyond its yield strength many times, and the same thing may happen during installation. Similar yielding can be permitted after installation by design, provided the resulting deformation has no adverse effect on the function of the structure.

Basing design solely on approximations for real stress does not always yield safe results. The collapse of some structures has been traced to a trigger action of neglected points of high stress concentrations in materials that are not ductile at these points. Ductile materials may fail in a brittle fashion if subjected to overload in three planes at the same time. Careful attention to such conditions will result in safer design and will eliminate grossly overdesigned structures that waste both material and money.

Plastic deformation, especially at key points, sometimes is the real measure of structural strength. For example, a crack, once started, may be propagated by almost infinite stress, because at the bottom of the crack the material cannot yield a finite amount in virtually zero distance. In a ductile material, the crack will continue until the splitting load is resisted elsewhere. Plasticity underlies current design to an extent not usually realized and offers promise of greater economy in future construction (Symposium—Plastic Strength of Structural Members 1955, Neal 1956).

EFFECTS OF COLD WORKING ON STRENGTH AND DUCTILITY

During pipe fabrication, the steel plates or sheets are often formed into the desired shape at room temperatures. Such cold-forming operations obviously cause inelastic deformation, because the steel retains its formed shape. To illustrate the general effects of such deformation on strength and ductility, the elemental behavior of a carbon-steel tension specimen subjected to plastic deformation and subsequent reloadings will be discussed. The behavior of actual cold-formed plates may be much more complex.

As illustrated in Figure 1-9, if a steel specimen of plate material is unloaded after being stressed into either the plastic or strain-hardening range, the unloading curve will follow a path parallel to the elastic portion of the stress–strain curve, and a residual strain or permanent set will remain after the load is removed.

If the specimen is promptly reloaded, it will follow the unloading curve to the stress–strain curve of the virgin (unstrained) material. If the amount of plastic deformation is less than that required for the onset of strain hardening, the yield strength of the plastically deformed steel will be approximately the same as that of the virgin material. However, if the amount of plastic deformation is sufficient to cause strain hardening, the yield strength of the steel will be increased. In either case, the tensile strength will remain the same, but the ductility measured from the point of reloading will be decreased. As indicated in Figure 1-9, the decrease in ductility is approximately equal to the amount of inelastic prestrain.

A steel specimen that has been strained into the strain-hardening range, unloaded, and allowed to age for several days at room temperature (or for a much shorter time at a moderately elevated temperature) will tend to follow the path indicated in Figure 1-10 during reloading (Dieter 1961). This phenomenon, known as strain aging, has the effect of increasing yield and tensile strength while decreasing ductility (Chajes, Britvec, and Winter 1963).

The effects of cold work on the strength and ductility of the structural steels can be eliminated largely by thermal stress relief, or annealing. Such treatment is not always possible; fortunately, it is not often necessary.

BRITTLE FRACTURE CONSIDERATIONS IN STRUCTURAL DESIGN

General Considerations

As temperature decreases, there generally is an increase in the yield stress, tensile strength, modulus of elasticity, and fatigue strength of the plate steels. In contrast, the ductility of these steels, as measured by reduction in area or by elongation under

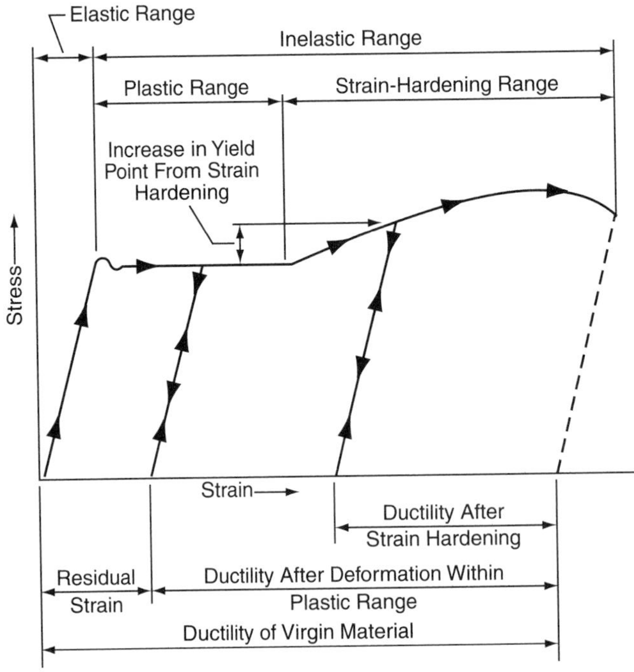

NOTE: Diagram is schematic and not to scale.

Source: Brockenbrough, R.L. & Johnston, B.G., U.S.S. Steel Design Manual. *ADUSS 27-3400-04. US Steel Corp. Pittsburgh, Pa. (Jan. 1981).*

Figure 1-9 Effects of strain hardening

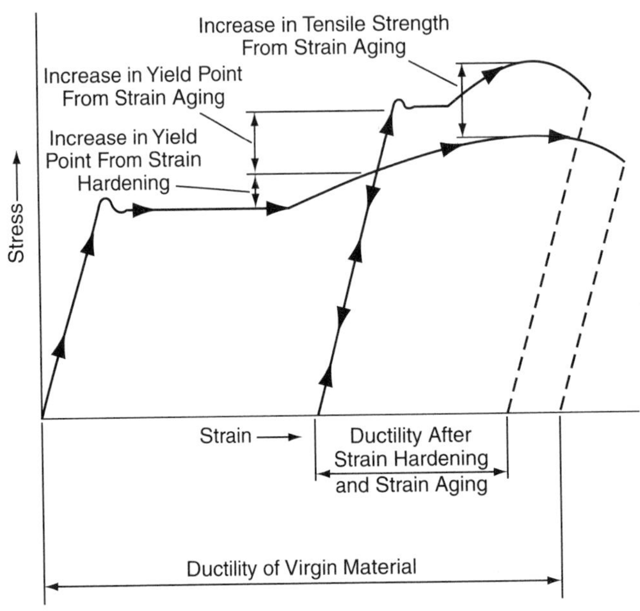

NOTE: Diagram is schematic and not to scale.

Source: Brockenbrough, R.L. & Johnston, B.G., U.S.S. Steel Design Manual. *ADUSS 27-3400-04. US Steel Corp. Pittsburgh, Pa. (Jan. 1981).*

Figure 1-10 Effects of strain aging

load, decreases with decreasing temperatures. Furthermore, there is a temperature below which a structural steel subjected to tensile stresses may fracture by cleavage with little or no plastic deformation, rather than by shear, which is usually preceded by a considerable amount of plastic deformation or yielding.*

Fracture that occurs by cleavage at a nominal tensile stress below the yield stress is referred to as *brittle fracture*. Generally, a brittle fracture can occur when there is an adverse combination of tensile stress, temperature strain rate, and geometrical discontinuity (such as a notch). Other design and fabrication factors may also have an important influence. Because of the interrelation of these effects, the exact combination of stress, temperature, notch, and other conditions that cause brittle fracture in a given structure cannot be readily calculated. Preventing brittle fracture often consists mainly of avoiding conditions that tend to cause brittle fracture and selecting a steel appropriate for the application. These factors are discussed in the following paragraphs. Parker (1957), the Welding Research Council (1957), Lightner and Vanderbeck (1956), Rolfe and Barsom (1977), and Barsom (1993) describe the subject in much more detail.

Fracture mechanics offers a more direct approach for prediction of crack propagation. For this analysis, it is assumed that an internal imperfection forming a crack is present in the structure. By linear-elastic stress analysis and laboratory tests on precracked specimens, the applied stress causing rapid crack propagation is related to the size of the imperfection. Fracture mechanics has become increasingly useful in developing a fracture-control plan and establishing, on a rational basis, the interrelated requirements of material selection, design stress level, fabrication, and inspection requirements (Barsom 1993).

Conditions Causing Brittle Fracture

Plastic deformation occurs only in the presence of shear stresses. Shear stresses are always present in a uniaxial or a biaxial state of stress. However, in a triaxial state of stress, the maximum shear stress approaches zero as the principal stresses approach a common value. As a result, under equal triaxial tensile stresses, failure occurs by cleavage rather than by shear. Consequently, triaxial tensile stresses tend to cause brittle fracture and should be avoided. As discussed in the following material, a triaxial state of stress can result from a uniaxial loading when notches or geometrical discontinuities are present.

If a transversely notched bar is subjected to a longitudinal tensile force, the stress concentration effect of the notch causes high longitudinal tensile stresses at the apex of the notch and lower longitudinal stresses in adjacent material. The lateral contraction in the width and thickness direction of the highly stressed material at the apex of the notch is restrained by the smaller lateral contraction of the lower stressed material. Therefore, in addition to the longitudinal tensile stresses, tensile stresses are created in the width and thickness directions, so that a triaxial state of stress is present near the apex of the notch.

The effect of a geometrical discontinuity in a structure is generally similar to, although not necessarily as severe as, the effect of the notch in the bar. Examples of geometrical discontinuities include poor design details (such as abrupt changes in cross section, attachment welds on components in tension, and square-cornered cutouts) and fabrication flaws (such as weld cracks, undercuts, arc strikes, and scars from chipping hammers).

*Shear and cleavage are used in the metallurgical sense (macroscopically) to denote different fracture mechanisms. Parker (1957), as well as most elementary textbooks on metallurgy, discusses these mechanisms.

Increased strain rates tend to increase the possibility of brittle behavior. Therefore, structures that are loaded at fast rates are more susceptible to brittle fracture. However, a rapid strain rate or impact load is not a required condition for a brittle fracture.

Cold work and the strain aging that normally follows generally increases the likelihood of brittle fractures. This behavior is usually attributed to the reduction in ductility. The effect of cold work occurring in cold-forming operations can be minimized by selecting a generous forming radius, therefore limiting the amount of strain. The amount of strain that can be tolerated depends on both the steel and the application. A more severe but quite localized type of cold work occurs at the edges of punched holes or at sheared edges. This effect can be essentially eliminated for holes by drilling instead of punching or by reaming after punching; for sheared edges, it can be eliminated by machining or grinding. Severe hammer blows may also produce enough cold work to locally reduce the toughness of the steel.

When tensile residual stresses are present, such as those resulting from welding, they increase any applied tensile stress, resulting in the actual tensile stress in the member being greater than the applied stress. Consequently, the likelihood of brittle fracture in a structure that contains high residual stresses may be minimized by a postweld heat treatment. The decision to use a postweld heat treatment should be made with assurance that the anticipated benefits are needed and will be realized, and that possible harmful effects can be tolerated. Many modern steels for welded construction are designed for use in the less costly as-welded condition when possible. The soundness and mechanical properties of welded joints in some steels may be adversely affected by a postweld heat treatment.

Welding may also contribute to brittle fracture by introducing notches and flaws into a structure and changing the microstructure of the base metal. Such detrimental effects can be minimized by properly designing welds, by selecting their appropriate location, and by using good welding practice. The proper electrode must be selected so that the weld metal will be as resistant to brittle fracture as the base metal.

Charpy V-Notch Impact Test

Some steels will sustain more adverse temperature, notching, and loading conditions without fracture than other steels. Numerous tests have been developed to evaluate and assign a numerical value determining the relative susceptibility of steels to brittle fracture. Each of these tests can establish with certainty only the relative susceptibility to brittle fracture under the particular conditions in the test; however, some tests provide a meaningful guide to the relative performance of steels in structures subjected to severe temperature and stress conditions. The most commonly used rating test, the Charpy V-notch impact test, is described in this section, and the interpretation of its results is discussed briefly. Parker (1957) and the Welding Research Council (1957) discuss in detail many other rating tests.

The Charpy V-notch impact test specifically evaluates notch toughness—the resistance to fracture in the presence of a notch—and is widely used as a guide to the performance of steels in structures susceptible to brittle fracture. In this test, a small rectangular bar, with a V-shaped notch of specified size at its mid-length, is supported at its ends as a beam and fractured by a blow from a swinging pendulum. The energy required to fracture the specimen (which can be calculated from the height to which the pendulum raises after breaking the specimen), or the appearance of the fracture surface is determined for a range of temperatures. The appearance of the fracture surface is usually expressed as the percentage of the surface that appears to have fractured by shear as indicated by a fibrous appearance. A shiny or crystalline appearance is associated with a cleavage fracture.

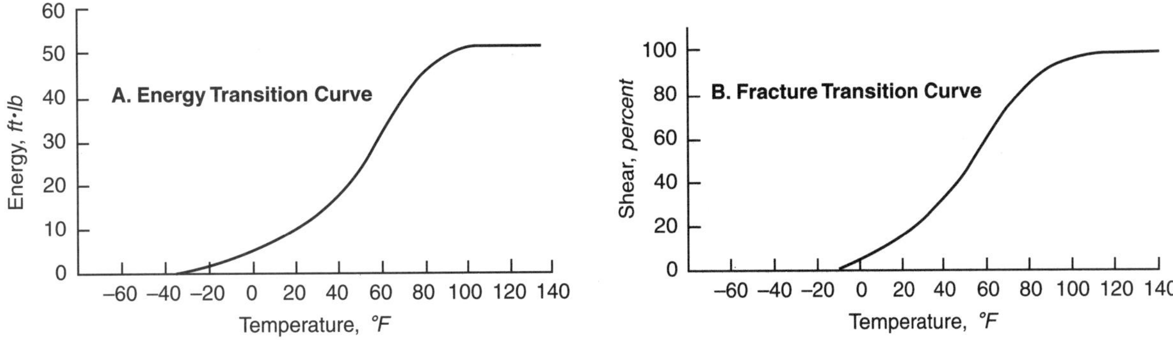

Source: Brockenbrough, R.L. & Johnston, B.G., U.S.S. Steel Design Manual. *ADUSS 27-3400-04. US Steel Corp. Pittsburgh, Pa. (Jan. 1981).*

NOTE: Curves are for carbon steel and are taken from the Welding Research Council (1957).

Figure 1-11 Transition curves obtained from Charpy V-notch impact tests

These data are used to plot curves of energy (see Figure 1-11) or percentage of shear fracture as a function of temperature. For the structural steels, the energy and percentage of shear fracture decrease from relatively high values to relatively low values with decreasing temperature. The temperature near the lower end of the energy-temperature curve, at which a selected value of energy is absorbed (often 15 ft•lb), is called the *ductility transition temperature*. The temperature at which the percentage of shear fracture decreases to 50 percent is often called the *fracture-appearance transition temperature* or *fracture transition temperature*. Both transition temperatures provide a rating of the brittle fracture resistance of various steels; the lower the transition temperature, the better the resistance to brittle fracture. The ductility transition temperature and the fracture transition temperature depend on many parameters (such as composition, thickness, and thermomechanical processing) and, therefore, can vary significantly for a given grade of steel.

Steel Selection

Requirements for notch toughness of steels used for specific applications can be determined through correlations with service performance. Fracture mechanics, when applied in conjunction with a thorough study of material properties, design, fabrication, inspection, erection, and service conditions, have been beneficial. In general, where a given steel has been used successfully for an extensive period in a given application, brittle fracture is not likely to occur in similar applications unless unusual temperature, notch, or stress conditions are present. Nevertheless, it is always desirable to avoid or minimize the previously cited adverse conditions that increase the susceptibility to brittle fracture.

GOOD PRACTICE

The ordinary water pipeline requires little stress calculation. The commonly used internal pressures for steel water pipe are given in Table 4-1 in chapter 4 and Table A-1 in appendix A. Suggested design stresses to resist other loadings are given as guides in various chapters on the different design subjects.

When designing the details of supports, wye branches, and other specials, especially for large pipe, the designer should review the data in chapter 13.

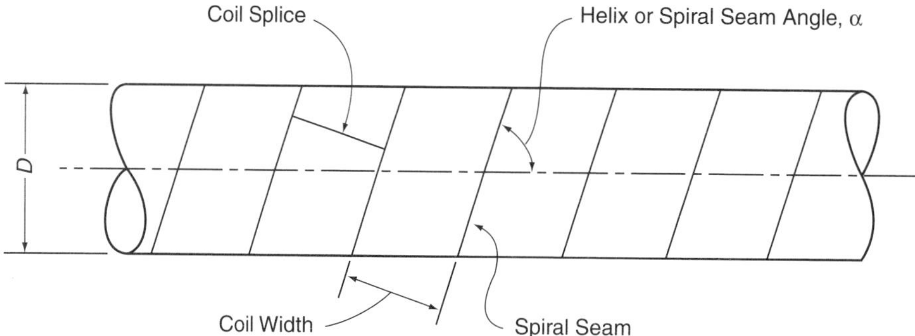

Figure 1-12 Spiral pipe weld seams

The concept of designing based on strain as well as stress will increase the knowledge of the behavior of steel and other materials in many cases where consideration of stress alone offers no reasonable explanation. The action and undesirable effects of stress raisers or stress concentrations—such as notches, threads, laps, and sudden changes in cross section—will be better understood. The steps involved in counteracting adverse effects become evident. Design using strain as well as stress will result in safer and more economical structures than if strain is ignored. Safe loads are a more important design consideration than safe stresses.

EVALUATION OF STRESSES IN SPIRAL-WELDED PIPE

Occasionally, it is necessary to evaluate stresses of spiral-welded pipe seams. The hoop or circumferential stress, σ_2, is the maximum stress that occurs in a pipeline loaded under internal pressure. Therefore, the weld located perpendicular to this direction of stressing is subjected to the greatest amount of loading. The more the seam angle deviates from the direction of the pipe axis, the more the normal stress acting perpendicular to the welding seam decreases when neglecting the longitudinal stresses (Sommer 1982). The relationship is shown in the following equation:

$$\sigma_n = \sigma_1(\cos^2\alpha) + \sigma_2(\sin^2\alpha) \tag{1-1}$$

Where:

σ_n = stress normal to the axis of the weld (psi)

σ_1 = longitudinal stress parallel to the longitudinal axis of the pipe (psi)

σ_2 = circumferential or hoop stress in pipe (psi)

α = angle of the spiral seam relative to longitudinal axis of the pipe in degrees (Figure 1-12)

REFERENCES

American Water Works Association. ANSI/AWWA Standard C200, *Steel Water Pipe—6 In. (150 mm) and Larger.* Latest edition. Denver, Colo.: American Water Works Association.

Barnard, R.E. 1950. Design of Steel Ring Flanges for Water Works Service—A Progress Report. *Jour. AWWA*, 42(10):931.

Barsom, J.M. 1993. *Welded Steel Water Pipe Fracture Toughness and Structural Performance*. Bull. 11-3. Cincinnati, Ohio: SPFA Steel Water Pipe.

Bethlehem Steel Co. 1946. *Bolt Tests—Tension Applied by Tightening Nut Versus Pure Tension*. Bethlehem, Pa.: Bethlehem Steel Co. (unpublished).

Brockenbrough, R.L., and B.G. Johnston. 1981. *USS Steel Design Manual*. ADUSS 27-3400-04. Pittsburgh, Pa.: US Steel Corp.

Cates, W.H. 1971. History of Steel Water Pipe, Its Fabrication and Design Development.

Chajes, A., S.J. Britvec, and G. Winter. 1963. Effects of Cold-Straining on Structural Sheet Steels. *Jour. of the Structural Div.*, Proc., ASCE, 89, No. ST2.

Dieter, G.E., Jr. 1961. *Mechanical Metallurgy*. New York: McGraw-Hill Book Co.

Elliot, G.A. 1922. The Use of Steel Pipe in Water Works. *Jour. AWWA*, 9(11):839.

Hinds, J. 1954. Notable Steel Pipe Installations. *Jour. AWWA*, 46(7):609.

Lightner, M.W., and R.W. Vanderbeck. 1956. *Factors Involved in Brittle Fracture. Regional Technical Meetings*. Washington, D.C.: American Iron and Steel Institute.

Neal, B.G. 1956. *The Plastic Methods of Structural Analysis*. New York: John Wiley & Sons.

Parker, E.R. 1957. *Brittle Behavior of Engineering Structures*. New York: John Wiley & Sons.

Rolfe, S.T., and J.M. Barsom. 1977. *Fracture and Fatigue Control in Structures—Applications of Fracture Mechanics*. Englewood Cliffs, N.J.: Prentice Hall, Inc.

Sommer, B. 1982. Spiral-Weld Pipe Meets High-Pressure Needs, *Oil and Gas Journal*, (Feb.):106–116.

Symposium—Plastic Strength of Structural Members. 1955. *Trans. ASCE*, Paper 2772.

Welding Research Council. 1957. *Control of Steel Construction to Avoid Brittle Failure*. New York: Welding Research Council.

This page intentionally blank.

Chapter 2

Manufacture and Testing

MANUFACTURE

Electric resistance welding and electric fusion welding are the most common methods used to convert flat-rolled steel bars, plates, sheets, and strips into tubular products.

Electric resistance welding (ERW) is performed without filler material. The flat strip, with edges previously trimmed to provide a clean, even surface for welding, is formed progressively into a tubular shape as it travels through a series of rolls. The forming is done cold. Welding is then effected by the application of heat and pressure. The welding heat for the tubular edges is generated by resistance to the flow of an electric current, which can be introduced through electrodes or by induction. Pressure rolls force the heated edges together to effect the weld. The squeezing action of the pressure rolls forming the weld causes some of the hot weld metal to extrude from the joint, forming a bead of weld flash both inside and outside the pipe. The flash is normally trimmed within tolerance limits while it is still hot from welding, using mechanical cutting tools contoured to the shape of the pipe (Figures 2-1 through 2-5).

Electric fusion welding (EFW) differs from ERW in that filler material is used and mechanical pressure is unnecessary to effect the weld. Pipe produced with this process can have straight or spiral seams. Straight-seam pipe is made from plate with edges planed parallel to each other and square with the ends. Curving the plate edges with crimping rolls is the first step of the forming process. This is followed by presses that form the plate first into a U-shaped trough and then into a full O-shaped tube. The O-shaped tube is then fed into a longitudinal seam-welding machine. Spiral-seam pipe is made from coiled strip or plate by a continuous process (Figure 2-6). An automatic machine unrolls the coil, prepares the edges for welding, and spirally forms the strip into a tubular shape. As the tube leaves the forming element, the edges are joined by fusion welding in the same submerged-arc process as is generally used in straight-seam pipe (Figure 2-7). The welded tube is cut to the desired length by an automatic cutoff device.

Pipe sizes manufactured using the fusion welding process are limited only by size limitations of individual pipe manufacturers. This process is especially suited for pipe in sizes larger than possible or feasible by other methods.

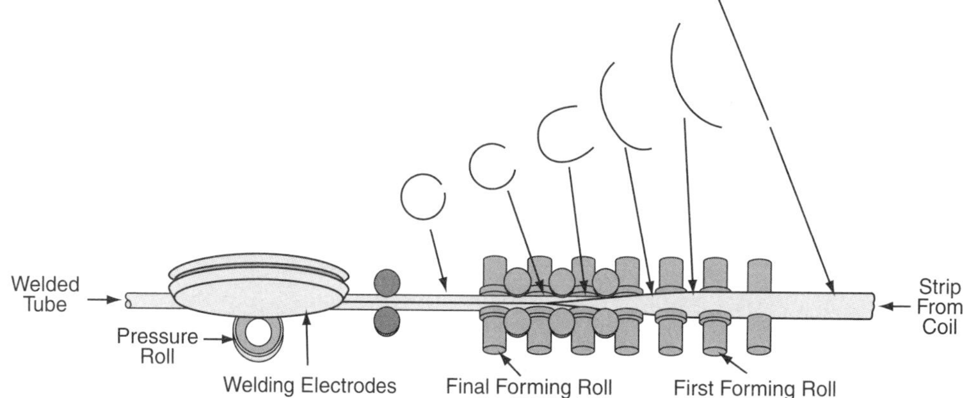

Source: Carbon Steel Pipe, Structural Tubing, Line Pipe and Oil Country Tubular Goods. Steel Products Manual. *American Iron and Steel Institute, Washington, D.C. (Apr. 1982).*

Figure 2-1 Schematic representation of the sequence of operations performed by a typical machine for making electric-resistance-welded tubes from steel strip

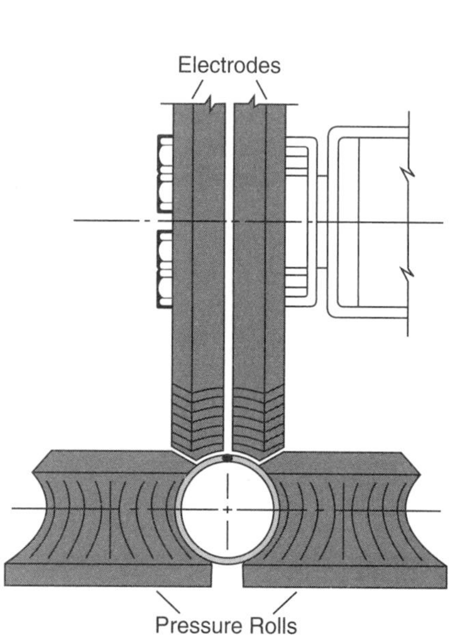

Application of pressure by rolls on both sides and beneath the electrodes forces the heated tube edges together to form a weld.

Source: See Figure 2-1.

Figure 2-2 Cross section through weld point

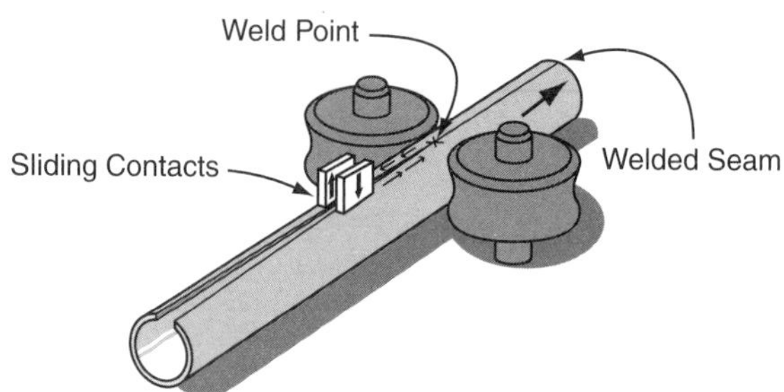

The current enters the tube via sliding contacts and flows along Vee edges to and from weld point.

Source: See Figure 2-1.

Figure 2-3 Electric resistance welding using high-frequency welding current

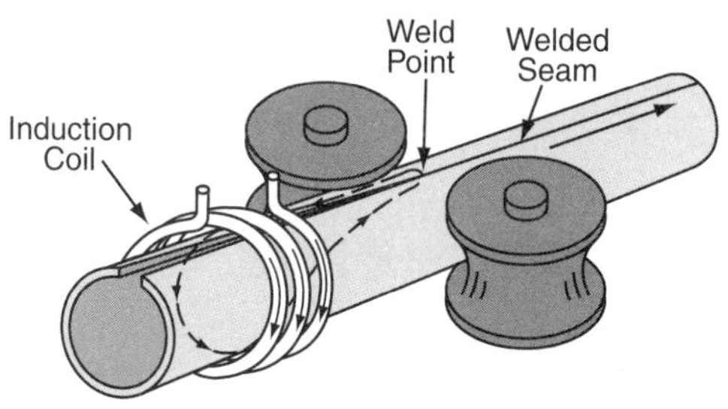

Eddy current flows around the back of the tube and along the edges to and from the weld point.

Source: See Figure 2-1.

Figure 2-4 Electric resistance welding by induction using high-frequency welding current

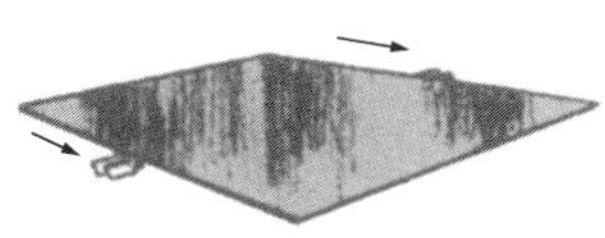

1.
Edge Planing—Submerged arc weld pipe begins as a flat rectangular steel plate from the plate mill. The first step in transforming it to pipe is planing the edges parallel to each other and square with the ends.

2.
Edge Crimping Rolls—Here the edges of the plate are curved to facilitate final forming of the pipe, reduce die wear, and produce greater uniformity at the seam edges when the plate is pressed to a cylindrical shape. The total surface of the plate, both sides edge to edge, is also inspected ultrasonically.

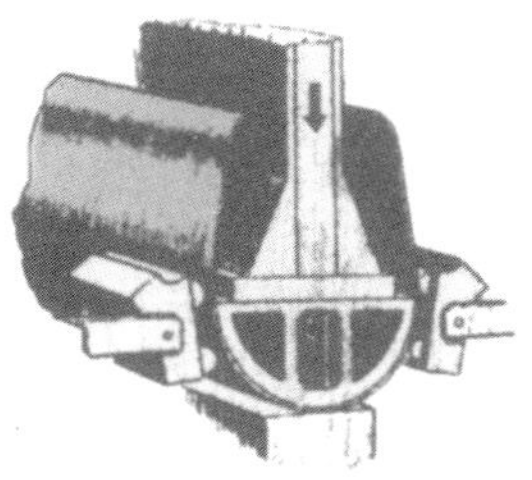

3.
U-ing Press—A semicircular ram descends on the plate, forcing it down between the rocker dies to form a U. The plate is slightly overbent to allow for spring-back.

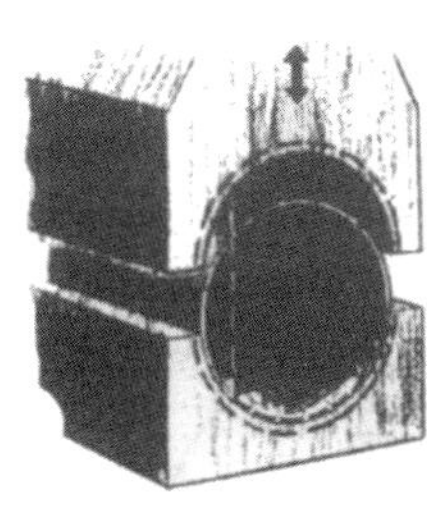

4.
O-ing Press—The U-shaped plate enters this press with the semicircular dies open. The top die, under hydraulic pressure, is forced down on the U, cold forming it to a cylindrical shape.

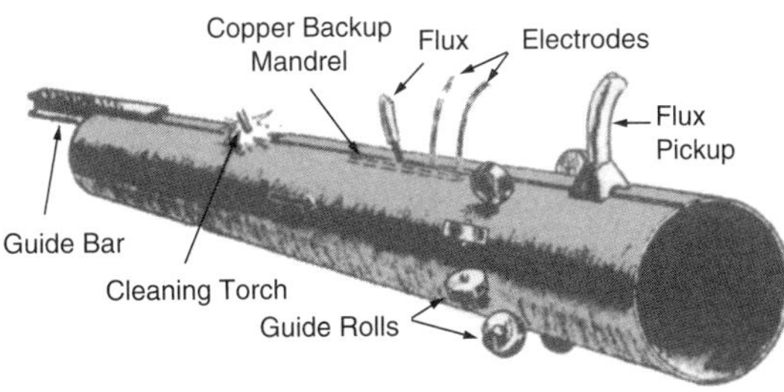

5.
Outside Welding—The O-formed plate is now fed into a longitudinal seam-welding machine in which the abutting edges are properly aligned, firmly pressed together, and welded by the submerged arc process. Two electrodes are used, and the weld is completed to within 3 in. of the pipe ends.

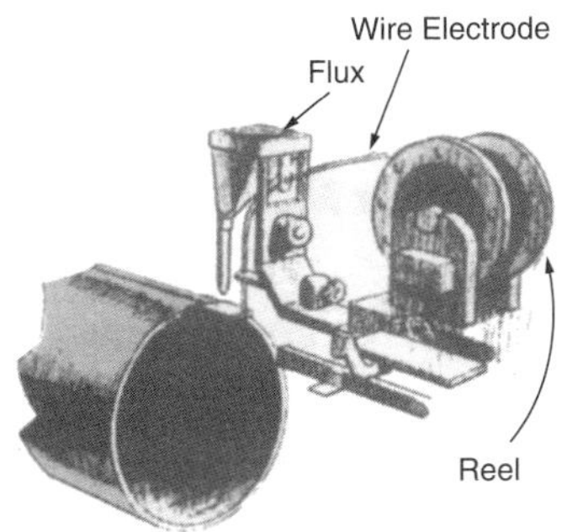

6.
End Welding—Here a 5-in. steel plate is attached to each end of the pipe at the seam, permitting the last few inches of the OD seam to be welded.

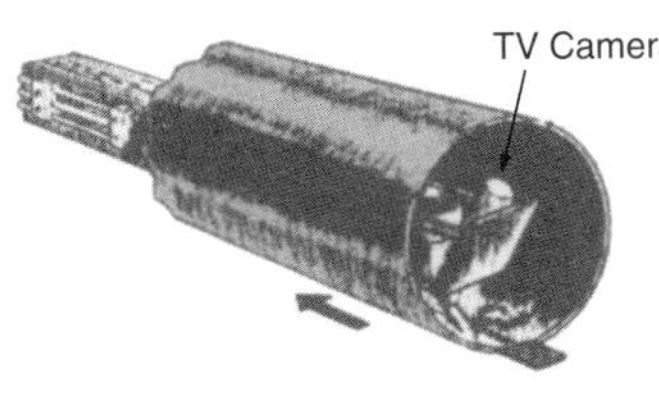

7.
Inside Welding—Here, the welding head and a small TV camera are mounted on a long cantilever boom; as the pipe is drawn over the welding boom, a TV screen at the operator's control board enables him to keep the welding exactly on the seam. Finishing up on the tab, the last few inches of the seam are welded. The small plates are then removed, and the completed weld is inspected inside and out.

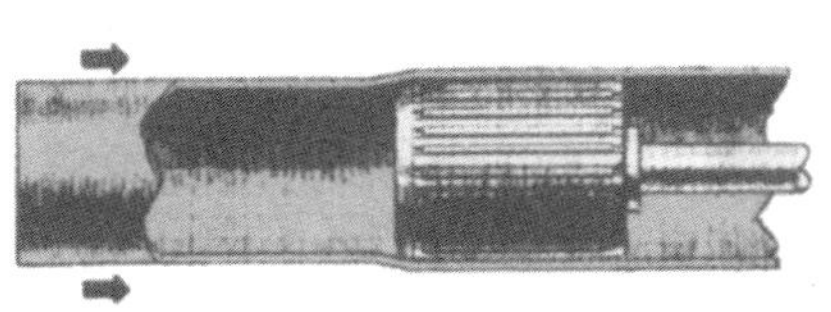

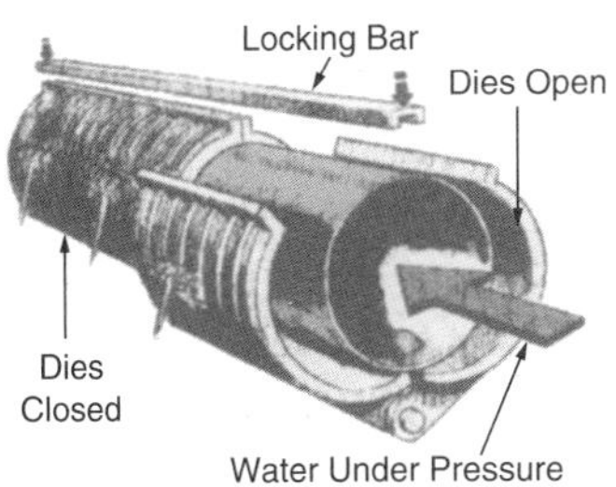

8.
Expanding and Testing—The pipe is either mechanically or hydrostatically expanded, depending on the mill location. In either case, accurate size and straightness and improved transverse yield strength are obtained by expansion.

Mechanical Expander—The pipe is mechanically expanded in 24-in. through 27-in. increments until half of the length is completed. The pipe rolls to a second expander die where the remaining half of the length is expanded. The pipe length then proceeds to a hydrostatic unit where a specified internal pressure is applied to test the weld for sweats or leaks.

Hydrostatic Expander—Both ends of the pipe are sealed by mandrels. The semicircular dies, slightly larger than the pipe OD, are closed, and the pipe is hydraulically expanded against the dies. The dies are opened and a specified internal pressure applied to test the weld for sweats or leaks.

Source: Carbon Steel Pipe, Structural Tubing, Line Pipe and Oil Country Tubular Goods. Steel Products Manual. *American Iron and Steel Institute, Washington, D.C. (Apr. 1982).*

Figure 2-5 Sequence of operations in a typical double submerged arc weld process

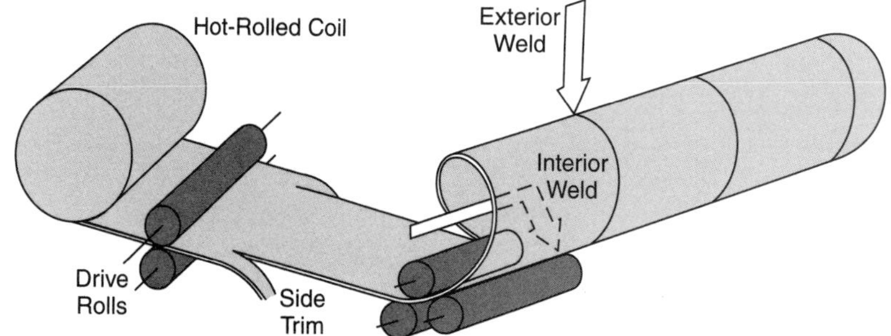

Figure 2-6 Schematic diagram of process for making spiral-seam pipe

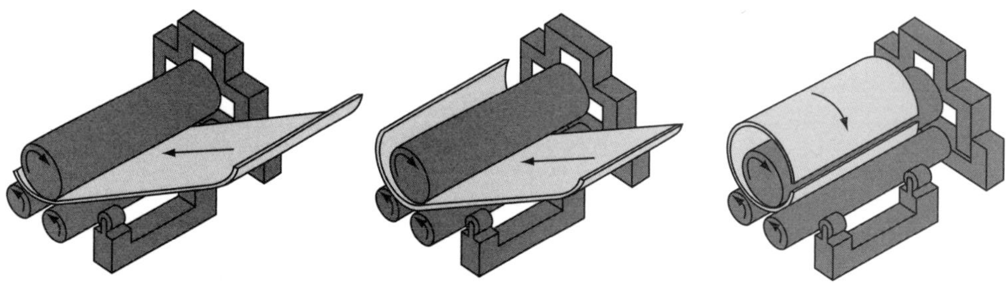

Figure 2-7 Schematic diagram for making plate pipe

TESTING

Tests of Chemical Properties

The various services to which steel pipe is used require a variety of chemical compositions to produce the necessary characteristics. The chemical compositions established in the AWWA steel pipe standards are appropriate for the needs of water utility applications. However, there are other steel materials that may be equally suitable.

Ladle analysis. Ladle analysis is the chemical analysis representative of the heat or blow of steel. This analysis is reported to the purchaser. Analysis results are determined by testing for specific elements, using a test ingot sample obtained from the first or middle part of the heat or blow during the pouring of the steel from the ladle.

Most steel melting operations obtain more than one ladle-test ingot sample from each heat or blow; often three or more are taken, representing the first, middle, and last portions of the heat or blow. Drillings taken from the first or middle sample are used in determining the ladle analysis because experience has shown that these locations most closely represent the chemical analysis of the entire heat or blow. The additional samples are used for a survey of uniformity and for control purposes.

Check analysis. Check analysis, as used in the steel industry, means analysis of the metal after it has been rolled or forged into semifinished or finished forms. Such an analysis is made either to verify the average composition of the heat, to verify the composition of a lot as represented by the ladle analysis, or to determine variations in the composition of a heat or lot. Check analysis is not used, as the term might imply, to confirm the accuracy of a previous result. Check analysis of known heats is justified only where a high degree of uniformity of composition is essential—for example, on material that is to be heat treated. Such analysis should rarely be necessary for water pipe, except to identify or confirm the assumed analysis of plates or pipe that have lost identity. The results of analyses representing different locations in the same piece, or taken from different pieces of a lot, may differ from each other and from the ladle analysis because of segregation. These permissible variations from the specified ranges or limits have been established in the applicable specification or by common practice. The variations are a natural phenomenon that must be recognized by inspectors. The methods of analysis commonly used are in accordance with the latest edition of ASTM A751, those approved by the National Bureau of Standards, or others of equivalent accuracy.

Tests of Physical Properties

The methods of testing the physical properties of steel pipe are established in ASTM A370 (1995a). The physical properties required are contained in AWWA C200, *Standard for Steel Water Pipe 6 In. (150 mm) and Larger* (latest edition), or are as otherwise specified by the purchaser.

Hydrostatic Test of Straight Pipe

Straight lengths of pressure pipe and tubing are customarily subjected to an internal hydrostatic pressure test. This operation is conducted as a part of the regular mill inspection procedure to detect defects. It is not intended to bear a direct relationship to bursting pressures, working pressures, or design data, although test pressures sometimes influence design pressures. AWWA C200 contains a formula for determining hydrostatic test.

Hydrostatic tests are performed at the pressure required by the standard during manufacture of the pipe. The requirements for hydrostatic testing in the presence of the purchaser's inspector involve additional handling, unless the inspector is present during the course of manufacture. The producer, on request, customarily furnishes a certificate confirming such testing.

Tests of Dimensional Properties

The diameter, length, wall thickness, straightness, and out-of-roundness of pipe are checked as part of the normal manufacturing procedure. Such dimensions are subject to the tolerances prescribed in the appropriate standards or specifications.

REFERENCES

American Society for Testing and Materials. 1995a. ASTM Standard A370-95, *Standard Test Methods and Definitions for Mechanical Testing of Steel Products*. Philadelphia, Pa.: American Society for Testing and Materials.

ASTM Standard A751, *Standard Test Methods, Practices, and Terminology for Chemical Analysis of Steel Products*. Latest edition. Philadelphia, Pa.: American Society for Testing and Materials.

American Water Works Association. ANSI/AWWA Standard C200, *Steel Water Pipe—6 In. (150 mm) and Larger*. Latest edition. Denver, Colo.: American Water Works Association.

This page intentionally blank.

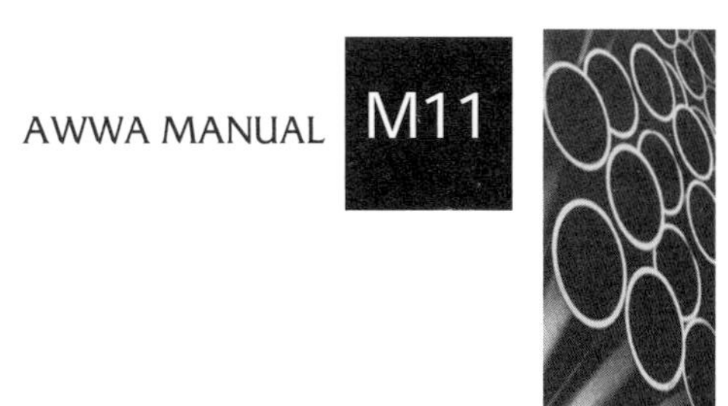

Chapter 3

Hydraulics of Pipelines

This chapter concentrates on the flow of water in transmission conduits. It is not intended to cover flows through the complicated networks of distribution systems. Because this manual is a guide to practice rather than a textbook, historical and theoretical development of the many hydraulic flow formulas has been omitted, as has discussion of universal or rational formulas.

The discussions and data in this chapter are therefore restricted to the three formulas most commonly used in water flow calculations in the Western Hemisphere.

FORMULAS

Hydraulic Symbols

The definition of hydraulic symbols used in this chapter are as follows:

A = area of pipe, ft^2 (m^2)
C = Hazen-Williams coefficient
d = nominal diameter of pipe, in.
D = inside diameter of pipe, in. (mm)
f = Darcy friction factor
g = acceleration of gravity, 32.2 ft/s^2 (9.81 m/s^2)
h_f = head loss [ft (m)] in pipe length L, ft (m)
H = head loss in 1,000 units of length of pipe, ft/1,000 ft (m/1,000 m)
K_s = Scobey constant
L = length of pipe, ft (m)
n = Manning coefficient
Q = discharge, ft^3/sec (m^3/sec)
r = hydraulic radius of pipe, ft (m)
$s = \frac{h_f}{L} = \frac{H}{1{,}000}$ = slope of hydraulic gradient
V = mean velocity, ft/sec (m/sec)

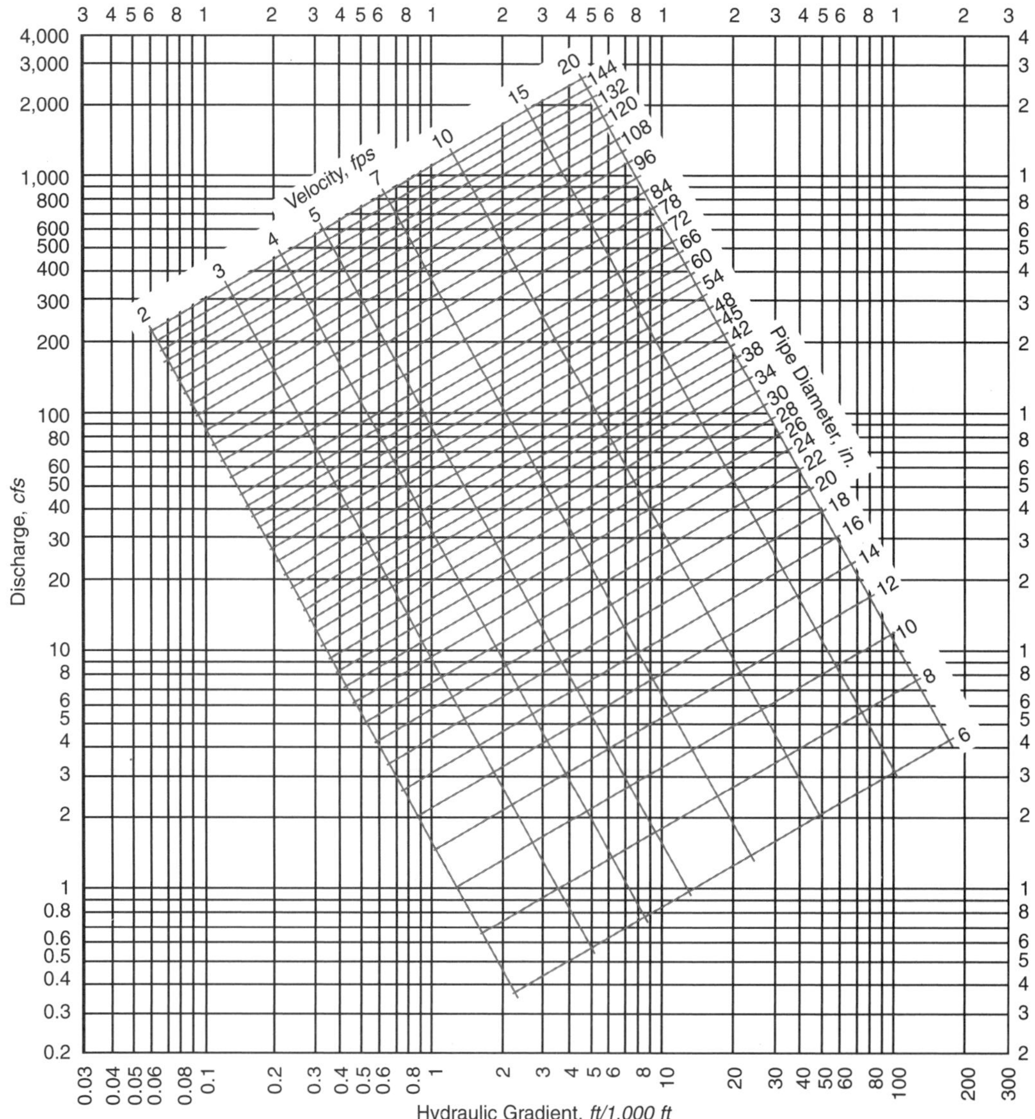

NOTE: To convert cfs to m³/sec, multiply by 0.02832; ft/1,000 ft = m/1,000 m; to convert in. to mm, multiply by 25.4; to convert fps to m/sec, multiply by 0.3048

Figure 3-1 Solution of the Hazen-Williams formula (based on $V = 1.318Cr^{0.63}s^{0.54}$ for $C = 150$)

Table 3-1 Multiplying factors corresponding to various values of *C* in Hazen-Williams formula*

Values of *C*	160	155	150	145	140	130	120	110	100	90	80	60	40
							Base *C* = 150						
Relative discharge and velocity for given loss of head	1.067	1.033	1.000	0.967	0.933	0.867	0.800	0.733	0.667	0.600	0.533	0.400	0.267
Relative loss of head for given discharge	0.887	0.941	1.000	1.065	1.136	1.297	1.511	1.775	2.117	2.573	3.199	5.447	11.533

Source: Barnard, R.E., Design Standards for Steel Water Pipe. Jour. AWWA, *40:1:24 (Jan. 1948).*

* Use with Figure 3-1.

The Hazen-Williams Formula

Probably the most popular formula in current use among waterworks engineers is the Hazen-Williams formula. This formula, first published in 1904, is

$$V = 1.318Cr^{0.63}s^{0.54} \tag{3-1}$$

Equivalent metric equation:

$$V = 0.849Cr^{0.63}s^{0.54} \tag{3-1M}$$

The head loss h_f may be calculated from

$$h_f = \frac{4.72Q^{1.852}L}{C^{1.852}D^{4.87}} \tag{3-2}$$

Equivalent metric equation:

$$h_f = \frac{10.65Q^{1.852}L}{C^{1.852}D^{4.87}} \tag{3-2M}$$

Tests have shown that the value of the Hazen-Williams roughness coefficient C is dependent not only on the surface roughness of the pipe interior but also on the diameter of the pipe. Flow measurements indicate that for pipe with smooth interior linings in good condition, the average value of $C = 104 + 0.17d$.

However, in consideration of long-term lining deterioration, slime buildup, etc., a higher value is recommended: that is $C = 130 + 0.16d$.

A graphical solution of the Hazen-Williams formula for $C = 150$ is presented in Figure 3-1 for pipe sizes 6 in. through 144 in. (150 mm through 3,600 mm). The multiplying factors in Table 3-1 provide a convenient means of changing the flow capacities shown in Figure 3-1 to the flows for other values of C.

The Scobey Formula

The Scobey formula for steel pipe, used perhaps more commonly in irrigation work than in the waterworks industry, is

$$V = \frac{D^{0.58}h_f^{0.526}}{K_s^{0.526}} \tag{3-3}$$

Equivalent metric equation:

$$V = 1.13\frac{D^{0.58}h_f^{0.526}}{K_s^{0.526}} \tag{3-3M}$$

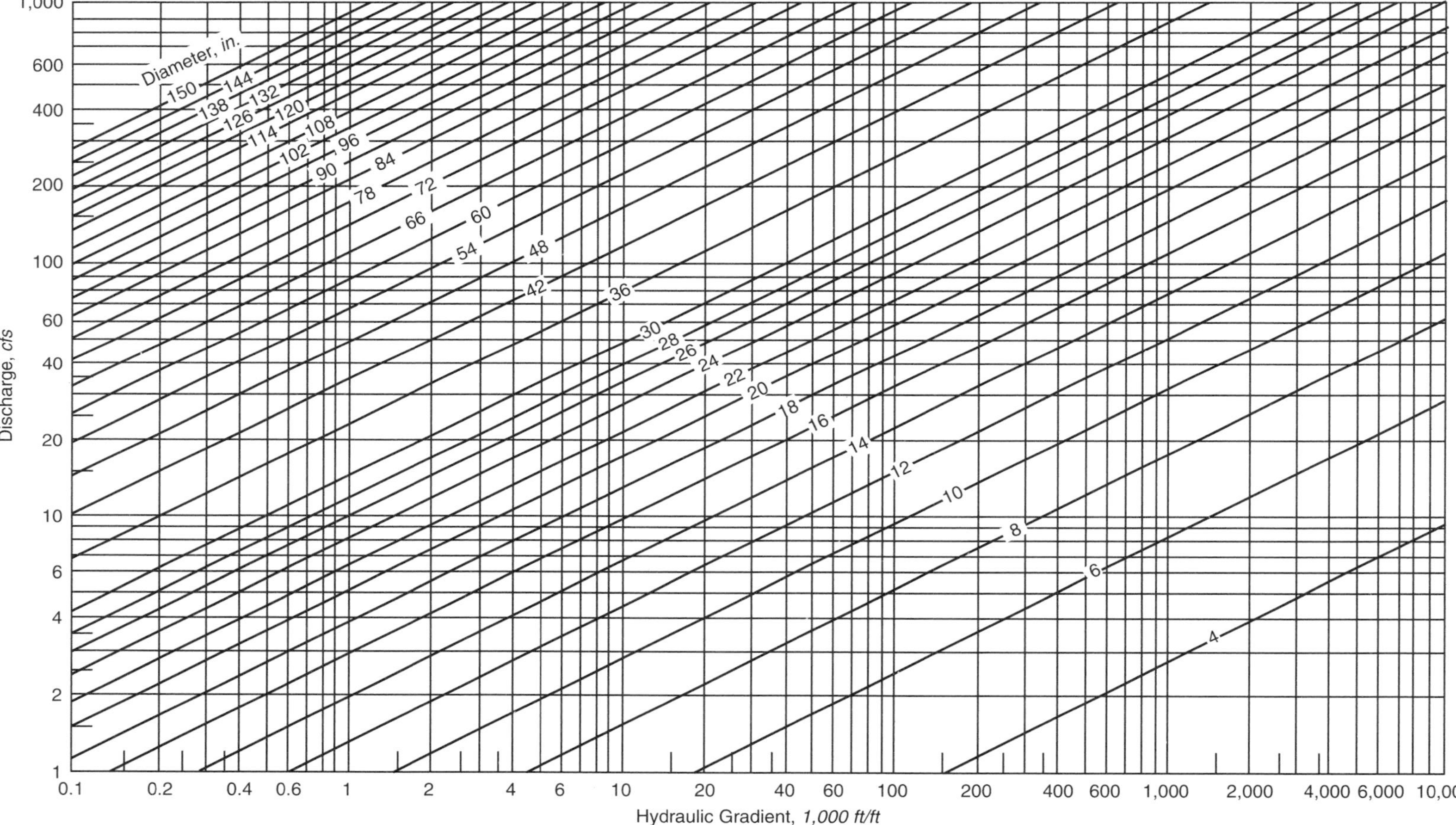

NOTE: To convert cfs to m³/sec, multiply by 0.02832; ft/1,000 ft = m/1,000 m; to convert in. to mm, multiply by 25.4; to convert fps to m/sec, multiply by 0.3048.

Figure 3-2 Solution of Scobey flow formula for K_s = 0.36 (See data in Table 3-2 for other K_s values)

Table 3-2 Multiplying factors for friction coefficient values—Base K_s = 0.36*

K_s value	0.32	0.34	0.36	0.38	0.40
Relative discharge	1.125	1.059	1.000	0.946	0.900

* Data for use with Figure 3-2.

or for determining head loss:

$$h_f = K_s \frac{V^{1.9}}{D^{1.1}} \tag{3-4}$$

Equivalent metric equation:

$$h_f = 0.789 K_s \frac{V^{1.9}}{D^{1.1}} \tag{3-4M}$$

The recommended K_s value for new bare steel pipe or pipe with linings conforming to current AWWA standards is 0.36. A graphical solution to the Scobey formula for $K_s = 0.36$ is shown in Figure 3-2. Multiplying factors for other friction coefficients are given in Table 3-2.

The Manning Formula

The Manning formula is

$$V = \frac{0.59}{n} D^{0.667} s^{0.5} \tag{3-5}$$

Equivalent metric equation:

$$V = \frac{0.397}{n} D^{0.667} s^{0.5} \tag{3-5M}$$

or

$$h_f = 2.87 n^2 \frac{LV^2}{D^{1.33}} \tag{3-6}$$

Equivalent metric equation:

$$h_f = 6.37 n^2 \frac{LV^2}{D^{1.33}} \tag{3-6M}$$

For design, an n value of 0.011 is recommended for steel pipe with linings conforming to current AWWA standards. A graphical solution to the Manning formula for n = 0.011 is shown in Figure 3-3. Multiplying factors for other values of n are given in Table 3-3.

CALCULATIONS

Computations for Flow Through Pipe

The quantity of water passing through any given pipe depends on the head (pressure) producing the flow, the diameter and length of the pipe, the condition of the pipe interior (smooth or rough), the number and abruptness of bends or elbows, and the presence of tees, branches, valves, and other appurtenances in the line.

The total head, or pressure, affecting flow may be divided into four parts: velocity head loss, entrance head loss, loss of head through friction, and minor losses caused

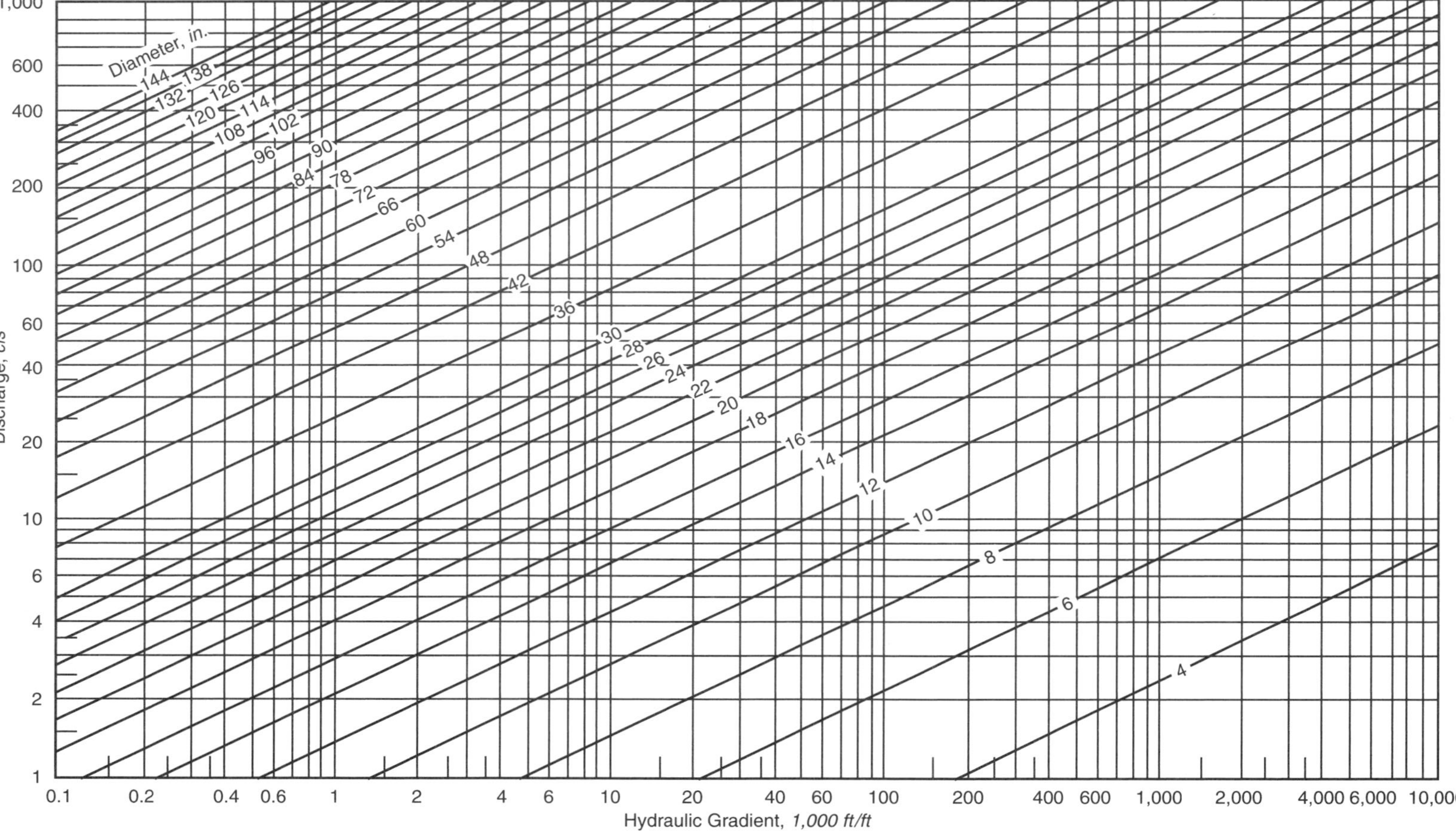

NOTE: To convert cfs to m^3/sec, multiply by 0.02832; ft/1,000 ft = m/1,000 m; to convert in. to mm, multiply by 25.4; to convert fps to m/sec, multiply by 0.3048.

Figure 3-3 Solution of Manning flow formula for n = 0.011

Table 3-3 Multiplying factors for friction coefficient values—Base n = 0.011*

n value	0.009	0.010	0.011	0.012	0.013
Relative discharge	1.222	1.100	1.000	0.917	0.846

* Data for use with Figure 3-3.

by elbows, fittings, and valves. Pressure (P) is related to total head (H) by the following formula:

$$P = 0.433H, \text{ in ft} \quad (3\text{-}7)$$

Equivalent metric equation:

$$P = 909.96H, \text{ in Pa} \quad (3\text{-}7M)$$

Velocity Head Loss ($V^2/2g$)

Velocity head loss is defined as the height a body must fall in a vacuum to acquire the velocity at which the water flows in the pipe. This loss is usually considered to be unrecoverable at the outlet. This head loss is calculated by the formula:

$$H = \frac{V^2}{2g} \quad (3\text{-}8)$$

Where:
- H = head loss in ft (m)
- V = velocity in ft/sec (m/sec)
- g = acceleration due to gravity or 32.2 ft/sec^2 (9.81 m/sec^2)

Entrance Head Loss

Entrance head loss is the head required to overcome the resistance at the entrance to the pipe; it is usually less than the velocity head. When the conditions are not specified, it is ordinarily considered equal to one-half the velocity head, assuming a sharp-edge entrance. Head losses for other than sharp-edge entrances may be found in treatises on hydraulics.

Loss of Head Through Friction

Friction head loss may be determined by one of the formulas that have been discussed previously. Data are given in this chapter to aid in solving the formulas.

Minor Losses Caused by Elbows, Fittings, and Valves

In long water lines, minor head losses caused by bends and fittings are occasionally ignored. In any given line, however, all losses should be considered so that no important factors will be overlooked. Minor losses should always be recognized when evaluating flow tests. Total head loss in long lines with low velocities, the sum of velocity head loss and entrance head loss may be relatively insignificant; in short lines with high velocities, this sum becomes very important. Ordinary tables and charts showing flow of water in pipe usually give only the friction head loss in straight pipe. In long water lines, this is typically the largest loss.

In the final solution to a flow problem, the sum of all losses must equal the available head, or pressure, producing the flow. The foregoing formulas determine H or V, and the volume of flow Q is found from

$$Q = AV \quad (3\text{-}9)$$

The information contained in Tables 3-4 through 3-9 will be useful when making hydraulic calculations.

Table 3-4 Slope conversions

Grade of Pipe *Percent*	Fall of Pipe in ft per ft *(m per m)*	Drop per 1,000 ft of Pipe *ft**	Drop per Mile of Pipe *ft**	Length of Pipe in 1-ft Drop *ft**
0.005	0.00005	0.05	0.264	20,000.00
0.01	0.0001	0.10	0.528	10,000.00
0.02	0.0002	0.20	1.056	5,000.00
0.03	0.0003	0.30	1.584	3,330.00
0.04	0.0004	0.40	2.112	2,500.00
0.05	0.0005	0.50	2.640	2,000.00
0.06	0.0006	0.0	3.168	1,666.67
0.07	0.0007	0.70	3.696	1,428.57
0.08	0.0008	0.80	4.224	1,250.00
0.09	0.0009	0.90	4.752	1,111.11
0.10	0.0010	1.00	5.280	1,000.00
0.20	0.0020	2.00	10.56	500.00
0.30	0.0030	3.00	15.84	333.33
0.40	0.0040	4.00	21.12	250.00
0.50	0.0050	5.00	26.40	200.00
0.60	0.0060	6.00	31.68	166.67
0.70	0.0070	7.00	36.96	142.86
0.80	0.0080	8.00	42.24	125.00
0.90	0.0090	9.00	47.52	111.11
1.00	0.0100	10.00	52.80	100.00
2.00	0.0200	20.00	105.60	50.00
3.00	0.0300	30.00	158.40	33.33
4.00	0.0400	40.00	211.20	25.00
5.00	0.0500	50.00	264.00	20.00
6.00	0.0600	60.00	316.80	16.67
7.00	0.0700	70.00	369.60	14.29
8.00	0.0800	80.00	422.40	12.5
9.00	0.0900	90.00	475.20	11.11
10.00	0.1000	100.00	528.00	10.00
12.00	0.1200	120.00	633.60	8.33

* NOTE: 1 ft = 0.3048 m.
1 km = 3,280 ft.

Source: Barnard, R.E., Design Standards for Steel Water Pipe. Jour. AWWA, 40:1:24 *(Jan. 1948).*

Table 3-5 Flow equivalents

gal/day	*liter/day**	*gal/min*	*m³/sec**	*cfs*	*gal/day*	*liter/day**	*gal/min*	*m³/sec**	*cfs*
1,000,000	378,500	694	0.044	1.55	36,000,000	13,626,000	25,000	1.577	55.73
2,000,000	757,000	1,389	0.088	3.09	37,000,000	14,004,500	25,694	1.620	57.28
3,000,000	1,135,500	2,083	0.131	4.64	38,000,000	14,383,000	26,389	1.664	58.82
4,000,000	1,514,000	2,778	0.175	6.19	39,000,000	14,761,500	27,083	1.708	60.37
5,000,000	1,892,500	3,472	0.219	7.74	40,000,000	15,140,000	27,778	1.752	61.92
6,000,000	2,271,000	4,167	0.263	9.28	42,000,000	15,897,000	29,167	1.839	65.02
7,000,000	2,649,500	4,861	0.307	10.83	44,000,000	16,654,000	30,556	1.927	68.11
8,000,000	3,028,000	5,556	0.350	12.38	46,000,000	17,411,000	31,944	2.014	71.21
9,000,000	3,406,500	6,250	0.394	13.93	48,000,000	18,168,000	33,333	2.102	74.31
10,000,000	3,785,000	6,944	0.438	15.48	50,000,000	18,925,000	34,722	2.190	77.40
11,000,000	4,163,500	7,639	0.482	17.02	52,000,000	19,682,000	36,111	2.277	80.50
12,000,000	4,542,000	8,333	0.526	18.57	54,000,000	20,439,000	37,500	2.365	83.60
13,000,000	4,920,500	9,028	0.569	20.12	56,000,000	21,196,000	38,889	2.452	86.69
14,000,000	5,299,000	9,722	0.613	21.67	58,000,000	21,953,000	40,278	2.540	89.79
15,000,000	5,677,500	10,417	0.657	23.22	60,000,000	22,710,000	41,667	2.628	92.88
16,000,000	6,056,000	11,111	0.701	24.77	62,000,000	23,467,000	43,056	2.715	95.98
17,000,000	6,434,500	11,806	0.744	26.31	64,000,000	24,224,000	44,444	2.803	99.08
18,000,000	6,813,000	12,500	0.788	27.86	66,000,000	24,981,000	45,833	2.890	102.17
19,000,000	7,191,500	13,194	0.832	29.41	68,000,000	25,738,000	47,222	2.978	105.27
20,000,000	7,570,000	13,889	0.876	30.96	70,000,000	26,495,000	48,611	3.066	108.37
21,000,000	7,948,500	14,583	0.920	32.51	72,000,000	27,252,000	50,000	3.153	111.46
22,000,000	8,327,000	15,278	0.963	34.05	74,000,000	28,009,000	51,389	3.241	114.56
23,000,000	8,705,500	15,972	1.007	35.60	76,000,000	28,766,000	52,778	3.328	117.65
24,000,000	9,084,000	16,667	1.051	37.15	78,000,000	29,523,000	54,167	3.416	120.75
25,000,000	9,462,500	17,361	1.095	38.70	80,000,000	30,280,000	55,556	3.503	123.85
26,000,000	9,841,000	18,056	1.139	40.25	82,000,000	31,037,000	56,944	3.591	126.94
27,000,000	10,219,500	18,750	1.182	41.80	84,000,000	31,794,000	58,333	3.679	130.04
28,000,000	10,598,000	19,444	1.226	43.34	86,000,000	32,551,000	59,722	3.766	133.14
29,000,000	10,976,500	20,139	1.270	44.89	88,000,000	33,308,000	61,111	3.854	136.23
30,000,000	11,355,000	20,833	1.314	46.44	90,000,000	34,065,000	62,500	3.941	139.33
31,000,000	11,733,500	21,528	1.358	47.99	92,000,000	34,822,000	63,889	4.029	142.43
32,000,000	12,112,000	22,222	1.401	49.54	94,000,000	35,579,000	65,278	4.117	145.52
33,000,000	12,490,500	22,917	1.445	51.08	96,000,000	36,336,000	66,667	4.204	148.62
34,000,000	12,869,000	23,611	1.489	52.63	98,000,000	37,093,000	68,056	4.292	151.71
35,000,000	13,247,500	24,306	1.533	54.18	100,000,000	37,850,000	69,444	4.379	154.81

Source: Barnard, R.E., Design Standards for Steel Water Pipe. Jour. AWWA, 40:1:24 *(Jan. 1948).*

* Metric equivalents added.

Table 3-6 Pressure (*psi*) for heads (*ft*)

	Additional Heads									
	0	+1	+2	+3	+4	+5	+6	+7	+8	+9
Head *ft*	Pressure *psi*									
0	—	0.43	0.87	1.30	1.73	2.16	2.60	3.03	3.46	3.90
10	4.33	4.76	5.20	5.63	6.06	6.49	6.93	7.36	7.79	8.23
20	8.66	9.09	9.53	9.96	10.39	10.82	11.26	11.69	12.12	12.56
30	12.99	13.42	13.86	14.29	14.72	15.15	15.59	16.02	16.45	16.89
40	17.32	17.75	18.19	18.62	19.05	19.48	19.92	20.35	20.78	21.22
50	21.65	22.08	22.52	22.95	23.38	23.81	24.25	24.68	25.11	25.55
60	25.98	26.41	26.85	27.28	27.71	28.14	28.58	29.01	29.44	29.88
70	30.31	30.74	31.18	31.61	32.04	32.47	32.91	33.34	33.77	34.21
80	34.64	35.07	35.51	35.94	36.37	36.80	37.24	37.67	38.10	38.54
90	38.97	39.40	39.84	40.27	40.70	41.13	41.57	42.00	42.43	42.87

Source: Barnard, R.E., Design Standards for Steel Water Pipe. Jour. AWWA, 40:1:24 *(Jan. 1948).*

Table 3-6M Pressure (*kPa*) for heads (*cm*)

	Additional Heads									
	0	+10	+20	+30	+40	+50	+60	+70	+80	+90
Head *cm*	Pressure *kPa*									
0	0.00	0.98	1.96	2.94	3.92	4.90	5.88	6.86	7.85	8.83
100	9.81	10.79	11.77	12.75	13.73	14.71	15.69	16.67	17.65	18.63
200	19.61	20.59	21.57	22.55	23.54	24.52	25.50	26.48	27.46	28.44
300	29.42	30.40	31.38	32.36	33.34	34.32	35.30	36.28	37.26	38.24
400	39.23	40.21	41.19	42.17	43.15	44.13	45.11	46.09	47.07	48.05
500	49.03	50.01	50.99	51.97	52.95	53.94	54.92	55.90	56.88	57.86
600	58.84	59.82	60.80	61.78	62.76	63.74	64.72	65.70	66.68	67.66
700	68.64	69.63	70.61	71.59	72.57	73.55	74.53	75.51	76.49	77.47
800	78.45	79.43	80.41	81.39	82.37	83.35	84.33	85.32	86.30	87.28
900	88.26	89.24	90.22	91.20	92.18	93.16	94.14	95.12	96.10	97.08

Source: Barnard, R.E., Design Standards for Steel Water Pipe. Jour. AWWA, 40:1:24 *(Jan. 1948).*

Table 3-7 Head (*ft*) for pressures (*psi*)

Pressure *psi*	Additional Heads, *ft* 0	+1	+2	+3	+4	+5	+6	+7	+8	+9
	Head *ft*									
0	—	2.3	4.6	6.9	9.2	11.5	13.9	16.2	18.5	20.8
10	23.1	25.4	27.7	30.0	32.3	34.6	36.9	39.3	41.6	43.9
20	46.2	48.5	50.8	53.1	55.4	57.7	60.0	62.4	64.7	67.0
30	69.3	71.6	73.9	76.2	78.5	80.8	83.1	85.4	87.8	90.1
40	92.4	94.7	97.0	99.3	101.6	103.9	106.2	108.5	110.8	113.2
50	115.5	117.8	120.1	122.4	124.7	127.0	129.3	131.6	133.9	136.3
60	138.6	140.9	143.2	145.5	147.8	150.1	152.4	154.7	157.0	159.3
70	161.7	164.0	166.3	168.6	170.9	173.2	175.5	177.8	180.1	182.4
80	184.8	187.1	189.4	191.7	194.0	196.3	198.6	200.9	203.2	205.5
90	207.9	210.2	212.5	214.8	217.1	219.4	221.7	224.0	226.3	228.6

Source: Barnard, R.E., Design Standards for Steel Water Pipe. Jour. AWWA, 40:1:24 *(Jan. 1948).*

Table 3-7M Head (*cm*) for pressures (*kPa*)

Pressure *kPa*	Additional Pressures, *kPa* 0	1	2	3	4	5	6	7	8	9
	Head *cm*									
0	0.00	10.20	20.39	30.59	40.79	50.99	61.18	71.38	81.58	91.78
10	101.97	112.17	122.37	132.57	142.76	152.96	163.16	173.36	183.55	193.75
20	203.95	214.15	224.34	234.54	244.74	254.94	265.13	275.33	285.53	295.73
30	305.92	316.12	326.32	336.52	346.71	356.91	367.11	377.31	387.50	397.70
40	407.90	418.10	428.29	438.49	448.69	458.88	469.08	479.28	489.48	499.67
50	509.87	520.07	530.27	540.46	550.66	560.86	571.06	581.25	591.45	601.65
60	611.85	622.04	632.24	642.44	652.64	662.83	673.03	683.23	693.43	703.62
70	713.82	724.02	734.22	744.41	754.61	764.81	775.01	785.20	795.40	805.60
80	815.80	825.99	836.19	846.39	856.59	866.78	876.98	887.18	897.37	907.57
90	917.77	927.97	938.16	948.36	958.56	968.76	978.95	989.15	999.35	1,009.55

Source: Barnard, R.E., Design Standards for Steel Water Pipe. Jour. AWWA, 40:1:24 *(Jan. 1948).*

Table 3-8 Pressures (*kPa*) for heads *ft* (*m*)

Head		Additional Heads, *ft* (*m*)									
		0	+1(.30)	+2(.61)	+3(.91)	+4(1.22)	+5(1.52)	+6(1.83)	+7(2.13)	+8(2.44)	+9(2.74)
		Pressure									
ft	(*m*)	*kPa*									
0	(0)	0	2.99	5.98	8.97	11.96	14.94	17.93	20.92	23.91	26.90
10	(3.05)	29.9	32.88	35.87	38.86	41.85	44.83	47.82	50.81	53.80	56.79
20	(6.10)	59.8	62.77	65.76	68.75	71.74	74.72	77.71	80.70	83.69	86.68
30	(9.14)	89.7	92.66	95.65	98.64	101.63	104.61	107.60	110.59	113.58	116.57
40	(12.19)	119.6	122.55	125.54	128.53	131.52	134.50	137.49	140.48	143.47	146.46
50	(15.24)	149.4	152.44	155.43	158.42	161.40	164.39	167.38	170.37	173.36	176.35
60	(18.29)	179.3	182.33	185.32	188.30	191.29	194.28	197.27	200.26	203.25	206.24
70	(21.34)	209.2	212.22	215.21	218.20	221.18	224.17	227.16	230.15	233.14	236.13
80	(24.38)	239.1	242.11	245.10	248.09	251.07	254.06	257.05	260.04	263.03	266.02
90	(27.43)	269.0	272.00	274.99	277.98	280.96	283.95	286.94	289.93	292.29	295.91

Source: Barnard, R.E., Design Standards for Steel Water Pipe. Jour. AWWA, 40:1:24 *(Jan. 1948).*

Table 3-9 Pressure equivalents

Mercury		Water				Mercury		Water			
in.	(*mm*)	*in.*	(*mm*)	*psi*	*kPa*	*in.*	(*mm*)	*in.*	(*mm*)	*psi*	*kPa*
1	(25.4)	13.6	(345.3)	0.49	3.39	13	(330.2)	176.7	(4,489.2)	6.38	44.03
2	(50.8)	27.2	(690.6)	0.98	6.77	14	(355.6)	190.3	(4,834.5)	6.88	47.41
3	(76.2)	40.8	(1,036.0)	1.47	10.16	15	(381.0)	203.9	(5,179.9)	7.37	50.80
4	(101.6)	54.4	(1,381.3)	1.96	13.55	16	(406.4)	217.5	(5,525.2)	7.86	54.18
5	(127.0)	68.0	(1,726.6)	2.46	16.93	17	(431.8)	231.1	(5,870.5)	8.35	57.57
6	(152.4)	81.6	(2,071.9)	2.95	20.32	18	(457.2)	244.7	(6,215.8)	8.84	60.95
7	(177.8)	95.2	(2,417.3)	3.44	23.70	20	(508.0)	271.9	(6,906.5)	9.82	67.73
8	(203.2)	108.8	(2,762.6)	3.93	27.09	22	(558.8)	299.1	(7,597.1)	10.8	74.50
9	(228.6)	122.4	(3,107.9)	4.42	30.48	24	(609.6)	326.3	(8,287.8)	11.7	81.27
10	(254.0)	136.0	(3,453.2)	4.91	33.86	26	(660.4)	353.5	(8,978.4)	12.7	88.05
11	(279.4)	149.5	(3,798.6)	5.40	37.25	28	(711.2)	380.7	(9,669.1)	13.7	94.82
12	(304.8)	163.1	(4,143.9)	5.89	40.64	30	(762.0)	407.9	(10,359.7)	14.7	101.59

Source: Barnard, R.E., Design Standards for Steel Water Pipe. Jour. AWWA, 40:1:24 *(Jan. 1948).*

Flow Through Fittings—Equivalent-Length Method

Experiments have shown that the head loss in bends, fittings, and valves is related to flow velocity and pipe diameter similar to that in straight pipe.

Consequently, it is possible to determine the length of a theoretical piece of straight pipe in which the head loss caused by friction would be the same as for a specific fitting. This method of equivalent lengths is recognized by several authorities (Crocker 1945; *Flow of Fluids Through Valves, Fittings, and Pipe* 1942). By developing the total equivalent length (piping plus bends, fittings, valves, etc.), the total head loss in a piping system can be determined.

The classical equation developed by Darcy-Weisbach for energy loss of flow in a pipeline is

$$H_L = f\left(\frac{L}{D}\right)\left(\frac{V^2}{2g}\right) \tag{3-10}$$

Equivalent metric equation:

$$H_L = 0.3048f\left(\frac{L}{D}\right)\left(\frac{V^2}{2g}\right) \tag{3-10M}$$

In the equation, H_L is the head loss caused by friction in the length of pipe L of inside diameter D for average velocity V. The friction factor f is a function of pipe roughness, velocity, pipe diameter, and fluid viscosity. Values for f have been developed by Moody (1944) and others (Figure 3-4). With a known f and L/D, the Darcy-Weisbach formula can be expressed:

$$H_L = K\left(\frac{V^2}{2g}\right) \tag{3-11}$$

Equivalent metric equation:

$$H_L = 0.3048K\left(\frac{V^2}{2g}\right) \tag{3-11M}$$

In this equation, K is the resistance coefficient. Figure 3-5 shows values for K based on a summary of experimental data.

Examples to determine head loss H_L for fittings and valves and equivalent pipe lengths using Figure 3-4 are as follows:

Assume:

D = 6 in. (152.4 mm) C = 100

Q = 450 gpm (1,704 L/m) V = 5.12 fps (1.56 m/sec)

Calculations:

Velocity head: $\frac{V^2}{2g} = \frac{5.12^2}{64.4} = 0.41$ ft $\frac{1.56^2}{2 \times 9.806 \text{ m/sec}^2} = 0.124$ m

1. 6-in. gate valve, fully open:

$K = 0.2$ $H_L = 0.2 \times 0.41$ ft $= 0.08$ ft 0.2×0.124 m $= 0.025$ m

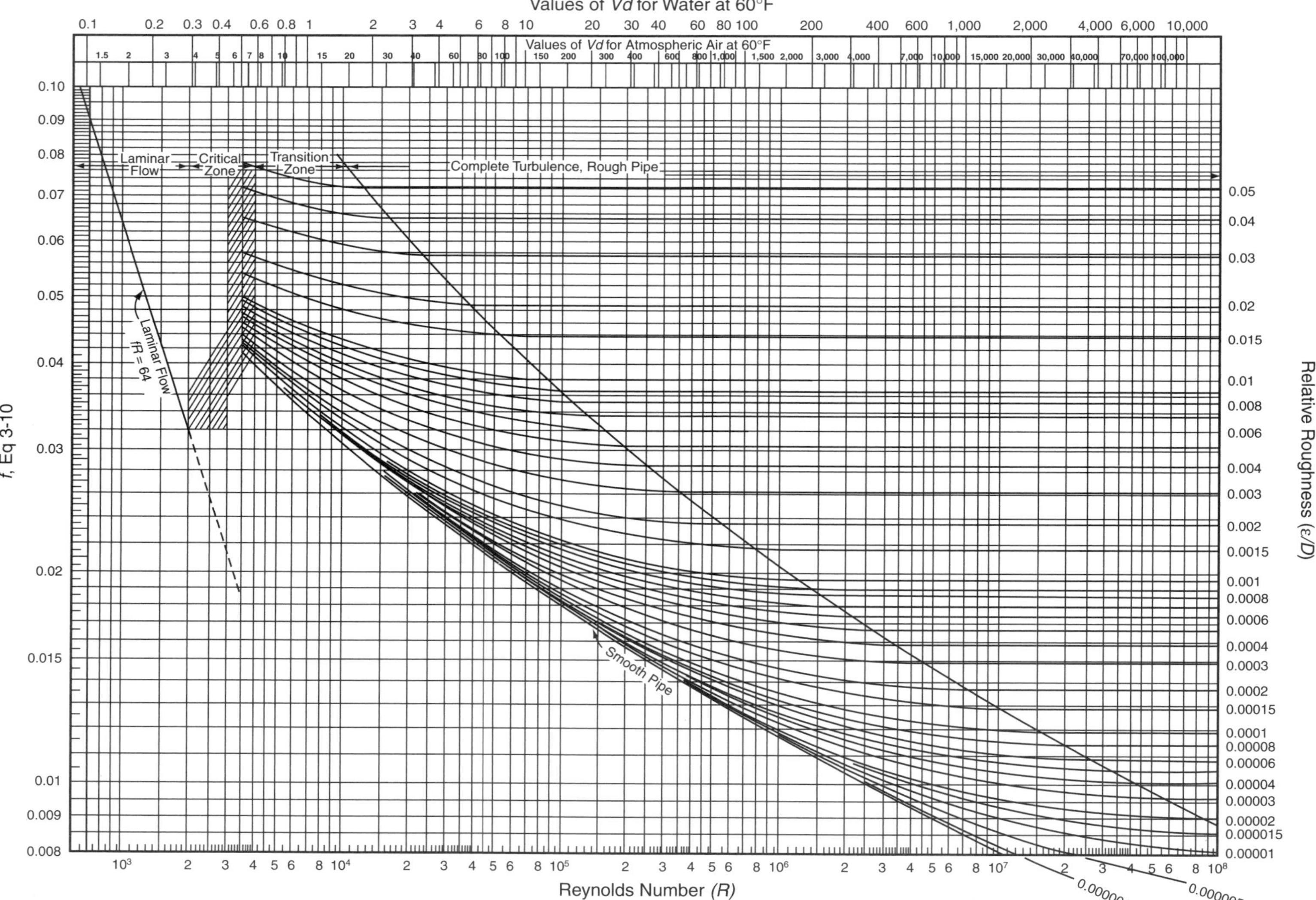

Source: Pipe Friction Manual. *Hydraulic Institute New York (1954).*

Figure 3-4 Moody diagram for friction in pipe

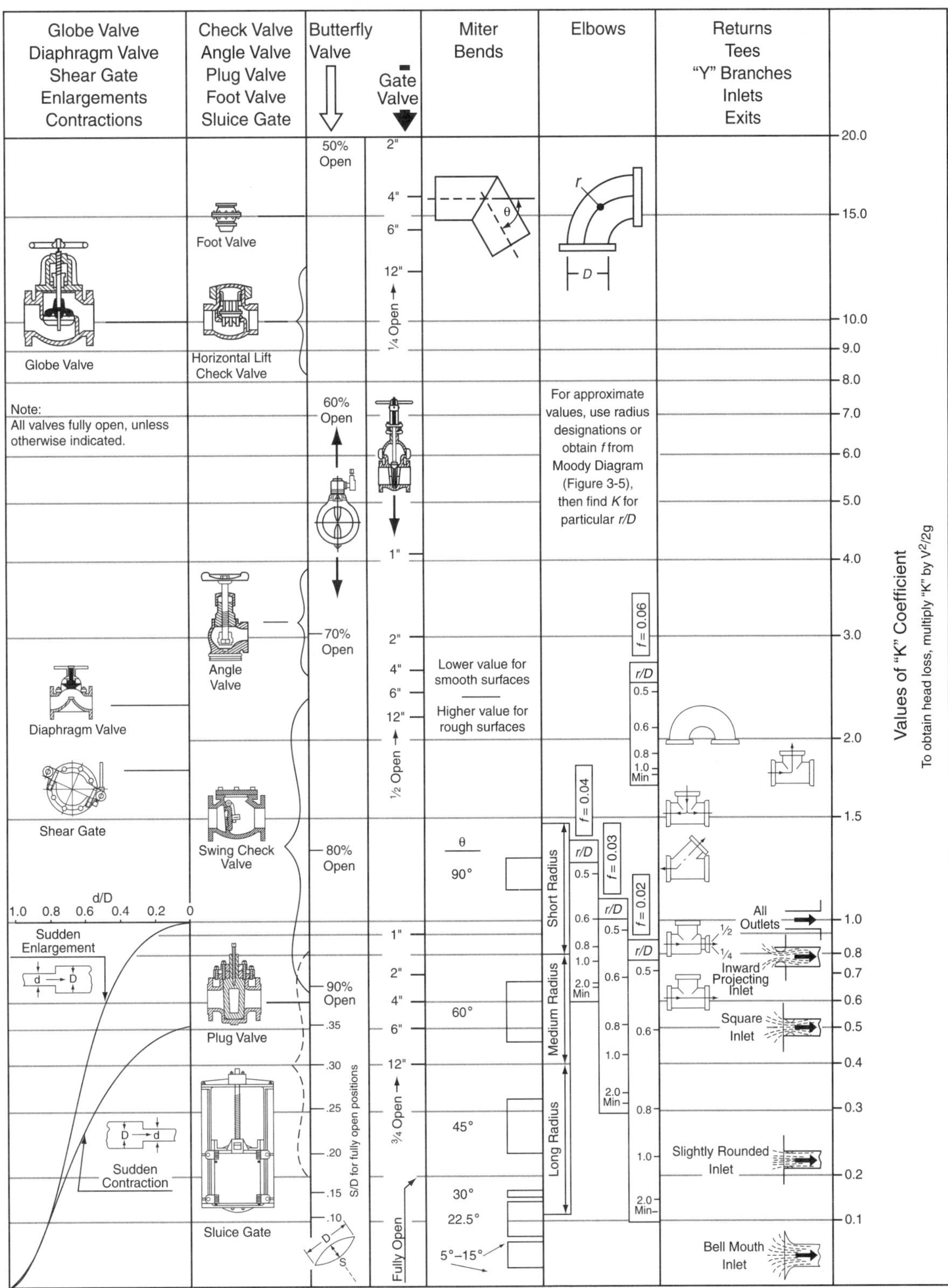

Source: John F. Lenard, President, Lenard Engineering, Inc.

Figure 3-5 Resistance coefficients of valves and fittings for fluid flows

2. 6-in. swing check valve, fully open:

   $K = 1.4$ $H_L = 1.4 \times 0.41$ ft $= 0.57$ ft 1.4×0.124 m $= 0.174$ m

3. Sudden enlargement from 6 in. to 8 in.:

   $d/D = 0.76$, K = 0.18 $H_L = 0.18 \times 0.41 = 0.07$ ft 0.18×0.124 m $= 0.022$ m

4. 6-in. elbow:

   $K = 0.6$ $H_L = 0.6 \times 0.41 = 0.25$ ft (0.6×0.124 m $= 0.074$ m)

   Total head loss 0.97 ft (0.295 m)

Using the Hazen-Williams formula, the equivalent pipe length for 6 in. pipe, $C = 100$ with an $H_L = 0.97$ ft (0.295 m), equals 35.3 ft (10.8 m).

ECONOMICAL DIAMETER OF PIPE

Hydraulic formulas give the relation between flow rate and head loss in pipes of various diameters and interior surface conditions. When a limited amount of head loss is available, the smallest diameter pipe that will deliver the required flow is selected. This results in the least construction cost. Where head is provided by pumping, a part of the cost is for energy to provide head to overcome friction. The cost for energy decreases as pipe diameter increases and friction losses decrease; however, the cost for the pipe increases. The objective is usually to minimize total cost (initial cost, operation, and maintenance) by selecting the pipe diameter that results in least life-cycle cost. Energy costs may prove to be the most significant cost. However, when making an assessment of future energy costs, present costs should be used when comparing all other costs.

Aqueducts

Economic studies of large aqueducts are frequently complicated by the desire to combine different means of carrying water—for example, through open conduits, pipe, and tunnels—in the same system. Hinds (1937a, b) demonstrated the use of graphical means in making such studies in the design of the Colorado River Aqueduct. The method of finding economical slopes described by Hinds had been used previously in the design of the Owens River Aqueduct of Los Angeles (Babbitt and Doland 1927, 1955) and the Catskill Aqueduct of New York (White 1913).

Penstocks

An economic study to determine penstock size generally requires that the annual penstock cost plus the value of power lost in friction be minimal. The annual penstock cost includes amortization of all related construction costs, operation and maintenance costs, and replacement reserve. A precise analytical evaluation, taking all factors into account, may be neither justified nor practical, because all variables in the problem are subject to varying degrees of uncertainty.

Methods used to determine the economic diameter for steel penstocks and pump lines can be found in the *ASCE Manual and Reports on Engineering Practice No. 79, Steel Penstocks* (1993), or the *Steel Plate Fabricators Engineering Data, Volume 4, Buried Steel Penstocks* (1992).

DISTRIBUTION SYSTEMS

Methods of determining economical sizes of pipe for distribution systems have been published (Lischer 1948).

AIR ENTRAINMENT AND RELEASE

Air entrained in flowing water tends to form bubbles at or near the high points in a pipeline. If not removed, such bubbles become serious obstacles to flow. The removal of air will prevent the formation of a hydraulic jump in a pipe at the end of these bubbles. Possible air entrainment and its removal must be considered and remedies applied if needed. The ability of the hydraulic jump to entrain the air and to carry it away by the flowing water has been investigated. Quantitative data have been published (Hall, Kalinske, and Robertson 1943) relating characteristics of the jump to the rate of air removal. Removal of air through air valves is discussed in chapter 9.

GOOD PRACTICE

Waterworks engineers should use the hydraulic-friction formulas with which they are most familiar and with which they have had experience. Three of the common conventional formulas have been discussed in this chapter. In any particular case, the results calculated using the different conventional formulas can be compared. Engineers should, however, recognize the increasing use of the rational or universal formulas, become familiar with them, and make check calculations using them. A practical coefficient value for the formulas should be conservatively selected.

The results of flow tests will generally be more useful if they are related to the rational concept of fluid flow. This entails giving more attention to relative surface roughness, water temperature, Reynolds numbers, and an analysis of test results aimed at incorporating them into the fluid-mechanics approach to flow determination.

REFERENCES

American Society of Civil Engineers. 1993. *ASCE Manual and Reports on Engineering Practice No. 79, Steel Penstocks.*

Babbitt, H.E., and J.J. Doland. 1927, 1955. *Water Supply Engineering*. New York: McGraw-Hill Book Co.

Barnard, R.E. 1948. Design Standards for Steel Water Pipe. *Jour. AWWA*, 40(1):24.

Buried Steel Penstocks, Steel Plate Engineering Data. 1992. Vol. 4. New York: AISI.

Crocker, S., ed. 1945. *Piping Handbook*, 4th ed. New York: McGraw-Hill Book Co.

Flow of Fluids Through Valves, Fittings, and Pipe. 1942. Tech. Paper 409. Chicago: Crane Co.

Hall, L.S., A.A. Kalinske, and J.M. Robertson. 1943. Entrainment of Air in Flowing Water—A Symposium. *Trans. ASCE*, 108:1393.

Hinds, J. 1937a. Economic Sizes of Pressure Conduits. *Engineering News Record*, 118:443.

Hinds, J. 1937b. Economic Water Conduit Size. *Engineering News Record*, 118:113.

Lischer, V.C. 1948. Determination of Economical Pipe Diameters in Distribution Systems. *Jour. AWWA*, 40(8):849.

Moody, L.F. 1944. *Friction Factors for Pipe Flow*. New York: American Society of Mechanical Engineers.

Voetsch, C., and M.H. Fresen. 1938. Economic Diameter of Steel Penstocks. *Trans. ASCE*, 103:89.

White, L. 1913. *Catskill Water Supply of New York, N.Y.* New York: John Wiley & Sons.

The following references are not cited in the text.

Aldrich, E.H. 1938. Solution of Transmission Problems of a Water System. *Trans. ASCE*, 103:1759.

American Society of Civil Engineers. *Pipeline Design for Water and Wastewater.* 1975. Report of the Task Committee on Engineering Practice in the Design of Pipelines. New York: American Society of Civil Engineers.

Bradley, J.N., and L.R. Thompson. 1951. *Friction Factors of Large Conduits Flowing Full*. Engineering Monograph 7. Denver, Colo.: US Bureau of Reclamation.

Capen, C.H. 1950. Trends in Coefficients of Large Pressure Pipes. *Jour. AWWA*, 42(9):860.

Cross, H. 1936. *Analysis of Flow in Networks of Conduits of Conductors*. Bull. 286. Urbana, Ill.: Univ. of Illinois, Engrg. Expt. Stn.

Davis, C.V., ed. 1952. *Handbook of Applied Hydraulics*, 2nd ed. New York: McGraw-Hill Book Co.

Farnsworth, G., Jr., and A. Rossano Jr. 1941. Application of the Hardy Cross Method to Distribution System Problems. *Jour. AWWA*, 33(2):224.

Hinds, J. 1946. Comparison of Formulas for Pipe Flow. *Jour. AWWA*, 38(11):1226.

King, H.W. 1954. *Handbook of Hydraulics*, 4th ed. New York: McGraw-Hill Book Co.

Pigott, R.J.S. 1950. Pressure Losses in Tubing, Pipe, and Fittings. *Trans. ASME*, 72:679.

Pipe Friction Manual. 1954. New York: Hydraulic Institute.

Chapter 4

Determination of Pipe Wall Thickness

The wall thickness of steel pipe is affected by many factors that will be discussed in this and succeeding chapters, including the following:

1. Internal pressure
   a. Maximum design pressure (chapter 4)
   b. Surge or water-hammer pressure (chapter 5)
2. External pressure
   a. Trench loading pressure (chapter 6)
   b. Earth-fill pressure (chapter 6)
   c. Uniform collapse pressure, atmospheric or hydraulic (chapter 4)
   d. Vacuum underground (chapter 6)
3. Special physical loading
   a. Pipe on saddle supports (chapter 7)
   b. Pipe on ring-girder supports (chapter 7)
   c. Joint design for additional stresses (chapter 13)
4. Practical requirements (chapter 4)
5. Practical design considerations for steel stresses with various coatings and linings (chapter 11)

The thickness selected should satisfy the most severe requirement.

INTERNAL PRESSURE

When designing for internal pressure, the minimum thickness of a cylinder should be selected to limit the circumferential tension stress to a certain level. This stress is termed *hoop stress.* The internal pressure used in design should be that to which the pipe may be subjected during its lifetime.

In a transmission pipeline, the pressure is measured by the distance between the pipe centerline and the hydraulic grade line. If there are in-line valves, the maximum pressure on the pipe between them with the valves closed will be measured by the

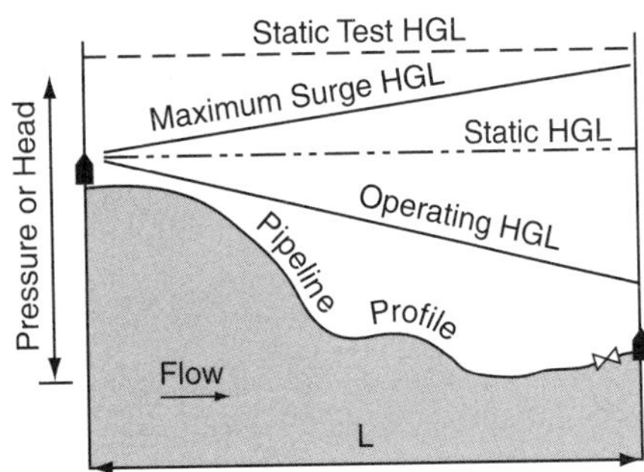

Figure 4-1 Relation of various heads or pressures for selection of design pressure (gravity flow)

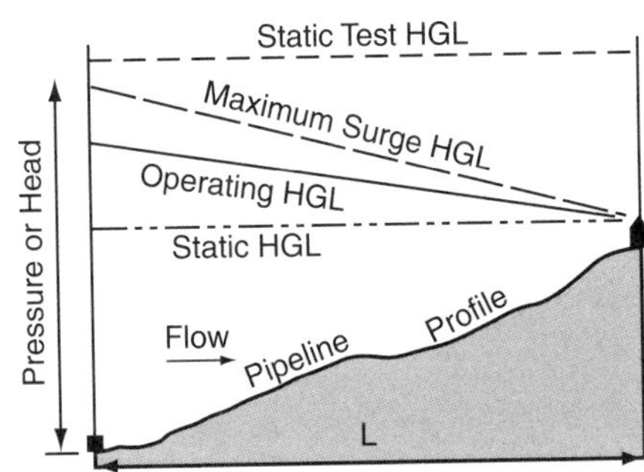

Figure 4-2 Relation of various heads or pressures for selection of design pressure (pumped flow)

distance between the pipe centerline and the elevation of the static level. Surge or water-hammer pressures must also be considered. These are discussed in chapter 5.

In a pump-discharge pipeline, the internal pressure is measured by the distance between the pipe and the hydraulic grade line created by the pumping operation. Pressure at the outlet and the loss caused by friction should be considered. If it is possible to impose a pressure equal to the shutoff head of the pumps, the pressure is measured between the pipe and the shutoff grade line. Figures 4-1 and 4-2 show typical pipeline and hydraulic grade profiles for gravity and pumped flow.

With pressure determined, the wall thickness is found using the Barlow formula, Eq 4-1:

$$t = \frac{pd}{2s} \tag{4-1}$$

Where:

t = minimum pipe wall thickness for the specified internal design pressure, in. (mm)

p = internal design pressure, psi (kPa)

d = outside diameter of pipe steel cylinder (not including coatings), in. (mm)

s = allowable design stress, psi (kPa)

ALLOWABLE TENSION STRESS IN STEEL

Tension Stress and Yield Strength

Modern steel technology has allowed increases in the allowable working stress for steel. This working stress is determined by the relation to the steel's yield strength rather than by its ultimate strength. A design stress equal to 50 percent of the specified minimum yield strength is often accepted for steel water pipe. Using the stated methods of stress analysis and proper quality control measures, these allowable design stresses are considered conservative for the usual water-transmission pipelines. Table 4-1 illustrates grades of steel used as a basis for working pressure and the design stress as compared to minimum yield point and minimum ultimate tensile strength for common grades of steel as referenced in AWWA C200, *Standard for Steel Water Pipe 6 In. (150 mm) and Larger* (latest edition).

Table A-1 gives the designer working pressures corresponding to 50 percent of the specified minimum-yield strength for several types of steel commonly used in waterworks pipelines. The designer is cautioned that the diameters and wall thicknesses

Table 4-1 Grades of steel used in AWWA C200 as basis for working pressures in Table A-1

Specifications for Fabricated Pipe		Design Stress 50%		Minimum Yield Point		Minimum Ultimate Tensile Strength	
		psi	*(MPa)*	*psi*	*(MPa)*	*psi*	*(MPa)*
		Steel Plate					
ASTM A36/A36M		18,000	(124.1)	36,000	(248.2)	58,000	(399.9)
ASTM A283/A283M	GR C	15,000	(103.4)	30,000	(206.8)	55,000	(379.2)
	GR D	16,500	(113.8)	33,000	(227.5)	60,000	(413.7)
ASTM A572/A572M	GR 42	21,000	(144.8)	42,000	(289.6)	60,000	(413.7)
	GR 50	25,000	(173.3)	50,000	(344.7)	65,000	(448.2)
		Steel Sheet—Coil or Flat					
ASTM A570/570M	GR 30	15,000	(103.4)	30,000	(206.8)	49,000	(337.8)
	GR 33	16,500	(113.8)	33,000	(227.5)	52,000	(358.5)
	GR 36	18,000	(124.1)	36,000	(248.2)	53,000	(365.4)
	GR 40	20,000	(137.9)	40,000	(275.8)	55,000	(379.2)
	GR 45	22,500	(155.1)	45,000	(310.3)	60,000	(413.7)
	GR 50	25,000	(172.4)	50,000	(344.7)	65,000	(448.2)
ASTM A607/A607M	GR 45	22,500	(155.1)	45,000	(310.2)	60,000	(413.7)
	GR 50	25,000	(172.4)	50,000	(344.7)	65,000	(448.2)
ASTM A907/A907M	GR 30	15,000	(103.4)	30,000	(206.8)	49,000	(337.8)
	GR 33	16,500	(113.8)	33,000	(227.5)	52,000	(358.5)
	GR 36	18,000	(124.1)	36,000	(248.2)	53,000	(365.4)
	GR 40	20,000	(137.9)	40,000	(275.8)	55,000	(379.2)
ASTM A935/A935M	GR 45	22,500	(155.1)	45,000	(310.2)	60,000	(413.7)
	GR 50	25,000	(172.4)	50,000	(344.7)	65,000	(448.2)
ASTM A936/A936M	GR 50	25,000	(172.4)	50,000	(344.7)	60,000	(413.7)
ASTM A53, A135 and A139	GR A	15,000	(103.4)	30,000	(206.8)	48,000	(330.9)
	GR B	17,500	(120.6)	35,000	(241.3)	60,000	(413.7)
ASTM A139	GR C	21,000	(144.8)	42,000	(289.6)	60,000	(413.7)
	GR D	23,000	(158.6)	46,000	(317.2)	60,000	(413.7)
	GR E	26,000	(179.3)	52,000	(385.5)	66,000	(455.1)

NOTES:

(1) ASTM standard specifications are subject to change. The latest edition should be used.

(2) Certain linings and coatings require that the maximum circumferential stress in the steel be limited to recommended maximum values for the prevention of detrimental damage. A commonly used criteria for cement–mortar coatings is to limit the maximum circumferential stress in the steel cylinder to 50 percent of the yield strength, with a limiting value of 18,000 psi (124.1 MPa) for design pressure.

(3) For pipelines over ½ in. (13 mm) in wall thickness, consider specifying a Charpy V-Notch (CVN) value of 25 lbf-ft (33.9 N-m) average at 30°F (–1°C), transverse specimen, heat-lot testing, for the steel. CVN specimens shall be prepared in accordance with ASTM A370. Heat lot testing is two coils per heat, outer wrap only. Filler material should meet the same toughness properties as the base metal.

listed in the table are for reference only and do not represent engineering or manufacturing limits. Modern steel-mill capabilities permit the manufacture of almost any diameter and wall thickness of pipe; in practice, however, most pipe manufacturers fabricate pipe to standard diameters. Pipe with thick linings, such as the cement–mortar lining specified in AWWA C205, *Standard for Cement–Mortar Protective Lining and Coating for Steel Water Pipe—4 in. (100 mm) and Larger—Shop Applied* (latest edition), and AWWA C602, *Standard for Cement–Mortar Lining of Water Pipelines—4 in. (100 mm) and Larger—In Place* (latest edition), is usually fabricated to the individual manufacturer's standard diameters to accommodate the required lining thicknesses. It is recommended that the pipe manufacturers be consulted before final selection of diameter and wall thicknesses.

Pressure Limits

High quality in the manufacture of both the pipe and the steel used in its manufacture is required by AWWA standards. Therefore, hoop stress may rise, within limits, above 50 percent of yield for transient loads. For steel pipe produced to meet AWWA standards, the increased hoop stress should be limited to 75 percent of the specified yield strength but should not exceed the mill test pressure.

CORROSION ALLOWANCE

At one time, it was a general practice to add a fixed, rule-of-thumb thickness to the pipe wall as a corrosion allowance. This was not an applicable solution in the waterworks field, where standards for coating and lining materials and procedures exist. The design should be made for the required wall-thickness pipe as determined by the loads imposed, then linings, coatings, and cathodic protection selected to provide the necessary corrosion protection.

EXTERNAL FLUID PRESSURE—UNIFORM AND RADIAL

The proper wall thickness must be selected to resist external loading imposed on the pipe. Such loading may take the form of outside pressure, either atmospheric or hydrostatic, both of which are uniform and act radially as collapsing forces. Buried pipe must be designed to resist earth pressure in trench or fill condition. These considerations are discussed in chapter 6.

Atmosphere or Fluid Environments

A general theory of collapse-resistance of steel pipe to uniform, radially acting forces has been developed (Timoshenko 1940). Any unreinforced tube longer than the critical length can be considered a tube of infinite length, as its collapsing pressure is independent of further increase in length. The following formula applies to such tubes:

$$P_c = \frac{2E_s}{1 - \nu_s^2}\left(\frac{t_s}{d_n}\right)^3 + \frac{2E_L}{1 - \nu_L^2}\left(\frac{t_L}{d_n}\right)^3 + \frac{2E_c}{1 - \nu_c^2}\left(\frac{t_c}{d_n}\right)^3 \qquad (4\text{-}2)$$

Where:

d_n = diameter to neutral axis of the shell (in.) (for thin pipes, the difference between inside diameter, outside diameter, and neutral-axis diameter is negligible)

t_s = steel thickness, in. (mm)

t_L = cement lining thickness, in. (mm)

t_c = cement coating thickness, in. (mm)
E_s = modulus of elasticity for steel, 30,000,000 psi (206.84 GPa)
E_c & E_L = modulus of elasticity for mortar coating and lining, 4,000,000 psi (27.579 GPa)
P_c = collapsing pressure, psi (Pa)
ν_s = Poisson's ratio for steel (usually taken as 0.30)
ν_c & ν_L = Poisson's ratio for mortar coating and lining (taken as 0.25)

Substituting the above values of E and υ for bare pipe:

$$P_c = 66,666,667 \left(\frac{t}{d_n}\right)^3 \tag{4-3}$$

Equivalent metric equation:

$$P_c = 459.650^5 \times 10^9 \left(\frac{t}{d_n}\right)^3 \tag{4-3M}$$

NOTE: Eq 4-2 and 4-3 do not include any theoretical safety factors, as vacuum is considered an absolute value. Eq 4-4 includes an allowance of approximately 30 percent to account for various factors, such as pipe deflection and wall thickness.

Applied Calculations

Circular cylindrical shells under external pressure may fail either by buckling or by yielding. Relatively thin-walled shells fail through instability or buckling under stresses that, on the average, are below the yield strength of the material normally encountered in the waterworks field. A number of theoretical and empirical formulas have been promulgated to provide for the effect of instability caused by collapsing. They include the formulas of Timoshenko (1940), Love, Roark (1965), Stewart, and Bryan.

Stewart developed two empirical equations for the collapsing pressures of steel pipes. The Stewart formula, which automatically accounts for wall thickness variations, out-of-roundness, and other manufacturing tolerances, follows:

For buckling failure, where t/d_n is 0.023 or less and P_c is 581 psi (4,006 kPa) or less:

$$P_c = 50,200,000 \left(\frac{t}{d_n}\right)^3 \tag{4-4}$$

Equivalent metric equation:

$$P_c = 346,116,801 \left(\frac{t}{d_n}\right)^3 \tag{4-4M}$$

This formula is considered more conservative than the previous formulas.

Eq 4-4 is predicated on the pipe being commercially round, made of steel with a minimum yield of at least 27,000 psi (186.2 MPa), and having a length of six diameters or more between reinforcing elements.

MINIMUM WALL THICKNESS

Minimum plate or sheet thicknesses for handling are often based on three formulas. They are as follows (Parmakian 1982):

$$t = \frac{D}{288} \text{ (pipe sizes up to 54 in. (1,350 mm) ID)} \quad (4\text{-}5)$$

$$t = \frac{D + 20}{400} \text{ (pipe sizes greater than 54 in. (1,350 mm) ID)} \quad (4\text{-}6)$$

$$t = \frac{D}{240} \text{ (for mortar-lined and flexible coated steel pipe)} \quad (4\text{-}7)$$

In no case shall the shell thickness be less than 14 gauge [0.0747 in. (1.90 mm)].

For pipe diameters smaller than 54 in. (1,350 mm), the use of Eq 4-5 (Pacific Gas and Electric formula) will result in a thinner pipe wall than the use of Eq 4-6 (US Bureau of Reclamation formula). For 54-in. (1,350 mm) and larger pipe, the opposite is true.

Additional consideration for handling should be given for certain linings and coatings.

GOOD PRACTICE

Internal pressure, external pressure, special physical loading, type of lining and coating, and other practical requirements determine wall thickness. It is good practice regarding internal pressure to use a working tensile stress of 50 percent of the yield-point stress under the influence of maximum design pressure. The stress of transitory surge pressures may be taken at 75 percent of the yield-point stress. The designer should, however, consider the effect of water hammer, pump shutoff head, or surge pressures in design. It is more positive and economical to select a proven coating or lining for protection against corrosion hazards than to add sacrificial wall thickness.

REFERENCES

American Society of Mechanical Engineers. *Rules for Construction of Unfired Pressure Vessels.* Sec. VIII, ASME Boiler and Pressure Vessel Code. New York: American Society of Mechanical Engineers.

American Water Works Association. ANSI/AWWA Standard C205, *Cement–Mortar Protective Lining and Coating for Steel Water Pipe—4 In. and Larger—Shop Applied.* Latest edition. Denver, Colo.: American Water Works Association.

American Water Works Association. ANSI/AWWA Standard C602, *Cement–Mortar Lining of Water Pipelines In Place—4 In. (100 mm) and Larger.* Latest edition. Denver, Colo.: American Water Works Association.

American Water Works Association. ANSI/AWWA Standard C200, *Steel Water Pipe–6 In. (150 mm) and Larger.* Latest edition. Denver, Colo.: American Water Works Association.

Parmakian, J. 1982. *Minimum Thickness for Handling Pipes, Water Power and Dam Construction.*

Roark, R.J. 1965. *Formulas for Stress and Strain*, 4th ed. New York: McGraw-Hill Book Co.

Timoshenko, S. 1940. *Strength of Materials.* Part II.D. New York: Van Nostrand Co.

US Bureau of Reclamation. *Welded Steel Penstocks.* Engineering Monograph 3. Denver, Colo.: US Bureau of Reclamation.

Chapter 5

Water Hammer and Pressure Surge

Water hammer is the result of a change in flow velocity in a closed conduit causing elastic waves to travel upstream and downstream from the point of origin. The elastic waves, in turn, cause increases or decreases in pressure as they travel along the line, and these pressure changes are variously referred to as water hammer, surge, or transient pressure.

The phenomenon of water hammer is extremely complex, and no attempt will be made to cover the subject in-depth in this manual. Only the fundamentals of elastic-wave theory and specific data pertaining to the properties of steel pipe will be discussed. For a more detailed understanding of water hammer, the references listed at the end of this chapter should be consulted.

BASIC RELATIONSHIPS

The following fundamental relationships in surge-wave theory determine the magnitude of the pressure rise and its distribution along a conduit. The pressure rise for instantaneous closure is directly proportional to the fluid velocity at cutoff and to the magnitude of the surge wave velocity. It is independent of the length of the conduit. Its value is

$$h = \frac{aV}{g} \tag{5-1}$$

or

$$P = \frac{aWV}{144g} = \left(\frac{a}{g}\right)\left(\frac{\text{sp gr}}{2.3}\right)V \tag{5-2}$$

Equivalent metric equation:

$$P = 9{,}835\left(\frac{a}{g}\right)(\text{sp gr})V \qquad (5\text{-}2M)$$

Where:

$$a = \frac{12}{\sqrt{\frac{W}{g}\left(\frac{1}{k} + \frac{d}{Et}\right)}} \qquad (5\text{-}3)$$

Equivalent metric equation:

$$a = \frac{0.319}{\sqrt{\frac{W}{g}\left(\frac{1}{k} + \frac{d}{Et}\right)}} \qquad (5\text{-}3M)$$

In the above equations:

a = wave velocity, fps (m/sec)
h = pressure rise above normal, ft (m) of water
p = pressure rise above normal, psi (Pa)
V = velocity of flow, ft/sec (m/sec)
W = specific weight of fluid, lb/ft^3 (kg/m^3)
sp gr = specific gravity of fluid (water = 1.0)
k = bulk modulus of compressibility of liquid, psi (Pa)
E = Young's modulus of elasticity for pipe wall material, psi (Pa)
d = inside diameter of conduit, in. (mm)
t = thickness of conduit wall, in. (mm)
g = acceleration due to gravity, 32.2 ft/sec^2 (9.81 m/sec^2)
L = length of conduit, ft (m)
$\frac{2L}{a}$ = critical time of conduit, sec
T = closing time, sec

For steel pipe, using k = 300,000 psi (2.07 GPa) and E = 30,000,000 psi (206.85 GPa), Eq 5-3 reduces to

$$a = \frac{4{,}660}{\sqrt{1 + \frac{1}{100}\left(\frac{d}{t}\right)}} \qquad (5\text{-}4)$$

Equivalent metric equation:

$$a = \frac{1{,}420.4}{\sqrt{1 + \frac{1}{100}\left(\frac{d}{t}\right)}} \qquad (5\text{-}4M)$$

Figure 5-1 gives values of pressure wave velocity for various pipe materials with d/t ratios up to 90. For steel pipe, higher ratios are frequently encountered in large sizes, and Table 5-1 gives computed values up to d/t = 400.

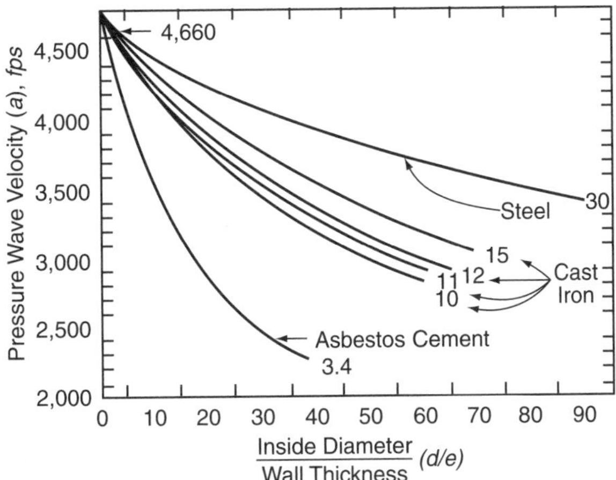

NOTE: The number at the end of each curve represents the modulus of elasticity (E) in 1,000,000-psi units for various pipe materials.

Figure 5-1 Surge wave velocity chart for water

Table 5-1 Velocity of pressure wave for steel pipe

Diameter/Thickness $\left(\frac{d}{t}\right)$	Wave Velocity a	
	ft/sec	(*m/sec*)
100	3,300	1,006
120	3,140	957
140	3,010	917
160	2,890	881
180	2,780	847
200	2,690	820
250	2,490	759
300	2,330	710
350	2,200	671
400	2,080	634

NOTE: Wave velocities shown are for bare steel pipe.

When the flow rate is changed in a time greater than zero but less than or equal to $2L/a$ seconds, the magnitude of the pressure rise is the same as with instantaneous closure, but the duration of the maximum value decreases as the time of closure approaches $2L/a$ seconds. Under these conditions, the pressure distribution along the pipeline varies as the time of closure varies. The pressure decreases uniformly along the line if closure is in $2L/a$ seconds. The maximum pressure at the control valve exists along the full length of the line with instantaneous closure and for slower rates, the pressure travels up the pipe a distance equal to L-(Ta/2), then decreases uniformly. The surge pressure distribution along the conduit is independent of the profile or ground contour of the line as long as the total pressure remains above the vapor pressure of the fluid.

For valve closing times greater than $2L/a$ seconds, the maximum pressure rise will be the maximum rate of change in flow with respect to time, dV/dT. Nonlinear closure rates of a valve can be investigated and the proper valve closing time determined to hold the maximum pressure rise to a desired limiting value. The effect of pumps and quick-closing check valves or control valves can be investigated using a graphical method or numerical method using a computer.

The profile of the conduit leading away from a pumping station may have a major influence on the surge conditions. When high points occur along the line, the surge hydraulic-grade elevation may fall below the pipe profile, causing negative pressures, perhaps as low as the vapor pressure of the fluid. If this occurs, the liquid column may be separated by a zone of vapor for a short time. Parting and rejoining of the liquid column can produce extremely high pressures and may cause failure of the conduit (Richards 1956).

The effect of friction can be accounted for in any surge problem. When friction losses are less than 5 percent of the normal static or working pressure, they can usually be neglected.

Accurate results of a surge analysis depend on knowing the various hydraulic and physical characteristics of the system. The velocity of the pressure wave is a fundamental factor in any surge study, as the surge pressures are directly proportional to its value. This velocity depends on the pipe diameter, wall thickness, material of the pipe walls, and the density and compressibility of the fluid in the pipe.

Determining the physical characteristics of the pipe material is straightforward. Young's modulus for steel lines can be taken at 30,000,000 psi (206.85 GPa), because it averages between 29,000,000 (199.95 GPa) and 31,000,000 psi (213.74 GPa), depending on the steel used. If the ratio of diameter to thickness is known, it is necessary to know only the density and the compressibility of the liquid within the pipe to determine the surge wave velocity a.

Within the range of ordinary operating temperatures for water, 32–100°F (0–38°C), and for pressures in the range of 0–1,000 psi (0–6.895 MPa), the specific gravity can be assumed to be 1.00. In the same range, the modulus of compressibility, or bulk modulus, has been measured and verified by field tests to be approximately 300,000 psi (2.07 GPa) with a variation of ±3 percent (Kerr, Kessler, and Gamet 1950).

CHECKLIST FOR PUMPING MAINS

A few factors can be checked to indicate whether surges of serious proportions will occur in any given system, once the physical, hydraulic, and operating characteristics are established. For most transmission mains supplied by motor-driven centrifugal pumps, the following 12 questions will indicate the seriousness of the surge problem (Kerr 1949, Water Hammer Control 1951):

1. Are there any high spots on the profile of the transmission main where the occurrence of a vacuum can cause a parting of the water column when a pump is cut off?
2. Is the length of the transmission main less than 20 times the head on the pumps (both values expressed in feet)?
3. Is the maximum velocity of flow in the transmission main in excess of 4.0 ft/sec (1.22 m/sec)?
4. Is the safety factor of the pipe less than 3.5 (related to ultimate strength) for normal operating pressures?
5. What is the natural decreasing rate of the water column if the pump is cut off? Will the column come to rest and reverse its direction of flow in less than the critical surge-wave time for the transmission main?

6. Will the check valve close in less than the critical time for the transmission main?
7. Are there any quick-closing automatic valves set to open or close in less than 5 sec?
8. Would the pump or its driving motor be damaged if allowed to run backward, reaching full speed?
9. Will the pump be tripped off before the discharge valve is fully closed?
10. Will the pump be started with the discharge gate valve open?
11. Are there booster stations on the system that depend on the operation of the main pumping station under consideration?
12. Are there any quick-closing automatic valves used in the pumping system that become inoperative with the failure of the pumping system pressure?

If the answer to any one of these questions is affirmative, there is a strong possibility that serious surges will occur. If the answer to two or more of the questions is affirmative, surges will probably occur with severity in proportion to the number of affirmative answers.

GENERAL STUDIES FOR WATER HAMMER CONTROL

Studies of surges can be performed during the design stage. Once the general layout of the system has been completed, the length, diameter, thickness, material, and capacity of the pipe, as well as the type and size of pumps, can be established. The normal operating pressures at various points in the system can be computed and the allowable maximum pressures fixed. Using this information, the margin for water hammer can be determined. The design should then be adjusted to provide either safety factors large enough to withstand such conditions as might be encountered or suitable remedial or control devices. The latter method is usually less costly. It is important to note that there is no single device that will correct all surge difficulties. Only by studying both normal operating conditions and possible emergency conditions can methods be determined to provide proper control.

General recommendations cannot be made on the type, size, and application of surge-control equipment for all plants. Several possible solutions should be considered for any individual installation, and the one selected should give the maximum protection for the least expenditure. Surges can often be reduced substantially by using bypasses around check valves, by cushioning check valves for the last 15–20 percent of the stroke, or by adopting a two-speed rate of valve stroke. Water hammer resulting from power failure to centrifugal pumps can sometimes be held to safe limits by providing flywheels or by allowing the pumps to run backward. Air-inlet valves may be needed, or it may be preferable to use a surge tank, a surge damper, or a hydropneumatic chamber. Under certain operating conditions, no devices will be required to hold the pressure rise within safe limits.

It is essential to coordinate all the elements of a system properly and to ascertain that operating practices conform to safety requirements. As changes take place in the system demand, it may be necessary to review and revise the surge conditions, particularly if the capacity is increased, additional pumpage or storage is added, or booster stations are planned.

If a complete surge study is made during design and the recommendations arising from it are followed, the system should operate without damage. The agreement between the theoretical analyses, properly applied, and the actual tests of installations is extremely close. When a surge study was not performed and dangerous conditions existed, invariably there were serious surges, and sometimes costly damage resulted. The time and effort spent on a surge study in advance of the final design is the least expensive means of preventing surges. The elastic-wave theory has been proved in actual practice, and design engineers should take the initiative in making surge studies and installing surge-control devices without waiting for serious failures to occur.

ALLOWANCE FOR WATER HAMMER

Many conditions have changed since the standard, rule-of-thumb empirical allowances for water hammer originated. Automatic stop, check, and throttling valves were not as widely used then as they are today. Valve closures measured in seconds and motor-driven centrifugal pumps were practically unknown. New types of pipe have since been introduced and used. Consequently, standard allowances for water hammer may not be applied universally to all types of installations. Also, these allowances may not provide full security under all circumstances. Potential water-hammer problems should be investigated in the design of pumping-station piping, force mains, and long transmission pipelines. Appropriate methods should be provided to reduce its effect to the minimum that is practicable or economical.

PRESSURE RISE CALCULATIONS

This manual does not cover an analysis of pressure rise in a complicated pipeline. However, basic data are provided for simple problems.

The pressure rise for instantaneous valve closure is given by Eq 5-1. Values of wave velocity a are presented in Figure 5-1 for diameter-thickness ratios of 90 and less and from Table 5-1 for higher ratios.

For solutions to more complex problems, reference can be made to the many publications available (see, for example Rich 1951, Parmakian 1963, Kinno 1968, and Streeter and Wylie 1967). Computer programs are available that include the effects of pipeline friction and give accurate results. There are several methods of reducing surges by the addition of devices or revising operating conditions, but these are outside the scope of this manual. Most of the available computer programs evaluate the various methods of reducing or controlling surges. (Streeter and Wylie 1967 describes some of these means.)

REFERENCES

Kerr, S.L. 1949. Minimizing Service Interruptions Due to Transmission Line Failures—Discussion. *Jour. AWWA*, 41(7):634.

Kerr, S.L. 1951. Water Hammer Control. *Jour. AWWA*, 43(12):985.

Kerr, S.L., L.H. Kessler, and M.B. Gamet. 1950. New Method for Bulk Modulus Determination. *Trans. ASME*, 72:1143.

Kinno, H. 1968. Water Hammer Control in Centrifugal Pump System. ASCE. *Jour. Hydraul. Div.* May.

Parmakian, J. 1963. *Water Hammer Analysis*. New York: Dover Publications.

Rich, G.R. 1951. *Hydraulic Transients* (Engineering Societies Monographs). New York: McGraw-Hill Book Co.

Richards, R.T. 1956. Water Column Separation in Pump Discharge Lines. *Trans. ASME*, 78:1297.

Streeter, V.L., and E.B. Wylie. 1967. *Hydraulic Transients*. New York: McGraw-Hill Book Co.

The following references are not cited in the text.

Alin, A.L. 1944. Penstock Surge Tank at Dennison Hydro Plant. *Civ. Eng.*, 14:296.

Allievi, L. 1925. *Theory of Water Hammer.* New York: American Society of Mechanical Engineers (out of print).

Angus, R.W. 1937. Water Hammer in Pipes, Including Those Supplied by Centrifugal Pumps; Graphical Treatment. *Proc. Inst. Mech. Engrs.*, 136:245.

Angus, R.W. 1939. Water Hammer Pressures in Compound and Branch Pipes. *Trans. ASCE*, 104:340.

Barnard, R.E. 1948. Design Standards for Steel Water Pipe. *Jour. AWWA*, 40(1):24.

Bennet, R. 1941. Water Hammer Correctives. *Water & Sew. Wks.*, 88:196.

Bergeron, L. 1961. *Water Hammer in Hydraulics and Wave Surge in Electricity.* New York: John Wiley & Sons.

Boerendans, W.L. 1939. Pressure Air Chambers in Centrifugal Pumping. *Jour. AWWA*, 31(11):1865.

Dawson, F.M., and A.A. Kalinske. 1939. Methods of Calculating Water Hammer Pressures. *Jour. AWWA*, 31(11):1835.

Effect of Valve Action on Water Hammer. 1960. *Jour. AWWA*, 52(1):65.

Evans, W.E., and C.C. Crawford. 1916. Charts and Designing Air Chambers for Pump Discharge Lines. *Proc. ASCE*, 79:57.

Fluid Engineering. 1976. In *Proc. of the Second International Conference on Pressure Surge*. London: British Hydraulic Res. Assoc.

Kerr, S.L. 1948. Practical Aspects of Water Hammer. *Jour. AWWA*, 40(6):599.

Parmakian, J. 1953. Pressure Surges at Large Pump Installations. *Trans. ASME*, 75:995.

Second Symposium on Water Hammer. 1937. *Trans. ASME*, 59:651.

Simin, O. 1904. Water Hammer. (Includes a digest of N. Joukovsky's study.) In *Proc. of the AWWA Annual Conference.* Denver, Colo.: American Water Works Association.

Standard Allowances for Water Hammer—Panel Discussion. 1952. *Jour. AWWA*, 44(11):977.

Stepanoff, A.J. 1949. Elements of Graphical Solution of Water Hammer Problems in Centrifugal Pump Systems. *Trans. ASME*, 71:515.

Streeter, V.L. 1972. Unsteady Flow Calculations by Numerical Methods. ASME. *Jour. of Basic Engrg.* June.

Surges in Pipe Lines—Oil and Water. 1950. *Trans. ASME*, 72:667.

Symposium on Water Hammer. 1949. New York: American Society of Mechanical Engineers.

Water Hammer Allowances in Pipe Design. 1958. Committee Report. *Jour. AWWA*, 50(3):340.

This page intentionally blank.

Chapter 6

External Loads

LOAD DETERMINATION

External loads on buried pipe are generally comprised of the weight of the backfill combined with live and impact loads. The Marston theory (1929) is generally used to determine the loads imposed on buried pipe by the soil surrounding it. This theory is applicable to both flexible and rigid pipes installed in a variety of conditions, including ditch and projecting conduit installations. Ditch conduits are structures installed and completely buried in narrow ditches in relatively passive or undisturbed soil. Projecting conduits are structures installed in shallow bedding with the top of the conduit projecting above the surface of the natural ground and then covered with the embankment. For purposes of calculating the external vertical loads on projecting conduits, the field conditions affecting the loads are conveniently grouped into four subclassifications. They are based on the magnitude of settlement of the interior prism* of soil relative to that of the exterior prism† and the height of the embankment in relation to the height at which settlements of the interior and exterior prisms of soil are equal (Spangler 1947).

Steel pipe is considered to be flexible, and the Marston theory provides a simple procedure for calculating external soil loads on flexible pipe. If the flexible pipe is buried in a ditch less than two times the width of the pipe, the load is computed as follows:

$$W_c = C_d w B_d^2 \left(\frac{B_c}{B_d}\right) \quad (6\text{-}1)$$

Where:

W_c = dead load on the conduit, in lb/lin ft (kg/m) of pipe

C_d = load coefficient based on H_c/B_d where H_c is the height of fill above conduit, and B_d is defined below.

w = unit weight of fill, in lb/ft³ (kg/m³)

B_d = width of trench at top of pipe in ft (m)

B_c = diameter of pipe in ft (m)

*The backfill prism directly above the pipe.

†The backfill prism between the trench walls and vertical lines drawn at the OD of the pipe.

If the pipe is buried in an embankment or wide trench, the load is computed as follows:

$$W_c = C_c w B_c^2 \tag{6-2}$$

Where:

C_c = coefficient for embankment conditions, a function of soil properties.

For flexible pipe, the settlement ratio (Spangler 1947) is assumed to be zero, in which case

$$C_c = \frac{H_c}{B_c} \tag{6-3}$$

Where:

H_c = height of fill above top of pipe in ft (m)

Then:

$$W_c = \frac{H_c}{B_c} w B_c^2 = w H_c B_c \tag{6-4}$$

The dead load calculation in Eq 6-4 is the weight of a prism of soil with a width equal to that of the pipe and a height equal to the depth of fill over the pipe. This prism load is convenient to calculate and is usually used for all installation conditions for both trench and embankment conditions. For use in the Iowa deflection formula, divide Eq 6-4 by 12 for US Customary units and by 1,000 for metric units.

In addition to supporting dead loads created by earth cover, buried pipelines can also be exposed to superimposed, concentrated, or distributed live loads. Concentrated live loads are generally caused by truck-wheel loads and railway-car loads. Distributed live loads are caused by surcharges, such as piles of material and temporary structures. The effect of live loads on a pipeline depends on the depth of cover over the pipe. A method for determining the live load using modified Boussinesq equations is presented by Handy (1982).

DEFLECTION DETERMINATION

The Iowa deflection formula was first proposed by M.G. Spangler (1941). It was later modified by Watkins and Spangler (1958) and has frequently been rearranged. In one of its most common forms, deflection is calculated as follows:

$$\Delta x = D_l \left(\frac{\mathrm{K} W r^3}{EI + 0.061 E' r^3} \right) \tag{6-5}$$

Where:

Δx = horizontal deflection of pipe, in in. (mm)
D_l = deflection lag factor* (1.0–1.5)

*Deflection lag factor, D_l, accounts for long-term deflection as a result of consolidation or settlement of backfill material at the sides of the pipe. For pressure pipe, D_l is 1.0, because long-term deflections are largely prevented by the supporting action of the internal hydrostatic pressure.

K = bedding constant (0.1)
W = load per unit of pipe length, in lb/lin in.
r = radius, in in. (mm)

EI = pipe wall stiffness

Where:

E = modulus of elasticity (30,000,000 psi [206,842,710 kPa] for steel and 4,000,000 psi [27,579,028 kPa] for cement mortar)
I = transverse moment of inertia per unit length of individual pipe wall components* = $t^3/12$, where t is in in. (mm)
E' = modulus of soil reaction in lbf/in^2 (kPa) (Tables 6-1 and 6-2).

Allowable pipe deflection for various lining and coating systems that are often accepted are

Mortar-lined and coated = 2 percent of pipe diameter†
Mortar-lined and flexible coated = 3 percent of pipe diameter
Flexible lined and coated = 5 percent of pipe diameter

In addition to other considerations, the allowable pipe deflection is also dependent on the type of jointing system being utilized.

Live-load effect, added to dead load when applicable, is generally based on AASHTO HS-20 truck loads or Cooper E-80 railroad loads as indicated in Table 6-3. These values are given in pounds per square foot and include a 50 percent impact factor. There is no live-load effect for HS-20 loads when the earth cover exceeds 8 ft (2.44 m) or for E-80 loads when the earth cover exceeds 30 ft (9.14 m).

Modulus of soil reaction E' is a measure of stiffness of the embedment material, which surrounds the pipe. This modulus is required for the calculation of deflection and critical buckling stress. E' is actually a hybrid modulus that has been introduced to eliminate the spring constant used in the original Iowa formula. It is the product of the modulus of passive resistance of the soil used in Spangler's early derivation and the radius of the pipe. It is not a pure material property.

Values of E' were originally determined by measuring deflections of actual installations of metal pipe and then back-calculating the effective soil reaction. Because E' is not a material property, it cannot be uniquely measured from a soil sample, therefore determining E' values for a given soil has historically presented a serious problem for designers.

*Under load, the individual elements—i.e., mortar lining, steel shell, and mortar coating—work together as laminated rings ($E_sI_s + E_1I_1 + E_cI_c$—shell, lining, and coating.) Structurally, the combined action of these elements increases the moment of inertia of the pipe section, above that of the shell alone, thus increasing its ability to resist loads. The pipe wall stiffness of these individual elements is additive.

†Mortar-lined and coated (AWWA C205) pipe deflection is based on a maximum mortar coating thickness of 1¼ in. (32 mm). Flexible pipe coatings include AWWA C209, C210, C213, C214, C215, and C222. Flexible pipe linings and coatings include AWWA C210, C213, C222, C224, and C225.

Table 6-1 Values* of modulus of soil reaction, *E′* (psi) based on depth of cover, type of soil, and relative compaction

	Depth of Cover		Standard AASHTO relative compaction‡							
			85%		90%		95%		100%	
Type of Soil†	*ft*	*(m)*	*psi*	*(kPa)*	*psi*	*(kPa)*	*psi*	*(kPa)*	*psi*	*(kPa)*
Fine-grained soils	2–5	(0.06–1.5)	500	(3,450)	700	(4,830)	1,000	(6,895)	1,500	(10,340)
with less than 25%	5–10	(1.5–3.1)	600	(4,140)	1,000	(6,895)	1,400	(9,655)	2,000	(13,790)
sand content (CL,	10–15	(3.1–4.6)	700	(4,830)	1,200	(8,275)	1,600	(11,030)	2,300	(15,860)
ML, CL-ML)	15–20	(4.6–6.1)	800	(5,520)	1,300	(8,965)	1,800	(12,410)	2,600	(17,930)
Coarse-grained soils	2–5	(0.06–1.5)	600	(4,140)	1,000	(6,895)	1,200	(8,275)	1,900	(13,100)
with fines (SM, SC)	5–10	(1.5–3.1)	900	(6,205)	1,400	(9,655)	1,800	(12,410)	2,700	(18,615)
	10–15	(3.1–4.6)	1,000	(6,895)	1,500	(10,340)	2,100	(14,480)	3,200	(22,065)
	15–20	(4.6–6.1)	1,100	(7,585)	1,600	(11,030)	2,400	(16,545)	3,700	(25,510)
Coarse-grained soils	2–5	(0.06–1.5)	700	(4,830)	1,000	(6,895)	1,600	(11,030)	2,500	(17,235)
with little or no fines	5–10	(1.5–3.1)	1,000	(6,895)	1,500	(10,340)	2,200	(15,170)	3,300	(22,750)
(SP, SM, GP, GW)	10–15	(3.1–4.6)	1,050	(7,240)	1,600	(11,030)	2,400	(16,545)	3,600	(24,820)
	15–20	(4.6–6.1)	1,100	(7,585)	1,700	(11,720)	2,500	(17,235)	3,800	(26,200)

* Hartley, James D. and Duncan, James M., "*E′* and its Variation with Depth." *Journal of Transportation,* Division of ASCE, Sept. 1987.

† Soil type symbols are from the Unified Classification System.

‡ *Soil compaction.* When specifying the amount of compaction required, it is very important to consider the degree of soil compaction that is economically obtainable in the field for a particular installation. The density and supporting strength of the native soil should be taken into account. The densification of the backfill envelope must include the haunches under the pipe to control both the horizontal and vertical pipe deflections. Specifying an unobtainable soil compaction value can result in inadequate support and injurious deflection. Therefore, a conservative assumption of the supporting capability of a soil is recommended, and good field inspection should be provided to verify that design assumptions are met.

Table 6-2 Unified soil classification

Symbol	Description
GW	Well-graded gravels, gravel-sand mixtures, little or no fines
GP	Poorly graded gravels, gravel-sand mixtures, little or no fines
GM	Silty gravels, poorly graded gravel-sand-silt mixtures
GC	Clayey gravels, poorly graded gravel-sand-clay mixtures
SW	Well-graded sands, gravelly sands, little or no fines
SP	Poorly graded sands, gravelly sands, little or no fines
SM	Silty sands, poorly graded sand-silt mixtures
SC	Clayey sands, poorly graded sand-clay mixtures
ML	Inorganic silts and very fine sand, silty or clayey fine sands
CL	Inorganic clays of low to medium plasticity
MH	Inorganic silts, micaceous or diatomaceous fine sandy or silty soils, elastic silts
CH	Inorganic clays of high plasticity, fat clays
OL	Organic silts and organic silt-clays of low plasticity
OH	Organic clays of medium to high plasticity
Pt	Peat and other highly organic soils

Source: Classification of Soils for Engineering Purposes. ASTM Standard D2487-69, *ASTM, Philadelphia, Pa. (1969).*

Table 6-3 Live-load effect

Highway HS-20 Loading*				Railroad E-80 Loading*			
Height of Cover		Load		Height of Cover		Load	
ft	*(m)*	*psf*	*(kg/m²)*	*ft*	*(m)*	*psf*	*(kg/m²)*
1	(0.30)	1,800	(8,788)	2	(0.61)	3,800	(18,553)
2	(0.61)	800	(3,906)	5	(1.52)	2,400	(11,718)
3	(0.91)	600	(2,929)	8	(2.44)	1,600	(7,812)
4	(1.22)	400	(1,953)	10	(3.05)	1,100	(5,371)
5	(1.52)	250	(1,221)	12	(3.66)	800	(3,906)
6	(1.83)	200	(976)	15	(4.57)	600	(2,929)
7	(2.13)	176	(859)	20	(6.10)	300	(1,465)
8	(2.44)	100	(488)	30	(9.14)	100	(488)

* Neglect live load when less than 100 psf; use dead load only.

To circumvent the problems inherent in working with the hybrid modulus E', the constrained soil modulus M_s (Krizek et al. 1971) has been used more frequently. The constrained modulus is a constitutive material property, which is measured as the slope of the secant of the stress–strain diagram obtained from a confined compression test of soil. It may also be calculated from Young's modulus E_s, and Poisson's ratio υ of the soil by

$$M_s = \frac{E_s(1-\upsilon)}{(1+\upsilon)(1-2\upsilon)} \tag{6-6}$$

The soil modulus can be determined from common consolidation tests, triaxial laboratory tests, or from field plate-bearing tests of the actual soil in which the pipe will be embedded.

Because M_s is taken as the secant modulus, it accounts in part for nonlinearities in stress–strain response of soil around the pipe. Determination of M_s is based on the actual load applied to a pipe. Decreasing the load results in a decreased value for M_s. Many researchers have studied the relationship between E' and M_s, with recommendations varying widely ($E' = 0.7$ to $1.5\ M_s$). This is understandable, because M_s is a "pure" soil property, whereas E' is empirical. It appears justified to assume the two to be the same, $E' = M_s$.

BUCKLING

Pipe embedded in soil may collapse or buckle from elastic instability resulting from loads and deformations. The summation of external loads should be equal to or less than the allowable buckling pressure. The allowable buckling pressure q_a may be determined by the following:

$$q_a = \left(\frac{1}{FS}\right)\left(32R_w B'E'\frac{EI}{D^3}\right)^{1/2} \tag{6-7}$$

Where:

q_a = allowable buckling pressure*, in psi (kPa)
FS = 2.0
D = diameter of pipe, in in. (mm)
R_w = water buoyancy factor
= $1 - 0.33(h_w/h)$, $0 \le h_w \le h$
h_w = height of water surface above top of pipe, in in. (mm)
h = height of ground surface above top of pipe, in in. (mm)
B' = empirical coefficient of elastic support (dimensionless)

$$= \frac{1}{1 + 4e^{(-0.065H)}}$$

Equivalent metric equation:

$$= \frac{1}{1 + 4e^{(-0.213H)}}$$

Where:

H = height of fill above pipe, in ft (m)
E' = modulus of soil reaction (see Table 6-1)
EI = pipe wall stiffness (see Eq 6-5)

Normal Pipe Installations

For determination of external loads in normal pipe installations, use the following equation:

$$\gamma_w h_w + R_w \frac{W_c}{D} + P_v \le q_a \tag{6-8}$$

Where:

h_w = height of water above conduit in in. (mm)
γ_w = specific weight of water = 0.0361 lb/cu in. (0.0098 kPa/mm^3)
P_v = internal vacuum pressure in psi (kPa) = atmospheric pressure less absolute pressure inside pipe, in psi (kPa)
W_c = vertical soil load on pipe per unit length, in lb/in. (kPa/mm)

In some situations, live loads should be considered as well. However, simultaneous application of live-load and internal-vacuum transients need not normally be considered. Therefore, if live loads are also considered, the buckling requirement is satisfied by

$$\gamma_w h_w + R_w \frac{W_c}{D} + \frac{W_L}{D} \le q_a \tag{6-9}$$

Where:

W_L = live load on the pipe per unit length, in lb/in. (kPa/mm)

---

*NOTE: Where internal vacuum occurs with cover depth less than 4 ft (1.2 m), but not less than 2 ft (0.6 m), care should be exercised. This is particularly important for large-diameter pipe. In no case shall cover depth be less than 2 ft (0.6 m) for pipe diameters less than 24 in. (600 mm), 3 ft (0.9 m) for pipe diameters 24 in. (600 mm) through 96 in. (2,400 mm), and 4 ft (1.2 m) for pipe over 96 in. (2,400 mm) in diameter.

EXTREME EXTERNAL LOADING CONDITIONS

An occasional need to calculate extreme external loading conditions arises—for example, to determine off-highway loading from heavy construction equipment. A convenient method of solution for such load determination using modified Boussinesq equations (Table 6-4) is presented by Handy (1982). As an example:

Assume:

Live load from a large loader:
Total weight = 127,000 lb (57,660 kg)
Weight on one set of dual wheels, P = 42,300 lb (19,200 kg)
Tire pattern is 44 in. × 24 in. (1,118 mm × 610 mm)

Calculation:

Using Figure 6-1 as reference:*

Tire pattern:

$$\frac{44}{12} \times \frac{24}{12} = 3.66 \times 2.0 = 7.33 \text{ sq ft (for dual wheels)}$$

$$\left(\frac{1{,}118}{1{,}000} \times \frac{610}{1{,}000} = 0.682 \text{ m}^2\right)$$

Surface pressure:

$$\frac{42{,}300}{7.33} = 5{,}768 \text{ psf}$$

$$\left(\frac{19{,}200 \text{ kg}}{0.682 \text{ m}^2} = 28{,}152 \text{ kg/m}^2\right)$$

If height of cover H is 2.0 ft, then

$$A = \frac{3.66}{2} = 1.83 \qquad B = \frac{2.0}{2} = 1.0$$

$$\left(A = \frac{1.118}{2} = 0.559 \text{ m}\right) \qquad \left(B = \frac{0.610}{2} = 0.305 \text{ m}\right)$$

$$m = \frac{A}{H} = 0.915 \qquad n = \frac{B}{H} = 0.5$$

$$\left(m = \frac{0.559}{0.61} = 0.916\right) \qquad \left(n = \frac{0.305}{0.61} = 0.5\right)$$

Coefficient from Table 6-4 = 0.117

$$P = 0.117(4)(5{,}768) = 2{,}700 \text{ psf}$$

$$(P = 0.117(4)(28{,}152) = 13{,}180 \text{ kg/m}^2)$$

---

*NOTE: The product of A and B is ¼ of the total tire pattern.

Table 6-4 Influence coefficients for rectangular areas

$m = A/H$ or $n = B/H$	$n = B/H$ or $m = A/H$								
	0.1	0.2	0.3	0.4	0.5	0.6	0.7	0.8	0.9
0.1	0.005	0.009	0.013	0.017	0.020	0.022	0.024	0.026	0.027
0.2	0.009	0.018	0.026	0.033	0.039	0.043	0.047	0.050	0.053
0.3	0.013	0.026	0.037	0.047	0.056	0.063	0.069	0.073	0.077
0.4	0.017	0.033	0.047	0.060	0.071	0.080	0.087	0.093	0.098
0.5	0.020	0.039	0.056	0.071	0.084	0.095	0.103	0.110	0.116
0.6	0.022	0.043	0.063	0.080	0.095	0.107	0.117	0.125	0.131
0.7	0.024	0.047	0.069	0.087	0.103	0.117	0.128	0.137	0.144
0.8	0.026	0.050	0.073	0.093	0.110	0.125	0.137	0.146	0.154
0.9	0.027	0.053	0.077	0.098	0.116	0.131	0.144	0.154	0.162
1.0	0.028	0.055	0.079	0.101	0.120	0.136	0.149	0.160	0.168
1.2	0.029	0.057	0.083	0.106	0.126	0.143	0.157	0.168	0.178
1.5	0.030	0.059	0.086	0.110	0.131	0.149	0.164	0.176	0.186
2.0	0.031	0.061	0.089	0.113	0.135	0.153	0.169	0.181	0.192
2.5	0.031	0.062	0.090	0.115	0.137	0.155	0.170	0.183	0.194
3.0	0.032	0.062	0.090	0.115	0.137	0.156	0.171	0.184	0.195
5.0	0.032	0.062	0.090	0.115	0.137	0.156	0.172	0.185	0.196
10.0	0.032	0.062	0.090	0.115	0.137	0.156	0.172	0.185	0.196
∞	0.032	0.062	0.090	0.115	0.137	0.156	0.172	0.185	0.196

	1.0	1.2	1.5	2.0	2.5	3.0	5.0	10.0	∞
0.1	0.028	0.029	0.030	0.031	0.031	0.032	0.032	0.032	0.032
0.2	0.055	0.057	0.059	0.061	0.062	0.062	0.062	0.062	0.062
0.3	0.079	0.083	0.086	0.089	0.090	0.090	0.090	0.090	0.090
0.4	0.101	0.106	0.110	0.113	0.115	0.115	0.115	0.115	0.115
0.5	0.120	0.126	0.131	0.135	0.137	0.137	0.137	0.137	0.137
0.6	0.136	0.143	0.149	0.153	0.155	0.156	0.156	0.156	0.156
0.7	0.149	0.157	0.164	0.169	0.170	0.171	0.172	0.172	0.172
0.8	0.160	0.168	0.176	0.181	0.183	0.184	0.185	0.185	0.185
0.9	0.168	0.178	0.186	0.192	0.194	0.195	0.196	0.196	0.196
1.0	0.175	0.185	0.193	0.200	0.202	0.203	0.204	0.205	0.205
1.2	0.185	0.196	0.205	0.212	0.215	0.216	0.217	0.218	0.218
1.5	0.193	0.205	0.215	0.223	0.226	0.228	0.229	0.230	0.230
2.0	0.200	0.212	0.223	0.232	0.236	0.238	0.239	0.240	0.240
2.5	0.202	0.215	0.226	0.236	0.240	0.242	0.244	0.244	0.244
3.0	0.203	0.216	0.228	0.238	0.242	0.244	0.246	0.247	0.247
5.0	0.204	0.217	0.229	0.239	0.244	0.246	0.249	0.249	0.249
10.0	0.205	0.218	0.230	0.240	0.244	0.247	0.249	0.250	0.250
∞	0.205	0.218	0.230	0.240	0.244	0.247	0.249	0.250	0.250

Source: Newmark, N.M., Simplified Computation of Vertical Pressures in Elastic Foundations. *Circ. 24. Engrg. Exp. Stn., Univ. of Illinois (1935).*

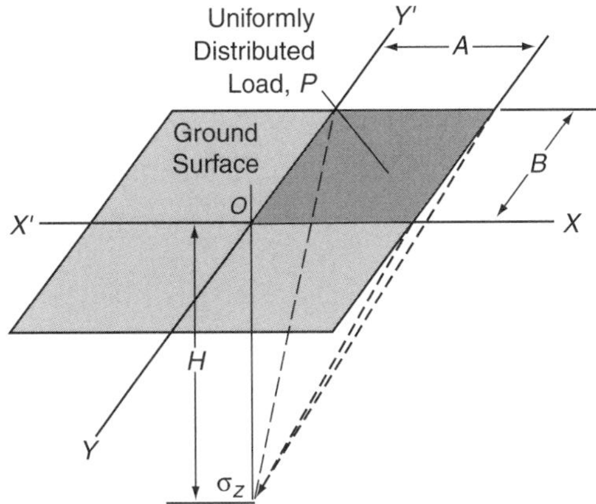

Source: Spangler, M.G. & Handy, R.L. Soil Engineering. *Harper and Row, New York (4th ed., 1982).*

Figure 6-1 Position of area

If height of cover is 3.0 ft (0.914 m), then

$$m = 0.610 \quad n = 0.333$$

Coefficient = 0.07

$$P = 1{,}615 \text{ psf}$$

$$(P = 0.7\ (4)(28{,}152) = 7{,}883 \text{ kg/m}^2)$$

Using the Iowa formula (Eq 6-5) to calculate deflection for 54-in. (1,372 mm) pipe and 60-in. (1,524 mm) pipe, wall thickness ¼ in. (6.35 mm) for each size, E' = 1,250 psi (8.618 MPa), D_l = 1.0, and soil weight of 120 pcf (1,922 kg/m^3), the results are

Total load (dead and live load):
2 ft (0.61 m) cover:

$$W_c = \left[(120)(2) + 2{,}700\right]\frac{2r}{144} = 40.8r$$

$$W_c = (1{,}922\,(2) + 13{,}180)\frac{2r}{1{,}000} = 34.0r$$

3 ft (0.914 m) cover:

$$W_c = \left[(120)(3) + 1{,}615\right]\frac{2r}{144} = 27.4r$$

$$W_c = (1{,}922\,(0.914) + 7{,}883)\frac{2r}{1{,}000} = 19.279r$$

Using Spangler's formula, deflection =

60 in. (1,524 mm),	2 ft (0.61 m) cover: = 1.58 in. (40.1 mm) = 2.6%
	3 ft (0.914 mm) cover: = 1.06 in. (26.9 mm) = 1.8%
54 in. (1,372 mm),	2 ft (0.61 m) cover: = 1.41 in. (35.8 mm) = 2.6%
	3 ft (0.914 m) cover: = 0.95 in. (24.1 mm) = 1.8%

COMPUTER PROGRAMS

Traditional procedures depending on weight and the elastic modulus of soil to determine pipe deflections and stress have been discussed. However, computer programs that permit a more rational determination in the design of pipe are now available from universities, consulting engineers, and manufacturers.

REFERENCES

Handy, R.L. 1982. *Soil Engineering*, 4th ed. New York: Harper & Row Publishers.

Howard, A. 1977. Modules of Soil Reaction Values for Buried Flexible Pipe. *Jour. Geotechnical Engr. Div.*—ASLE.

Krizek, R.J., R.A. Parmelee, J.N. Kay, and H.A. El Naggar. 1971. *Structural Analysis and Design of Pipe*. Report 116. HCHRP.

Marston, A. 1929. The Theory of External Loads on Closed Conduits in the Light of the Latest Experiments. In *Proc. of the Ninth Annual Meeting Highway Res. Board.*

Spangler, M.G. 1947. Underground Conduits—An Appraisal of Modern Research. *Proc. ASCE*. June.

———. *The Structural Design of Flexible Pipe Culverts*. 1941. Bull. 153. Ames, Iowa: Iowa State College.

Watkins, R.K., and M.G. Spangler. 1958. Some Characteristics of the Modulus of Passive Resistance of Soil: A Study in Similitude. *Highway Research Board Proc.*, 37:576.

The following references are not cited in the text.

Barnard, R.E. 1948. Design Standards for Steel Water Pipe. *Jour. AWWA*, 40:24.

———. 1955. *Behavior of Flexible Steel Pipe Under Embankments and in Trenches. Bull.* Middletown, Ohio: Armco Drainage & Metal Products, Inc.

———. 1957. Design and Deflection Control of Buried Steel Pipe Supporting Earth Loads and Live Loads. *Proc. ASTM*, 57:1233.

Braune, Cain, and Janda. 1929. Earth Pressure Experiments on Culvert Pipe. *Public Roads*, 10:9.

Burmister, D.M. 1951. *The Importance of Natural Controlling Conditions Upon Triaxial Compression Test Conditions.* Special Tech. Pub. 106. Philadelphia, Pa.: American Society for Testing and Materials.

Housel, W.S. 1951. *Interpretation of Triaxial Compression Tests on Granular Soils.* Special Tech. Pub. 106. Philadelphia, Pa.: American Society for Testing and Materials.

Luscher, U. 1966. Buckling of Soil Surrounded Tubes. *Jour. Soil Mechanics and Foundations Div.—ASCE*, November.

Proctor, R.R. 1933. Design and Construction of Rolled-Earth Dams. *Engineering News Record*, 111:245.

———. 1948. An Approximate Method for Predicting the Settlement of Foundations and Footings. In *Second International Conference on Soil Mechanics & Foundation Engr.* The Hague, Netherlands.

Proudfit, D.P. 1963. Performance of Large-Diameter Steel Pipe at St. Paul. *Jour. AWWA*, 55(3):303.

Reitz, H.M. 1956. Soil Mechanics and Backfilling Practices. *Jour. AWWA*, 48(12):1497.

Report on Steel Pipelines for Underground Water Service. 1936. Special Investigation 888. Chicago: Underwriters' Labs., Inc.

Sowers, G.F. 1956. Trench Excavation and Backfilling. *Jour. AWWA*, 48(7):854.

Spangler, M.G. 1948. Underground Conduits—An Appraisal of Modern Research. *Trans. ASCE*, 113:316.

———. 1951–1952. *Protective Casings for Pipelines*. Engineering Reports 11. Ames, Iowa: Iowa State College.

Spangler, M.G., and D.L. Phillips. 1955. *Deflections of Timber-Strutted Corrugated-Metal Pipe Culverts Under Earth Fills.* Bull. 102. Highway Research Board; Pub. 350. Washington, D.C.: National Academy of Sciences–National Research Council.

Terzaghi, K. 1943. *Theoretical Soil Mechanics.* New York: John Wiley & Sons.

Wagner, A.A. 1951. *Shear Characteristics of Remolded Earth Materials*. Special Tech. Pub. 106. Philadelphia, Pa.: American Society for Testing and Materials.

Wiggin, T.H., M.L. Enger, and W.J. Schlick. 1939. A Proposed New Method for Determining Barrel Thicknesses of Cast-Iron Pipe, *Jour. AWWA*, 31:811.

Chapter 7

Supports for Pipe

Pipe is supported in various ways, depending on size, circumstances, and economics. Small pipe within buildings may be held by adjustable hangers or by brackets or attached to building members. When subjected to temperature changes causing considerable longitudinal movement, steel pipe is frequently set on concave rollers. Data on adjustable hangers and rollers have been published (Roark 1954).

Pipe acting as a self-supporting bridge may rest on suitably padded concrete saddles (Figures 7-1 and 7-2) or may be supported by ring girders or flange rings welded to the pipe (Figures 7-3 through 7-6 and 7-15). The kind of support selected may be determined by installation conditions or by economics. The pipe cost is usually lower, and there is more flexibility in field erection when saddles can be used. Longer clear spans may be possible using ring-girder construction.

SADDLE SUPPORTS

There has been very little uniformity in the design or spacing of saddle supports. Spans have gradually increased, because experience has proved that such increases were safe and practical. In general, the ordinary theory of flexure applies when a circular pipe is supported at intervals, is held circular at and between the supports, and is completely filled. If the pipe is only partially filled and the cross section at points between supports becomes out-of-round, the maximum fiber stress is considerably greater than indicated by the ordinary flexure formula, being highest for the half-filled condition (Schorer 1933).

In the case of a pipe carrying internal pressure where the ends are fully restrained, the Poisson-ratio effect of the hoop stress, which produces lateral tension, must be added to the flexural stress to obtain the total beam stress. Excessive deflection should be avoided when the pipe acts as a beam. A maximum deflection of $1/360$ of the span is recommended. This is the same recommendation used for beams carrying plastered ceilings.

Saddle supports cause local stresses both longitudinally and circumferentially in unstiffened, comparatively thin-wall pipe at the tips and edges of the supports. The highest local stresses are circumferential bending stresses at the saddle tips. Stresses vary with the load, the diameter-wall thickness ratio, and the angle of contact with

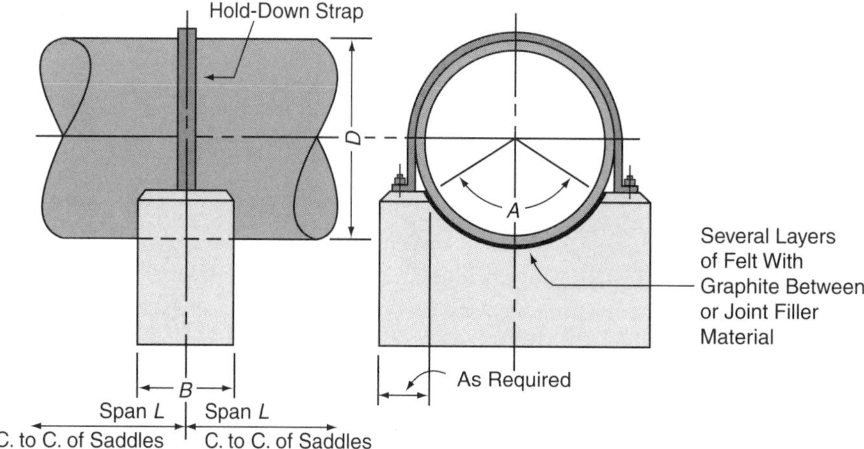

Pipe acting as a self-supporting bridge may rest on suitably padded concrete saddles

Source: Barnard, R.E., Design Standards for Steel Water Pipe, Jour. AWWA, *40:1:24 (Jan. 1948).*

Figure 7-1 Details of concrete saddle

Figure 7-2 Saddle supports for 78-in. pipe

the pipe. In practice, the contact angle varies from 90° to 120°. The difficulty encountered with 180° contact angles has been eliminated by reducing the angles to 120°. For equal load, the stresses are less for a large contact angle than for a small one, and interestingly, their intensity is practically independent of the width of the saddle (Dimension B, Figure 7-1). The width of the saddle may, therefore, be that which is most desirable for good pier design.

Because saddle supports cause critical points of stress in the metal adjacent to the saddle edges, it is frequently more economical to increase the wall thickness of the pipe when it is overstressed than to provide stiffening rings. This is especially true where pipe sizes are 36 in. (900 mm) in diameter and smaller. Even a small increase in wall thickness has a great stiffening effect. The whole length of the span may be thickened, or only a length at the saddle support—equal to about two pipe diameters plus saddle width—need be thickened.

When pipe lengths resting on saddles are joined by flanges or mechanical couplings, the strength and position of the joints must be able to safely resist bending and shear

Figure 7-3 Ring girders provide support for 54-in. diameter pipe

The rings are supporting a 54-in. diameter pipe laid on a slope.

Figure 7-4 Expansion joints between stiffener rings

This block anchors a 66-in. diameter pipe against longitudinal movement.

Figure 7-5 Anchor block

forces while remaining tight. Ordinarily it is advisable to place joints at, or as near as practicable to, the point of zero bending moment in the span or spans. Manufacturers of mechanical joints should be consulted regarding the use of their joints on self-supporting pipe spans.

The pipe should be held in each saddle by a steel hold-down strap bolted to the concrete. Secure anchorages must be provided at intervals in multiple-span installations.

Research* has shown that, for pipelines supported by saddles, secondary stresses at the supports are large enough to create critical conditions only near the saddle tips. The highest stress is the circumferential bending stress, which tends to decrease as the internal pressure increases. Therefore, the critical condition is usually with the pipe full but at zero pressure. This stress can be calculated from

$$S_{cs} = k\frac{P}{t^2}\log_e\left(\frac{R}{t}\right) \quad (7\text{-}1)$$

Where:

S_{cs} = local bending stress at saddle (psi)
k = 0.02 – 0.00012 (A-90) (contact angle factor)
A = contact angle (degrees) (See Figure 7-1)
P = total saddle reaction (lb)
R = pipe radius (in.)
t = pipe wall thickness (in.)

If a longitudinal stress exists near the saddle tips, such as a thermal stress or the beam bending stress at that depth on the pipe, designate its calculated value as S_{1s}. Then calculate the effective stress, S_e:

$$S_e = (S_{cs}^2 + S_{ls}^2 - S_{cs}S_{ls})^{1/2} \quad (7\text{-}2)$$

This stress (S_e) must not exceed the yield point. It is not necessary to apply a safety factor because tests have shown that, because this is a very localized condition, the resulting design will have a safety factor of approximately two.

The bending stress when the pipe is under pressure is calculated by multiplying S_{cs} by a reduction factor (RF) calculated from

$$RF = \frac{(\tanh A)}{A} \quad (7\text{-}3)$$

Where:

$A = 1.1\ (R/t)\ (S_h/E)^{1/2}$
S_h = hoop stress (psi)
E = modulus of elasticity (psi) (30,000,000 for steel)
tanh denotes hyperbolic tangent

The hoop stress equals the sum of the membrane stress caused by pressure (usually tension) and the membrane stress at the tip of the cradle caused by the supported load (usually compression). It must be added to the reduced bending stress to get the total circumferential stress. It is usually not necessary to make this calculation because the zero

*R.D. Stokes, "Stresses in Steel Pipelines at Saddle Support." *Civil Engineering Transactions,* October 1965, The Institution of Engineers, Australia.

pressure condition controls the design. The constant of 1.1 in the reduction factor was experimentally calibrated for a 150° saddle and is considered reasonable for 120° saddle.

As with all support systems, the maximum beam bending stress for the pipe span must be calculated and limited to a suitable allowable stress. It is usually not necessary to add the beam bending stress at the bottom of the pipe at the support (e.g., at an intermediate support in a continuous span arrangement) to a secondary saddle stress, as was sometimes done in past procedures, because Stokes has shown that these stresses are much smaller than those given in Eq 7-1. As mentioned previously, if the pipe is under pressure and the ends are restrained, the Poisson's ratio effect of the hoop stress ($0.30S_h$) must be added to the beam flexure stress. The total longitudinal stress (S_t) is calculated as

$$S_t = S_f + 0.30S_h \tag{7-4}$$

Example:

42-in. diameter by 5⁄16 in. wall pipe,
A283 Grade C steel ($F_y = 30{,}000$ psi),
40,000 lb total radial reaction on 120° saddle,
longitudinal stress of 3,000 psi compression
(thermal plus bending at saddle tips; $D/4$ below center of pipe).

$$k = 0.02 - 0.0012(120 - 90) = -0.0164$$

$$S_{cs} = 0.0164 \times \frac{40{,}000}{(0.3125)^2}\log_e\left(\frac{21}{0.3125}\right) = -28{,}300 \text{ psi}$$

$$S_{ls} = -3{,}000 \text{ psi} \tag{7-5}$$

$$S_e = 1{,}000\left[28.3^2 + 3.0^2 - (-3.0 \times -28.3)\right]^{1/2} = 26{,}926 \text{ psi}$$

$$26{,}926 < 30{,}000$$

Beam stresses must still be checked by Eq 7-4.

The flexure stress S_f should be calculated in the usual manner. In single spans, this stress is maximum at the center between supports and may be quite small over the support if flexible joints are used at the pipe ends. In multiple-span cases, the flexure stress in rigidly joined pipe will be that determined by continuous beam theory.

For pipe with diameters of 6 to 144 in. (150 to 3,600 mm), Table 7-1 gives practical, safe spans that are conservative for pipes supporting their weight plus that of the contained water. Other live loads such as earthquake, wind, or the like should also be calculated. Data for calculating spans for large pipe on saddles have been published (Schorer 1933).

PIPE DEFLECTION AS BEAM

In the design of free spans of pipe, the theoretical deflection should be determined in order to judge flexibility or ascertain that the deflection does not exceed an acceptable upper limit. Freely supported pipe sometimes must be laid so that it will drain fully and not pool water between supports. The allowable deflection or sag between supports must be found to determine the necessary grade.

Table 7-1 Practical safe spans for simply supported pipe in 120° contact saddles*

Nominal Size *in.*†	Wall Thickness *in.*									
	3/16	1/4	5/16	3/8	7/16	1/2	5/8	3/4	7/8	1
	Span L *ft*									
6	36	40	44							
8	38	42	45							
10	39	43	46							
12	40	44	47							
14	40	44	47							
16	41	45	48							
18	41	46	49	52						
20	42	46	50	53						
22	42	46	51	54						
24	42	48	52	55	58	60				
26	43	48	52	56	59	61				
28	43	48	53	56	59	62				
30	43	49	53	57	60	63				
32	44	49	54	57	61	64				
34	44	49	54	58	61	64				
36	44	50	54	58	62	65	70			
38	44	50	55	59	62	65	70			
40	44	50	55	59	63	66	71			
42	44	50	55	59	63	66	72			
45		51	55	60	63	67	72			
48		51	56	60	64	67	73	78		
51		51	56	60	64	68	74	79		
54		51	56	61	65	68	74	79		
57		51	57	61	65	69	75	80		
60		51	57	61	65	69	75	80		
63		52	57	62	66	69	76	81		
66		52	57	62	66	70	76	81	86	90
72		52	58	62	66	70	77	82	87	92
78			58	62	67	71	77	83	88	93
84			58	63	67	71	78	84	89	94
90			58	63	67	71	78	84	90	94
96			58	63	68	72	79	85	90	95
102			58	63	68	72	79	85	91	96
108				64	68	72	80	86	91	96
114				64	68	73	80	86	92	97
120					69	73	80	87	92	98
126					69	73	81	87	93	98
132					69	73	81	87	93	98
138					69	73	81	88	94	99
144					69	74	81	88	94	99

NOTE: This table is based on bending stresses; localized stresses at supports must be analyzed.

* After Cates (1950): d and t are pipe diameter and thickness respectively in in. (mm), and L is in ft (m); fiber stress = 8,000 psi (55.16 MPa), loaded by dead weight of pipe plus container water.

† To convert nominal in. to nominal mm, multiply by 25.0; to convert in. to mm, multiply by 25.4; to convert ft to m, multiply by 0.3048.

In any given case, the deflection is influenced by the conditions of installation. The pipe may be a single span or may be continuous over several supports. The ends may act as though free or fixed. In addition to its own weight and that of the water, the pipe may carry the weight of insulation or other uniform load. Concentrated loads, such as valves, other appurtenances, or fittings, may be present between supports.

The maximum theoretical deflection for a simple span can be determined using

$$y = 22.5\frac{WL^3}{EI} \tag{7-6}$$

Equivalent metric equation:

$$y = \frac{5\times10^6}{384}\times\frac{WL^3}{EI} \tag{7-6M}$$

Where:

y = maximum deflection at center of span, in. (mm)
W = total load on span, lbf (N)
L = length of span, ft (m)
E = modulus of elasticity, psi (GPa); 30,000,000 psi (206.84 GPa) for steel pipe
I = moment of inertia of pipe, in.^{4} (mm^4) (values of I are given in Table 7-6 on page 103)

Except for some changes in unit designation, this is the standard textbook formula for uniformly distributed load and free ends. It can be used for concentrated loads at the center of the span, and it can be applied to other end conditions by applying a correction factor described later in this chapter.

Tests conducted to determine the deflection of horizontal standard-weight pipelines filled with water (Roark 1954) have indicated that with pipe larger than 2 in. (76 mm) and supported at intervals greater than 10 ft (3.0 m), the deflection is less than that determined theoretically for a uniformly loaded pipe fixed at both ends. The actual deflection of smaller pipe approached the theoretical deflection for free ends.

METHODS OF CALCULATION

The following methods of calculating deflection are based on the formulas commonly found in textbooks for the cases given. Maximum deflection in a given case can be calculated by first assuming that the load is uniformly distributed and the ends are free. This is case 1 below. Later this result can be modified if the load is concentrated or the ends are fixed (cases 2, 3, and 4 below). The deflection for case 1 may be calculated using Eq 7-3. Note that in cases 1 and 2 the load W is the total uniformly distributed load on the span, but in cases 3 and 4 it is the load concentrated at the center of the span.

The four most commonly encountered conditions, with their corresponding deflection factors, are

- *Case 1:* If the load W is uniformly distributed and the ends are free, the deflection is calculated using Eq 7-3.
- *Case 2:* If the load W is uniformly distributed but the ends are fixed, the deflection is 0.2 times that for case 1.
- *Case 3:* If the load W is concentrated at the center and the ends are free, the deflection is 1.6 times that for case 1.

- *Case 4:* If the load W is concentrated at the center and the ends are fixed, the deflection is 0.4 times that for case 1.

The deflections caused by different loads are additive. Therefore, if a uniformly loaded pipe span contains a concentrated load, the calculated deflection for the latter is added to that for the uniform load, and the total sag in the pipe is the sum of the two deflections.

GRADIENT OF SUPPORTED PIPELINES TO PREVENT POCKETING

If intermittently supported pipelines are to drain freely, they must contain no sag pockets. To eliminate pockets, each downstream support level must be lower than its upstream neighbor by an amount that depends on the sag of the pipe between them. A practical average gradient of support elevations to meet this requirement may be found by using the following formula (Wilson and Newmark 1933):

$$G = \frac{4y}{L} \tag{7-7}$$

Where:

G = gradient, in./ft (mm/m)
L = span ft (m)
y = deflection from pipe dead load without weight of water, in. (mm)

The elevation of one end should be higher than the other by an amount equal to four times the deflection calculated at mid-span of the pipe.

Example: If the deflection of an insulated, 20-in. (500 mm) OD, 0.375-in. (9.5-mm) wall thickness pipe carrying steam is 0.4 in. (10.2 mm) in a simple, free-ended 50-ft (15.2-m) span, what should be the grade of a series of 50-ft spans to allow drainage?

Solution:

$$G = \frac{4(0.4)}{50} = 0.032 \text{ in./ft} = 2.67 \text{ mm/m} \tag{7-8}$$

It has been suggested (Roark 1954) that, in the interest of satisfactory operation, the calculated theoretical deflection should be doubled when determining the slope of the pipeline gradient. If this were done in the preceding example, the grade used would be 0.064 in./ft (5.33 mm/m).

The difference in elevation between a downstream support and its upstream neighbor must be four times the theoretical deflection of the pipe between them to establish the grade according to Eq 7-4. The elevation difference is eight times the deflection if the recommendation in the preceding paragraph is followed.

SPAN LENGTHS AND STRESSES

For a field-welded pipeline the span length is between 60 to 120 ft (18.3 to 36.6 m). The length to be used in any particular situation is dependent on economics. Longer spans result in fewer piers and the attendant monetary savings, but they may require substantially heavier support rings or ring girders and greater shell thicknesses over the supports and at mid-span, which could materially offset any savings on the decreased number of piers. These factors, together with required distances between anchor points (changes in direction or slope) dictated by field conditions and the prudent selection of optimum plate widths, will influence the determination of span length. All of these factors are considered in making preliminary layouts, which will lead to the selection of a final layout.

Span lengths for pipelines having sleeve-type, coupled field joints may be limited to a maximum length of 40 ft (12.2 m) between couplings. This limitation is based on the maximum allowable longitudinal movement within a sleeve-type coupling of 3/8 in. (9.5 mm), as set by coupling manufacturers, and is the maximum movement anticipated for a 40-ft (12.2 m) length of pipe allowing 100°F (55.6°C) temperature change.

Stresses considered between supports are

1. Longitudinal stresses caused by beam bending.
2. Longitudinal stresses caused by longitudinal movement under temperature changes and internal pressure.
3. Circumferential (hoop) stress as a result of internal pressure.
4. Equivalent stress based on the Hencky–Mises theory of failure.

Stresses considered at supports are

1. Circumferential stresses in supporting ring girder as a result of bending and direct stresses and tensile stress due to internal pressure.
2. Circumferential stresses in support rings at saddle supports.
3. Longitudinal stresses in the shell at support caused by beam bending, and stresses in the shell as a result of longitudinal movement of the shell under temperature changes and internal pressure.
4. Bending stresses imposed by the rigid ring girder.
5. Equivalent stress based on the Hencky–Mises theory of failure.

For a pipeline laid out as a continuous beam between an anchor and an expansion joint, several combinations of span lengths should be studied to determine the optimum span lengths for existing conditions. Span lengths between supports and length of cantilevered sections adjacent to the expansion joint should be proportioned so that the longitudinal bending moment at the supports is equal to or approaches the moment for a fixed-end beam, $M = \pm WL^2/12$. More importantly, the slope or deflection of the free end of the cantilevered section should be equal to that of the free end of the adjoining cantilevered section so that the connecting expansion joint will operate freely. The moments, reactions at the supports, and bending stresses are readily computed for any point along the continuous beam. Combined with these longitudinal stresses are the stresses due to longitudinal forces imposed on the shell in overcoming the forces of friction at the supports and expansion joints. The latter force is usually considered to be 500 lbf per circumferential ft (7,297 N/circumferential m). The stresses from these frictional forces are small but are combined with the longitudinal bending stresses when considering the combination of longitudinal and circumferential stresses. The circumferential or *pr/t* stress in the shell between supports is computed and combined with the longitudinal stress in accordance with the Hencky–Mises theory (Figure 7-7)—

$$S_e^2 = S_x^2 - S_x S_y + S_y^2 \tag{7-9}$$

—in which S_e equals equivalent stress and S_x and S_y are principal stresses. The equivalent stress is not permitted to exceed 33 percent of tensile strength for the normal condition. This analysis may frequently result in variable plate thickness between supports. Thicker shell plates may be required at supports and thinner plates in the region of inflection points and increasing plate thickness at the mid-span.

A pipeline installed aboveground is supported either on concrete saddles or on piers, and in the latter case, ring girders or support rings are provided to transfer the beam reactions through rocker assemblies, roller assemblies, or bearing plates to the concrete piers.

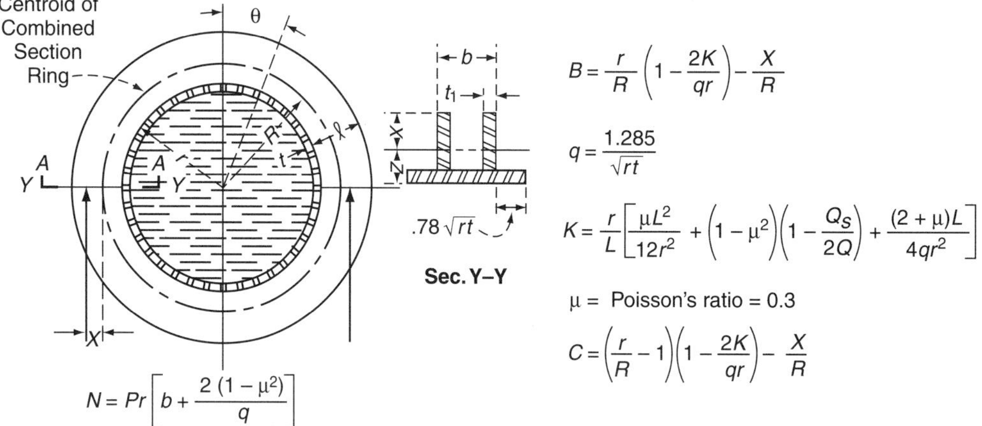

A = area combined ring section in square inches = $t(b + 1.56\sqrt{rt}) + 2\ell t_1$
I = moment of inertia of Section Y–Y
L = length of one span in inches
M = bending moment in the ring, inch pounds
N = tension due to internal pressure, on section of ring, in pounds
P = pressure head in pounds per square inch
Q_S = weight of pipe shell in one span in pounds
Q = combined weight pipe shell and water in one span, in pounds
S = radial shear stress in stiffener ring, in pounds
T = direct stress in the ring exclusive of N, in pounds

Total stress in outer fiber of ring = $\frac{T}{A} - M\frac{X}{I} + \frac{N}{A}$

Total stress in inner fiber of ring = $\frac{T}{A} + M\frac{Z}{I} + \frac{N}{A}$

	$T = Q(K_1 + BK_2)$		$M = Q(RK_3 + XK_4)$		$S = Q(K_5 + CK_6)$	
θ°	K_1	K_2	K_3	K_4	K_5	K_6
0°	−.238732	+.318310	+.011267	−.068310	0	0
15°	−.241384	+.307464	+.008618	−.057464	+.019651	+.082385
30°	−.248415	+.275664	+.001585	−.025665	+.032380	+.159155
45°	−.257198	+.225079	−.007198	+.024921	+.032117	+.225079
60°	−.263704	+.159155	−.013704	+.090845	+.014417	+.275664
75°	−.263023	+.082385	−.013023	+.167616	−.022945	+.307463
90°−	−.250000	0	0	+.250000	−.079577	+.318310
90°+	+.250000	0	0	−.250000	−.079577	+.318310
105°	+.263023	−.082385	+.013023	−.167616	−.022945	+.307463
120°	+.263704	−.159155	+.013704	−.090845	+.014417	+.275664
135°	+.257198	−.225079	+.007198	−.024921	+.032117	+.225079
150°	+.248415	−.275664	−.001585	+.025665	+.032380	+.159155
165°	+.241384	−.307464	−.008618	+.057464	+.019651	+.082385
180°	+.238732	−.318310	−.011267	+.068310	0	0

+ = Tension, − = Compression

US Bureau of Reclamation (1944)—Boulder Canyon Project—Final Reports, Part 5—Bulletin 5, Tables 3, 4, & 6.

Figure 7-6 Stiffener ring coefficients

RING GIRDERS

A satisfactory and rational design for ring girder construction was presented by Herman Schorer (1933) and fully described in Roark (1954) and AISI (1983), in which the maximum allowable design stress was given as 10,000 psi (68.95 MPa). Higher maximum allowable design stresses can be used when the ring girder analysis is based on the more comprehensive formulas and coefficients published in the Boulder Canyon Project, Final Reports, USBR Bulletin No. 5, Part 5, "Penstock Analysis & Stiffener Ring Design" (US Bureau of Reclamation 1944). Figure 7-6 is used to compute the circumferential stresses that result from bending, direct stress, and tensile stress. Longitudinal stresses at the supports include beam-bending stresses, stresses caused by frictional forces at the supports and expansion joints, and bending caused by the restraint of the pressurized pipe shell by the ring girder. Although the bending stress is only a local stress in the shell, which sharply decreases with increasing distance from the support, it is added to the other longitudinal stresses. Bending stress resulting from ring restraint, Shorer (1933), is

$$S_b = 1.82\frac{(A_r - ct)}{(A_r + 1.56t\sqrt{rt})}\frac{pr}{t} \tag{7-10}$$

Where:

A_r = cross-section area of ring, in.^{2} (see Figure 7-8)
c = width of girder, in in.
t = shell thicknesses, in in.
r = inside radius of shell, in in.
p = design pressure under normal conditions, psi

Circumferential and longitudinal stresses in the ring section are combined in accordance with the Hencky–Mises theory (Figure 7-7), and the maximum equivalent shell stress is limited to 33 percent of tensile strength for normal conditions. There are a number of other useful design references on the subject. (Foster 1949a, b; Bier 1940, 1949; Barnard 1948; Crocker 1954; Timoshenko 1936).

Intermittent Conditions

In addition to designing for normal conditions, certain intermittent conditions should be considered and stresses computed. These conditions involve stresses that occur when the pipeline is partially full during filling and draining operations and stresses are induced by earthquake forces under maximum normal operating conditions. Stresses occurring in a partially filled conduit are a maximum when the conduit is half full. The stresses in the ring girder supports may be satisfactorily determined by combining the values for tangential bending moments and the resulting stresses with the corresponding direct compressive and tensile stresses in accordance with Figures 7-8, 7-9, 7-10, 7-11, and 7-12. These stresses are summarized in Figure 7-13. Transverse earthquake loading should be considered based on local seismic records. These transverse loads are usually in the range of 0.10 to 0.50 of the gravity loads. Stiffener ring coefficients that may be used in determining ring girder stresses are shown in Figures 7-9, 7-10, 7-11, and 7-12.

Ring Girder Design Example: Determine plate thickness and support ring properties for a 96-in. (2,400-mm) diameter penstock supported at intervals on rockers. The span length is 80 ft (24.38 m), and the design pressure is 150 psi (1,034.2 kPa).

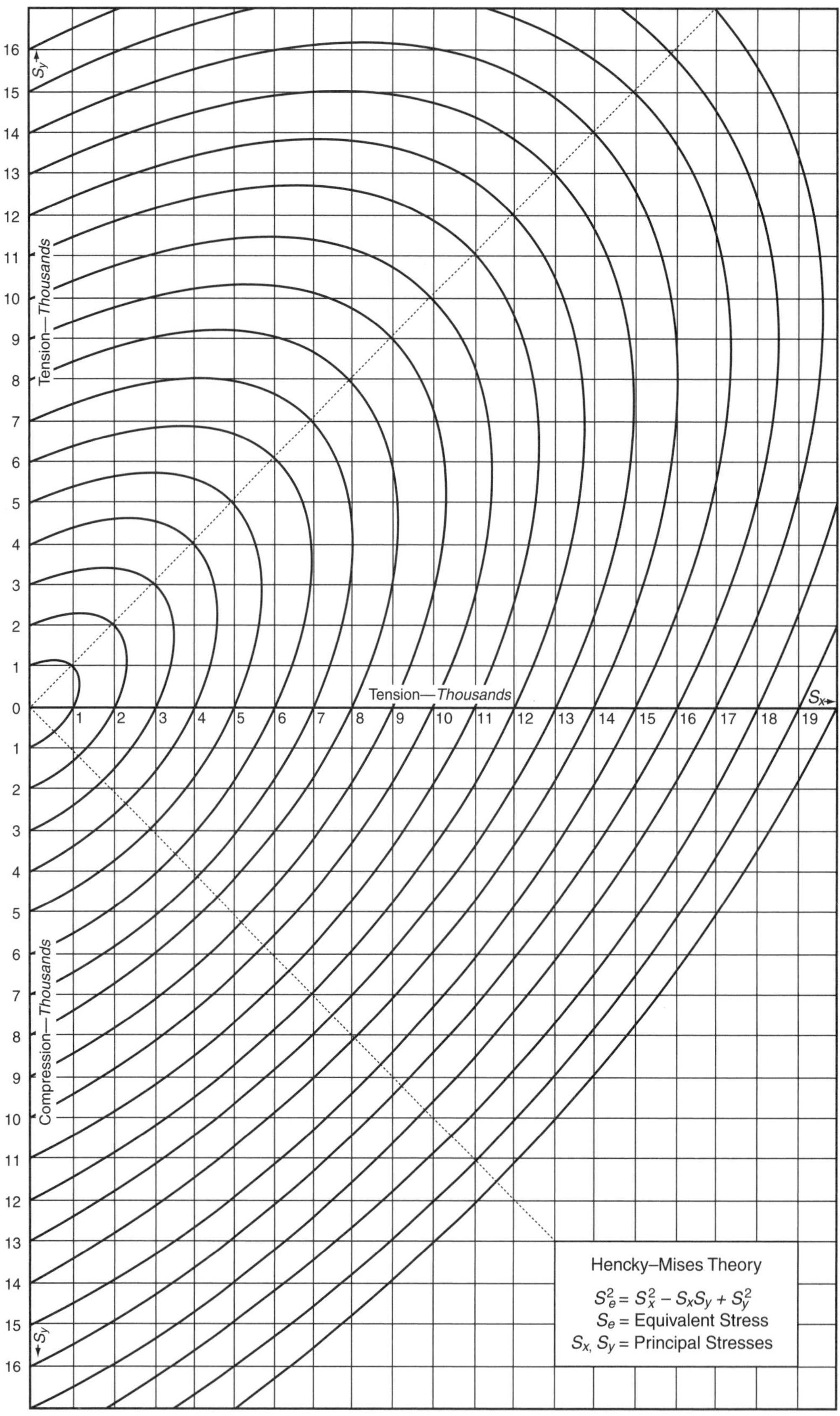

Figure 7-7 Equivalent stress diagram—Hencky–Mises theory

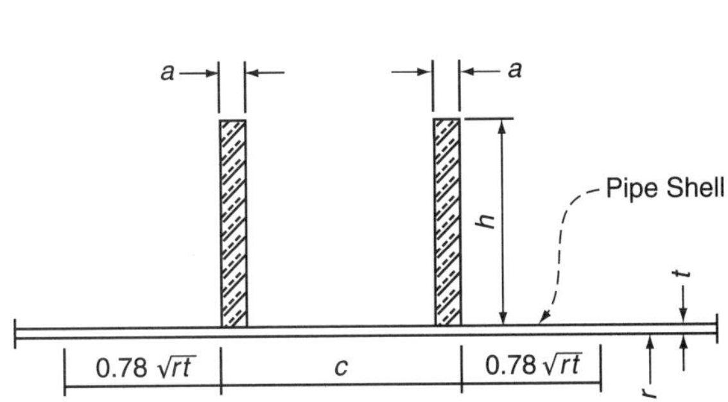

Figure 7-8 Bending stress in pipe shell with ring restraint

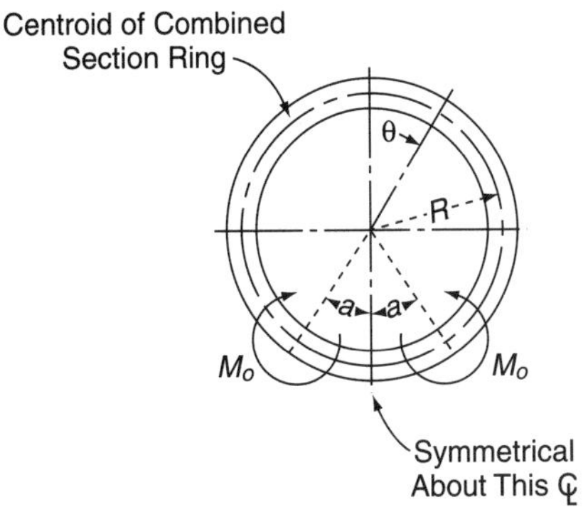

Figure 7-9 Stiffener ring coefficients, equal and opposite couples

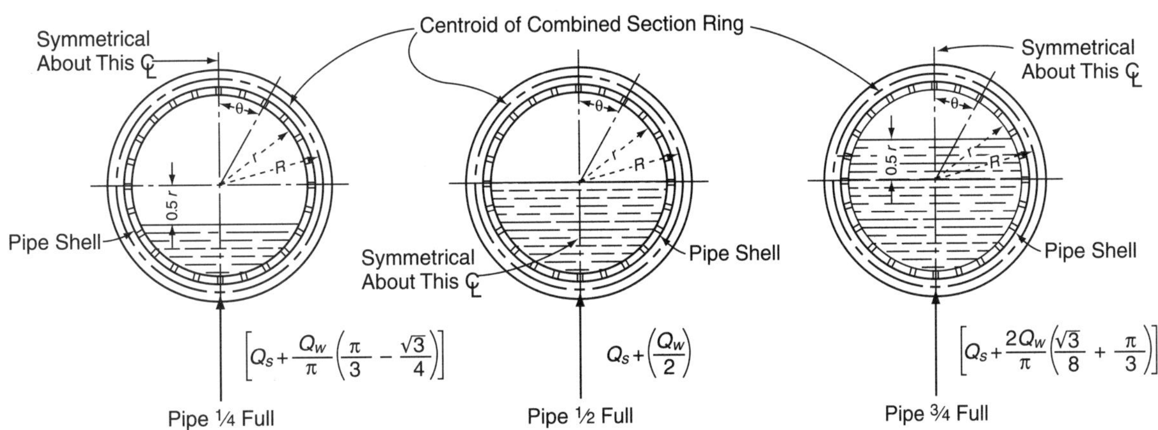

Figure 7-10 Stiffener ring stresses for partially filled pipe

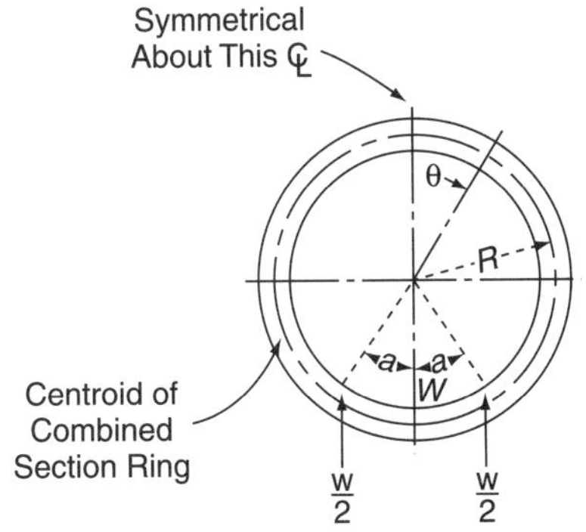

Figure 7-11 Stiffener ring coefficients, radial load supported by two reactions

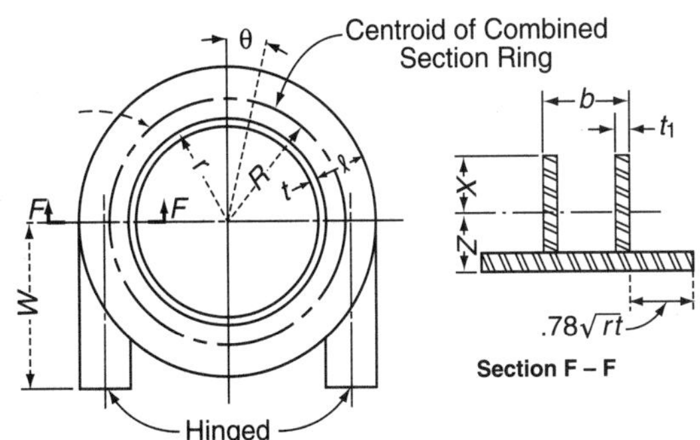

Figure 7-12 Stiffener ring coefficients—transverse earthquake

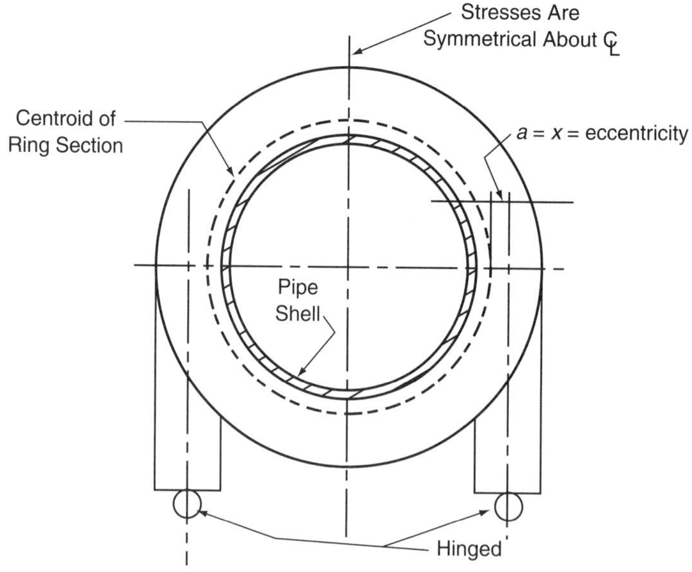

Loading Condition	Combination of Solutions*
Pipe just full or under pressure	Figure 7-10
Pipe half full	Fig. 7-10 Load $\pm Q_s + \frac{Q_w}{2}; W = Q_s + \frac{Q_w}{2}; M_o = \frac{\left(Q_s + \frac{Q_w}{2}\right)d}{2}$ $2a = 180°$
Transverse earthquake	Figure 7-12

*US Bureau of Reclamation (1944).

Figure 7-13 Combination of solutions

A. Nomenclature:

D = inside diameter, in. (mm)
r = inside radius, in. (mm)
R = radius of ring centroid, in. (mm)
t = shell thickness, in. (mm)
w' = weight of water = 62.4 lb/ft^3 (1 kg/m^3)
Q_w = weight of water in one span, lb (kg)
Q_s = weight of steel in one span, lb (kg)
Q = total weight of steel and water on one span, lb (kg)
w = combined weight of steel and water per linear in. (mm), lb/lin in. (kg/lin mm)
L = length of one span, in. (mm)
x = distance from end support, in. (mm)
$M_A = M_B$ = longitudinal bending moment at supports, in.-lbf (Nm)
M_l = longitudinal bending moment at mid-span, in.-lbf (Nm)
M_x = longitudinal bending moment at "x" distance from end supports, in.-lbf (Nm)
I/c = section modulus of shell, in.^{3} (mm^3)

S_x = circumferential stress in pipe shell between supports caused by internal pressure, psi (Pa)
S_y = longitudinal stress in pipe shell as a result of beam bending, psi (Pa)
M_L = maximum allowable longitudinal bending moment for a given plate thickness, in.-lbf (Nm)
S'_y = maximum allowable bending stress for a given plate thickness, psi (Pa)
S_L = maximum longitudinal stress, which when combined with the circumferential stress caused by internal pressure, the S_e, does not exceed S_A (see Figure 7-7), psi (Pa)
S_e = equivalent stress (see Figure 7-7), psi (Pa)
S_A = allowable stress, psi (Pa)
f_s = coefficient of friction at supports
n' = number of support piers
f' = frictional force at expansion joint
= approximately 500 lbf/circumferential ft (730 N circumferential m)
F_1 = longitudinal stress due to pressure on exposed pipe end at expansion joint, psi (Pa)
F_2 = longitudinal stress caused by frictional force at supports, psi (Pa)
F_3 = longitudinal stress caused by f', psi (Pa)
b = distance defined in Figure 7-12
c = distance defined in Figure 7-8
A_p = cross-sectional area of pipe, in.^{2} (m^2)
A = area of combined ring section, in.^{2} (m^2)
A_1 = area of single ring, in.^{2} (m^2)
A_2 = area of equivalent flange, in.^{2} (m^2)
I = moment of inertia of combined ring section, in.^{4} (mm^4)
N = tensile stress in ring section caused by internal pressure, psi (Pa)
$P = p$ = design pressure, psi (Pa)
T = direct stress in ring section exclusive of N, psi (Pa)
S_1 = ring bending stress in inside fiber, psi (Pa)
S_1' = ring bending stress in outside fiber, psi (Pa)
S_2 = total circumferential ring stress in inside fiber, psi (Pa)
S_2' = total circumferential ring stress in outside fiber, psi (Pa)
S_b = stress in pipe shell caused by ring restraint, psi (Pa)
S_3 = total longitudinal stress in pipe shell at support ring, psi (Pa)

B. Determine stresses and plate thicknesses between supports.

Basic data:

Material: ASTM A516, Gr. 60
S_A = 20,000 psi (1/3 of tensile strength)
D = 96 in.
r = 48 in.
t = 7/16 in. (assumed)
$p = P$ = 150 psi, including water hammer
L = 960 in.
w' = 62.4 lb/ft^3
n' = 2
f' = 500 lb/circumferential ft
f_s = 0.15

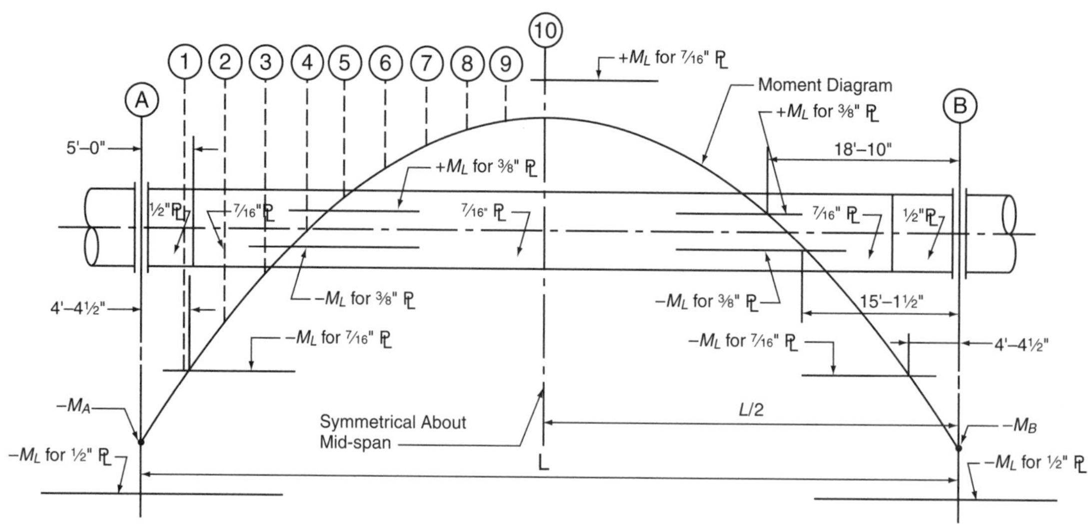

Figure 7-14 Stresses, moments, and plate thickness

Step 1:

Spans have been selected in laying out the penstock so that the beam bending moments at supports are given by the equation:

$$M_A = \frac{wL^2}{12}$$

Determine the moment curve and plot as shown in Figure 7-14.

$$Q_w = \frac{\pi D^2 L w'}{4(144)12} = \frac{\pi(96)^2(960)62.4}{4(144)12} = 250{,}925 \text{ lb}$$

$$Q_s = 0.89DtL = 0.89(96)(0.4375)960 = 35{,}885 \text{ lb}$$

$$Q = Q_w + Q_s = 250{,}925 + 35{,}885 = 286{,}810 \text{ lb}$$

$$w = \frac{Q}{L} = \frac{286{,}810}{960} = 299 \text{ lb per lin in.}$$

$$\frac{wL^2}{12} = \frac{299(960)^2}{12} = 22{,}963{,}200 \text{ in.-lb}$$

Moment calculations at x distances may be summarized as shown in Table 7-2. The circumferential stress between supports caused by internal pressure is

$$S_x = \frac{pr}{t}$$

for t = 3/8 in.; S_x = 150(48)/0.375 = +19,200 psi
t = 7/16 in.; S_x = 150(48)/0.4375 = +16,500 psi
t = 1/2 in.; S_x = 150(48)/0.5 = +14,400 psi
t = 9/16 in.; S_x = 150(48)/0.5625 = +12,800 psi

Table 7-2 Summary of moment calculations

x	$M_x = \frac{w}{12}\left(6Lx - L^2 - 6x^2\right)$	M_x in.-lb
0	$-\frac{wL^2}{12}$	–22,963,200
0.05L	$-0.0715\frac{wL^2}{12}$	–16,418,700
0.10L	$-0.460\frac{wL^2}{12}$	–10,563,100
0.15L	$-0.235\frac{wL^2}{12}$	–5,396,400
0.20L	$-0.040\frac{wL^2}{12}$	–918,500
0.25L	$+0.125\frac{wL^2}{12}$	+2,870,400
0.30L	$+0.260\frac{wL^2}{12}$	+5,970,400
0.35L	$+0.365\frac{wL^2}{12}$	+8,381,600
0.40L	$+0.440\frac{wL^2}{12}$	+10,103,800
0.45L	$+0.480\frac{wL^2}{12}$	+11,022,300
0.50L	$+0.500\frac{wL^2}{12}$	+11,481,600

NOTE: Values of M_x for x are symmetrical about mid-span.

With above values of S_x and $S_e = S_A = 20{,}000$ psi, determine from Figure 7-7 the maximum S'_L for the above plate thicknesses.

for $t = 3/8$ in.; $S_L = -1{,}600$ psi
$t = 7/16$ in.; $S_L = -5{,}800$ psi
$t = 1/2$ in.; $S_L = -8{,}500$ psi
$t = 9/16$ in.; $S_L = -10{,}200$ psi

Calculate F_1, F_2, and F_3:

$$F_1 = pt \qquad F_2 = \frac{f_s Q_n'}{\pi(D+t)t} \qquad F_3 = \frac{500D}{12(D+t)t}$$

t	F_1	F_2	F_3
3/8	–55	–760	–110
7/16	–65	–650	–95
1/2	–75	–570	–85
9/16	–85	–500	–75

NOTE: (–) sign denotes compression.

From: $-S'_y = -S_L + F_1 + F_2 + F_3$ and $I/c = \pi r^2 t$,

For $t = 3/8$ in.; $-S_y = +1{,}600 - 55 - 760 - 110 = 675$
$t = 7/16$ in.; $-S_y = +5{,}800 - 65 - 650 - 95 = 4{,}990$
$t = 1/2$ in.; $-S_y = +8{,}500 - 75 - 570 - 85 = 7{,}780$
$t = 9/16$ in.; $-S_y = +10{,}200 - 85 - 500 - 75 = 9{,}540$

The moment of inertia of a given plate thickness is obtained from the equation

$$\frac{I}{c} = \pi r^2 t$$

for $t = 3/8$ in.; $I/c = \pi(48)^2 0.375 = 2{,}714$
for $t = 7/16$ in.; $I/c = \pi(48)^2 0.4375 = 3{,}167$
for $t = 1/2$ in.; $I/c = \pi(48)^2 0.5 = 3{,}619$
for $t = 9/16$ in. $I/c = \pi(48)^2 0.5625 = 4{,}072$

From $M_L = \pm S'_y(I/c)$:

for $t = 3/8$ in.; $M_L = \pm 675(2{,}714) = \pm\ 1{,}834{,}650$
for $t = 7/16$ in.; $M_L = \pm 4{,}990(3{,}167) = \pm 15{,}803{,}330$
for $t = 1/2$ in.; $M_L = \pm 7{,}780(3{,}619) = \pm 28{,}155{,}820$
for $t = 9/16$ in.; $M_L = \pm 9{,}540(4{,}072) = \pm 38{,}838{,}750$

Values of M_L may be plotted as shown in Figure 7-14 and the intersect with the moment curve determined graphically to obtain points along the penstock where plate thicknesses vary. These points of intersection with the moment curve may be calculated as shown in Step 2.

Step 2:

Calculate x distances for maximum allowable moments for 3/8 in. and 1/2 in. plates (the moment curves for 1/2 in. and 9/16 in. plates do not intersect the moment diagram).

$$M_x = \frac{w}{12}(+6Lx - L^2 - 6x^2)$$

Where:

$w = 299$
$L = 960$

Then:

$$x^2 - 960x + 153{,}600 = -0.0067M_x$$

For 3⁄8 in. plate, $M_x = \pm 1{,}834{,}650$:

Then, for $M_x = +1{,}834{,}650$:

$x^2 - 960x + 153{,}600 = -0.0067(1{,}834{,}650) = -12{,}270$

$x^2 - 960x + 165{,}870 = 0$

$$x = \frac{960 \pm \sqrt{960^2 - 4(165{,}870)}}{2} = \frac{960 \pm 508}{2}$$

$x = 226$ in. or 18 ft 10 in. and 734 in. or 61 ft 2 in.

For $M_x = -1{,}834{,}650$:

$x^2 - 960x + 153{,}600 = +12{,}270$

$x^2 - 960x + 141{,}330 = 0$

$$x = \frac{960 \pm \sqrt{960^2 - 4(141{,}330)}}{2} = \frac{960 \pm 597}{2}$$

$x = 181.5$ in. or 15 ft 1½ in. and 778.5 in. or 64 ft 10½ in.

For 7⁄16 in. plate and for $M_x = \pm 15{,}803{,}330$:

$x^2 - 960x + 153{,}600 = -0.0067(-15{,}803{,}330)$

$x^2 - 960x + 47{,}718 = 0$

$$x = \frac{960 \times \sqrt{960^2 - 4(47{,}718)}}{2} = \frac{960 \pm 855}{2}$$

$x = 52.5$ in. or 4 ft 4½ in. and 907.5 in. or 75 ft 7½ in.

Note: The positive (+) M_x for 7⁄16 in. plate does not intersect the moment curve.

Step 3:

In summary, plate thicknesses of 7⁄16 in. and ½ in. were selected for reaches along the penstock that were consistent with the maximum allowable beam bending moment for a given thickness. From Figure 7-14, it should be noted that ½ in. plate is required over the support, but 7⁄16-in. plate is suitable along the remainder of the span.

C. Stiffener ring calculations:

Step 1:

Determine properties of assumed ring section.

Areas:

$2A_1 = 2(bd) = 2(0.75)(6) = 9.00$

$$A_2 = (1.56\sqrt{rt} + c)t = (1.56\sqrt{48(0.5)} + 9)0.5 = 8.32$$

Total area $= 2A_1 + A_2 = 17.32$ in.

Locate centroid:

$$2A_1\left(\frac{d}{2}+t\right) = 9.00(3.5) = 31.50$$

$$(A_2)\frac{t}{2} = 8.32(0.25) = 2.08$$

$$z = \frac{2A_1\left(\frac{d}{2}+t\right)+A_2\left(\frac{t}{2}\right)}{2A_1+A_2} = \frac{33.58}{17.32} = 1.94$$

$$x = 6.5 - z = 6.5 - 1.94 = 4.56$$

Calculate I of section about the centroid.

$$I_1 = 2\left[\frac{bd^3}{12}+A_1\left(\frac{d}{2}+t-z\right)^2\right] = 2\left[\frac{0.75(6)^3}{12}+0.75(6)\left(\frac{6}{2}+0.5-1.94\right)^2\right]$$

$$= 2(13.5+10.95) = 48.90 \text{ in.}^4$$

$$I_2 = \frac{(1.56\sqrt{rt}+c)t^3}{12}+A_2\left(z-\frac{t}{2}\right)^2 = \frac{(16.64)(0.5)^3}{12}+8.32(1.94-0.25)^2$$

$$= 0.17+23.76 = 23.93 \text{ in.}^4$$

$$I = I_1+I_2 = 48.90+23.93 = 72.83 \text{ in.}^4$$

$$\text{then } \frac{I}{z} = \frac{72.83}{1.94} = 37.54 \text{ in.}^3 \text{ and } \frac{I}{x} = \frac{72.83}{4.56} = 15.97 \text{ in.}^3$$

Step 2:
Determine circumferential ring stresses (Refer to Figure 7-6).

$$q = \frac{1.285}{\sqrt{rt}} = \frac{1.285}{\sqrt{48\times 0.5}} = 0.26$$

$$K = \frac{r}{L}\left[\frac{\mu L^2}{12r^2}+(1-\mu^2)\left(1-\frac{Q_s}{2Q}\right)+\frac{(2+\mu)L}{4qr^2}\right]$$

$$= \frac{48}{960}\left[\frac{0.3(960)^2}{12(48)^2}+(1-0.09)1-\left(\frac{35{,}880}{2(286{,}810)}\right)+\frac{2.3(960)}{(4)0.26(48)^2}\right]$$

$$K = 0.05(10+0.85+0.92) = 0.59$$

$$B = \frac{r}{R}\left(1-\frac{2K}{qr}\right)-\frac{X}{R}; \text{ where } X = 0.04R; \frac{X}{R} = 0.04$$

$$= \frac{48}{49.94}\left[1-\frac{2(0.59)}{0.26(48)}\right]-0.04 = 0.83$$

$$\frac{N}{A} = \frac{pr}{A}\left[b+\frac{2(1-\mu^2)}{q}\right] = \frac{150(48)}{17.32}\left[9+\frac{2(0.91)}{0.26}\right] = +6{,}650$$

Calculations for $\frac{T}{A}$, (refer to Figure 7-6).

For $\theta = 0°$:

$$\frac{T}{A} = \frac{Q}{A}(K_1 + BK_2) = \frac{286,810}{17.32}[-0.238732 + 0.83(0.318310)] = +420$$

For $\theta = 90°$ (–):

$$\frac{T}{A} = \frac{Q}{A}(K_1 + BK_2) = \frac{286,810}{17.32}(-0.25 + 0) = -4,140$$

For $\theta = 90°$(+):

$\frac{T}{A} = \frac{T}{A}$ for $\theta = 90°$ (–), except the sign is reversed = +4,140

For $\theta = 180°$:

$\frac{T}{A} = \frac{T}{A}$ for $\theta = 0°$, except the sign is reversed = –420

Ring bending moment computations (see Figure 7-6).

For $\theta = 0°$:

$x = 0.04R = 0.04(49.94) = 1.998$

$M = Q(RK_3 + xK_4) = 286,810\ [49.94(+0.011267) - 1.998(.068310)] = +122,235$

For $\theta = 90°$(–):

$M = Q(RK_3 + XK_4) = 286,810[50.06(0) + 1.998\ (0.25000)] = +143,260$

For $\theta = -90°$(+):

$M = M$ for $\theta = 90°$(–), except the sign is reversed = –143,260

For $\theta = 180°$:

$M = M$ for $\theta = 0°$, except the sign is reversed = –122,235

Step 3:

Tabulate values of $\frac{N}{A}, \frac{T}{A}$ and M in Table 7-3A, Figure 7-14, and calculate and tabulate S_1, S_2, and S'_2.

From section C, Step 1:

$I = 72.83$, $z = 1.94$, and $x = 4.56$

Then:

$$\frac{I}{z} = \frac{72.83}{1.94} + 37.54$$

$$\frac{I}{x} = \frac{72.83}{4.56} = 15.97$$

Table 7-3 Stresses at support ring

Table A

θ	N/A psi	T/A psi	M in.-lb	$I/\bar{z}$ in.^{3}	$I/\bar{x}$ in.^{3}	Bending Stress psi		Total Stress psi	
						S_1	S'_1	S_2	S'_2
0°	+6,650	+420	+122,000	37.54	15.97	+3,256	–7,654	+10,326	–584
90°–	+6,650	–4,140	+145,125	37.54	15.97	+3,816	–8,970	+6,326	–6,460
90°+	+6,650	+4,140	–145,125	37.54	15.97	–3,816	+8,970	+6,974	+19,760
180°	+6,650	–420	–122,000	37.54	15.97	–3,256	+7,654	+2,974	+13,884

Table B

θ	F_1 psi	F_2 psi	F_3 psi	S_b psi	S_y psi	S_3 Total Long. Stress–psi	S_e Equiv. Stress
0°	–75	–570	–85	+13,600	+6,350	+19,220	16,600
90°–	–75	–570	–85	+13,600	0	+12,870	11,100
90°+	–75	–570	–85	+13,600	0	+12,870	11,100
180°	–75	–570	–85	+13,600	+13,600	+6,520	5,600

+M produces tension(+) on inside fiber, S_1

– = compression

+ = tension

For θ = 0°:

$$S_1 = +\frac{122{,}235}{37.54} = +3{,}256$$

$$S'_1 = +\frac{122{,}235}{15.97} = -7{,}654$$

For θ = 90°(–):

$$S_1 = +\frac{143{,}260}{37.54} = +3{,}816$$

$$S'_1 = -\frac{143{,}260}{15.97} = -8{,}970$$

For θ = 90°(+):

$S_1 = S_1$ for θ = 90°(–), except sign is reversed = –3,816

$S'_1 = S_1$ for θ = 90°(–), except sign is reversed = +8,970

For θ = 180°:

$S_1 = S_1$ for θ = 0°, except sign is reversed = –3,256

$S'_1 = S'_1$ for θ = 0°, except sign is reversed = +7,654

For $\theta = 0°$:

$$S_2 = \frac{N}{A} + \frac{T}{A} + S_1 = +6{,}650 + 420 + 3{,}256 = +10{,}326$$

$$S'_2 = \frac{N}{A} + \frac{T}{A} + S'_1 = +6{,}650 + 420 + 7{,}654 = -584$$

For $\theta = 90°(-)$:

$$S_2 = \frac{N}{A} + \frac{T}{A} + S_1 = +6{,}650 - 4{,}140 + 3{,}816 = +6{,}326$$

$$S'_2 = \frac{N}{A} + \frac{T}{A} + S'_1 = +6{,}650 - 4{,}140 - 8{,}970 = -6{,}460$$

For $\theta = 90°(+)$:

$$S_2 = \frac{N}{A} + \frac{T}{A} + S_1 = +6{,}650 + 4{,}140 - 3{,}816 = +6{,}974$$

$$S'_2 = \frac{N}{A} + \frac{T}{A} + S'_1 = +6{,}650 + 4{,}140 + 8{,}970 = +19{,}760$$

For $\theta = 180°$:

$$S_2 = \frac{N}{A} + \frac{T}{A} + S_1 = +6{,}650 - 420 - 3{,}256 = +2{,}974$$

$$S'_2 = \frac{N}{A} + \frac{T}{A} + S'_1 = +6{,}650 - 420 + 7{,}654 = +13{,}884$$

Step 4:
Longitudinal stress computations.

Ring restraint stress—(see Figure 7-6 and ring properties, Step 1).

$A = 2ah + ct = 2(0.75)6 + 9(0.5) = 13.50$

$$1.56t\sqrt{rt} = 1.56(0.5)\sqrt{48(0.5)} = 3.82$$

$p = 150$

$$S_b = 1.82\left[\frac{(A - ct)}{A + 1.56\sqrt{rt}}\frac{pr}{t}\right] = 1.82\left[\frac{9.0}{13.5 + 3.82}\frac{150(48)}{0.5}\right] = +13{,}600$$

Beam bending at support ring:

$M_A = M_B = -22{,}963{,}200$ in.-lb; from section B.

$$Sy = \frac{M_A}{\left(\frac{I}{c}\right)} = \frac{22{,}963{,}200}{3{,}619} = \pm 6{,}350$$

Longitudinal stresses in the pipe shell caused by frictional forces at piers and expansion joint and hydrostatic pressure on the exposed end of pipe in expansion joint were determined in section B and are

$F_1 = -75$ $\quad F_2 = -570$ $\quad F_3 = -85$

Then:

For $\theta = 0°$,

$$S_3 = F_1 + F_2 + F_3 + S_b + S_y = -75 - 570 - 85 + 13{,}600 + 6{,}350 = +19{,}220$$

For $\theta = 90°(-)$,

$$S_3 = F_1 + F_2 + F_3 + S_b + S_y = -75 - 570 - 85 + 13{,}600 + 0 = +12{,}870$$

For $\theta = 90°\ (+)$,

S_3 = same as for $\theta = 90°\ (-) = +\ 12{,}870$

For $\theta = 180°$,

$$S_3 = F_1 + F_2 + F_3 + S_b + S_y = -75 - 570 - 85 + 13{,}600 - 6{,}350 = +\ 6{,}520$$

Using Figure 7-7 and S_3 and S_2 in Table 7-3, obtain S_e.

For $\theta = 0°$,

$S_2 = +10{,}326$

$S_3 = +19{,}220$

$S_e = 16{,}600$

For $\theta = 90°(-)$,

$S_2 = +6{,}326$

$S_3 = +12{,}870$

$S_e = 11{,}100$

For $\theta = 90°(+)$,

$S_2 = +6{,}974$

$S_3 = +12{,}870$

$S_e = 11{,}100$

For $\theta = 180°$,

$S_2 = +2{,}974$

$S_3 = +6{,}520$

$S_e = 5{,}600$

Step 5:

Circumferential, longitudinal, and equivalent stresses are tabulated in Table 7-3. Note that S'_2 at = 90°(+) = 19,760 psi, which is less than the allowable of 20,000 psi.

D. Determine stiffener ring stresses for half-full condition.

Basic data: Same as part B

Step 1:

From Figure 7-9, determine ring stresses for $2\alpha = 180°$ and tabulate resulting values in Table 7-4 for values of = 0°, 45°, 90°(–), 90°(+), 135°, and 180°.

$R = 49.94$

$Q_s = 35{,}885$

$Q_w = 250{,}925$

$a = 0.04R = 0.04\,(49.94) = 2.00$

$$M_o = (Q_s + 0.5Q_w)\frac{a}{2} = (35{,}885 + 125{,}463)\frac{2.00}{2} = 161{,}350$$

$$\frac{M_o}{R} = \frac{161{,}350}{49.94} = 3{,}231$$

$I = 72.83$

$x = 4.56$

$z = 1.94$

$A = 17.32$

σ_3 = Total stress in outer fiber = $\frac{T}{A} = \frac{Mx}{I}$

σ_3' = Total stress in inner fiber = $\frac{T}{A} = \frac{Mx}{I}$

For $\theta = 0°$,

$$M = M_0K_1 = 161{,}350(-0.136620) = -22{,}043$$

$$T = \frac{M_o}{R}K_2 = 3{,}231(-0.636620) = 2{,}057$$

$$\sigma_3 = \frac{T}{A} - \frac{Mx}{I} = \frac{-2{,}057}{17.32} - \frac{(-22{,}043)4.56}{72.83} = -119 + 1{,}380 = +1{,}261$$

$$\sigma_3' = \frac{T}{A} - \frac{Mz}{I} = \frac{-2{,}057}{17.32} - \frac{(-22{,}043)1.94}{72.83} = -119 - 587 = -706$$

For $\theta = 45°$,

$$M = M_0K_1 = 161{,}350\,(+0.049842) = +8{,}042$$

$$T = \frac{M_0}{R}K_2 = 3{,}231(-0.450158) = -1{,}454$$

$$\sigma_3 = \frac{T}{A} - \frac{Mx}{I} = \frac{-1{,}454}{17.32} - \frac{8{,}042(4.56)}{72.83} = -84 + 504 = -588$$

$$\sigma_3' = \frac{T}{A} - \frac{Mz}{I} = \frac{-2{,}057}{17.32} + \frac{8{,}042(1.94)}{72.83} = -84 + 214 = +130$$

For $\theta = 90°(-)$,

$$M = M_0K_1 = 161{,}350(+0.5) = +80{,}675$$

$$T = \frac{M_0}{R}K_2 = 3{,}231(0) = 0$$

$$\sigma_3 = \frac{T}{A} - \frac{Mx}{I} = 0 - \frac{80{,}675(4.56)}{72.83} = 0 - 5{,}051 = -5{,}051$$

$$\sigma_3' = \frac{T}{A} + \frac{Mz}{I} = 0 + \frac{80{,}675(1.94)}{72.83} = 0 + 2{,}149 = +2{,}149$$

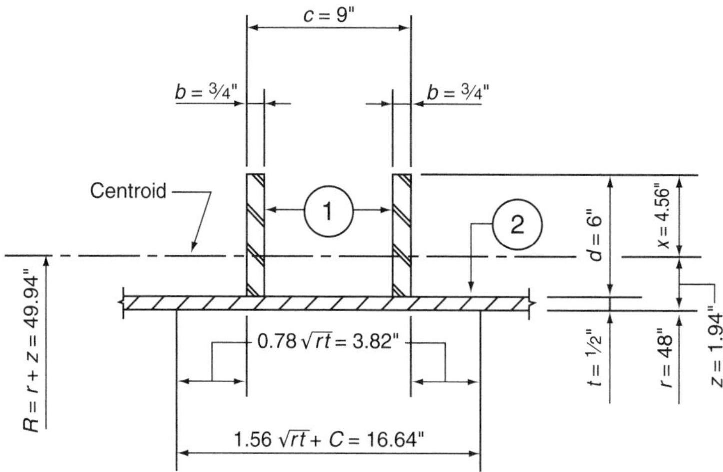

Figure 7-15 Detail of assumed ring section

For θ = 90°(+), 135°, and 180°, σ_3 and σ'_3 at these points are the same as at points θ = 90° (–), 45°, and 0°, except their signs are reversed as follows:

For θ = 90°(+):

$\sigma_3 = +5{,}051$

$\sigma'_3 = -2{,}149$

For θ = 135°:

$\sigma_3 = +588$

$\sigma'_3 = -130$

For θ = 180°:

$\sigma_3 = -1{,}261$

$\sigma'_3 = +706$

Step 2:

From Figure 7-15, determine ring stresses for 2α = 180° and tabulate resulting values in Table 7-5 for θ = 0°, 45°, 90°(–), 90°(+), 135°, and 180°.

$$W = Q_s + \frac{Q_w}{2} = 35{,}885 + \frac{250{,}925}{2} = 161{,}348$$

$R = 49.94$

$A = 17.32$

$I = 72.83$

$x = 4.56$

$z = 1.94$

$$\sigma_4 = \text{Total stress in outer fiber} = \frac{T}{A} - \frac{Mx}{I}$$

$$\sigma'_4 = \text{Total stress in outer fiber} = \frac{T}{A} + \frac{Mz}{I}$$

For $\theta = 0°$:

$M = WRK_1 = 161{,}350(49.94)(-0.068310) = -550{,}430$

$T = WK_2 = 161{,}350(-0.159155) = -25{,}680$

$$\sigma_4 = \frac{T}{A} - \frac{Mx}{I} = \frac{-25{,}680}{17.32} - \frac{(-550{,}430)4.56}{72.83} = -1{,}483 + 34{,}463 = +32{,}980$$

$$\sigma_4' = \frac{T}{A} - \frac{Mz}{I} = -\frac{-25{,}680}{17.32} + \frac{(-550{,}430)1.94}{72.83} = 1{,}483 - 14{,}662 = -16{,}145$$

For $\theta = 45°$:

$M = WRK_1 = 161{,}350(49.94)\,(-0.021694) = -174{,}806$

$T = WK_2 = 161{,}350(-0.112{,}540) = -18{,}158$

$$\sigma_4 = \frac{T}{A} - \frac{Mx}{I} = -\frac{-18{,}158}{17.32} - \frac{(-174{,}806)4.56}{72.83} = -1{,}048 + 10{,}945 = +9{,}897$$

$$\sigma_4' = \frac{T}{A} + \frac{Mz}{I} = -\,1{,}048 + \frac{(-174{,}806)1.94}{72.83} = -1{,}048 - 4{,}656 = -5{,}704$$

For $\theta = 90°(-)$:

$M = WRK_1 = 161{,}350(49.94)(+0.090845) = +732{,}013$

$T = WK_2 = 161{,}350(0) = 0$

$$\sigma_4 = \frac{T}{A} - \frac{Mx}{I} = 0 - \frac{(+732{,}013)4.56}{72.83} = 0 - 45{,}832 = -45{,}832$$

$$\sigma_4' = \frac{T}{A} - \frac{Mz}{I} = 0 + \frac{(+732{,}013)1.94}{72.83} = 0 + 19{,}499 = +19{,}499$$

For $\theta = 90°(+)$:

$M = WRK_1 = 8{,}057{,}819(+0.090845) = +732{,}013$

$T = WK_2 = 161{,}350(+0.5) = +80{,}675$

$$\sigma_4 = \frac{T}{A} - \frac{Mx}{I} = -\frac{80{,}675}{17.32} - \frac{(+732{,}013)4.56}{72.83} = +4{,}658 - 45{,}832 = -41{,}174$$

$$\sigma_4' = \frac{T}{A} + \frac{Mz}{I} = +4{,}658 + \frac{(+732{,}013)1.94}{72.83} = +4{,}658 + 19{,}499 = +24{,}157$$

For $\theta = 135°$:

$M = WRK_1 = 8{,}057{,}819(+0.056938) = +458{,}796$

$T = WK_2 = 161{,}350(+0.466{,}093) = +75{,}204$

$$\sigma_4 = \frac{T}{A} - \frac{Mx}{I} = -\frac{75{,}204}{17.32} - \frac{(+458{,}796)4.56}{72.83} = +4{,}342 - 28{,}725 = -24{,}383$$

$$\sigma_4' = \frac{T}{A} + \frac{Mz}{I} = +4{,}342 + \frac{(+458{,}796)1.94}{72.83} = +4{,}342 + 12{,}221 = +16{,}563$$

For $\theta = 180°$:

$M = WRK_1 = 8{,}057{,}819(-0.25) = -2{,}014{,}455$

$T = WK_2 = 161{,}350(+0.159155) = +25{,}680$

$$\sigma_4 = \frac{T}{A} - \frac{Mx}{I} = -\frac{25{,}680}{17.32} - \frac{(-2{,}014{,}455)4.56}{72.83} = +1{,}483 + 126{,}128 = +127{,}611$$

$$\sigma_4' = \frac{T}{A} + \frac{Mz}{I} = +1{,}483 + \frac{(-2{,}014{,}455)1.94}{72.83} = +1{,}483 + 53{,}660 = -52{,}177$$

Step 3:

From Figure 7-11 determine ring stresses for $2\alpha = 180°$ and tabulate resulting values in Table 7-5 for $\theta = 0°, 45°, 90°(-), 90°(+), 135°, 180°$. The following data are applicable to all of the calculations: $L = 960$; $L^2 = 921{,}600$; $\mu = 0.3$; $\mu^2 = 0.09$; $(1 - \mu^2) = 0.91$; $r = 48$; $r^2 = 2{,}304$; $t = ½$; $t^2 = 0.25$; $c = 9.0$; $Q_w = 250{,}925$; and $Q_s = 35{,}885$.

$$Q = \sqrt[4]{\frac{3(1-\mu^2)}{r^2t^2}} = \sqrt[4]{\frac{3(0.91)}{(2{,}304)0.25}} = 0.2624$$

$$Q_1'' = Q_s + \frac{Q_w}{2} = 35{,}885 + \frac{250{,}925}{2} = 161{,}350$$

$$Q_2'' = \frac{2Q_w}{3(\pi)^2} = \frac{2(250{,}925)}{3(3.1416)^2} = 16{,}949$$

$$Q_4'' = \frac{2}{15(\pi)^2}Q_w = \frac{2(250{,}925)}{15(3.1416)^2} = 3{,}390$$

$$Q_0'' = \frac{Q_w}{\pi^2} = \frac{250{,}925}{(3.1416)^2} = 25{,}424$$

$$G_2 = 1 + \frac{3(R-r)}{R}\left(1 - \frac{2K_z}{Lq}\right) = 1 + \frac{3(1.94)}{49.94}\left(1 - \frac{2(44.56)}{960(0.2624)}\right)$$

$$= 1 + 0.1165(0.6462) = 1.075$$

$$K_2 = \frac{\mu L^2}{3r^2} + (1-\mu^2) + \frac{(2+\mu)L}{qr^2} = \frac{0.3(921{,}600)}{3(2{,}304)} + 0.91 + \frac{2.3(960)}{0.2624(2{,}304)}$$

$$= 40 + 0.91 + 3.65 = 44.56$$

$$G_4 = 1 + \frac{15(R-r)}{R}\left(1 - \frac{2K_4}{Lq}\right) = 1 + \frac{15(1.94)}{49.94}\left(1 - \frac{2(175.52)}{960(0.2624)}\right)$$

$$= 1 + 0.5827(1 - 1.3935) = 1 - 0.23 = 0.077$$

$$K_4 = \frac{4\mu L^2}{3r^2} + (1-\mu^2) + \frac{4(2+\mu)L}{qr^2} = \frac{4(0.3)(921{,}600)}{3(2{,}304)} + 0.91 + \frac{4(2.3)(960)}{0.2624(2{,}304)}$$

$$= 160 + 0.91 + 14.61 = 175.52$$

$$K_1'' = \frac{r}{L}\left[\frac{\mu L^2}{12r^2} + (1-\mu^2)\left(1 - \frac{Q_s}{2Q_1''}\right) + \frac{(2+\mu)L}{4qr^2}\right]$$

$$= \frac{48}{960}\left[\frac{0.3(921{,}600)}{12(2{,}304)} + 0.91\left(1 - \frac{35{,}885}{2(161{,}350)}\right) + \frac{2.3(960)}{4(0.2624)2{,}304}\right]$$

$$= 0.05(10 + 0.81 + 0.91) = 0.59$$

$$\frac{Q_1''R}{\pi} = \frac{161{,}350(49.94)}{\pi} = 2{,}564{,}883$$

$$\frac{Q_2''RG_2}{3} = \frac{16{,}949(49.94)1.075}{3} = 303{,}305$$

$$\frac{Q_4''RG_4}{15} = \frac{3{,}390(49.94)0.77}{15} = 8{,}691$$

$$\frac{Q_1''}{4\pi} = \frac{161{,}350}{4\pi} = 12{,}340$$

$$\frac{Q_1''r}{\pi R}\left(1 - \frac{2K''_1}{qr}\right) = \frac{161{,}350(48)}{\pi(49.94)}\left(1 - \frac{2(0.59)}{(0.2624)48}\right) = 44{,}739$$

$$\frac{Q_o''}{L}\left[c + \frac{2}{q}(1-\mu^2)\right] = \frac{25{,}424}{960}\left[9.0 + \frac{2(0.91)}{0.2624}\right] = 421$$

$$Q_2''\left(\frac{4}{3} - \frac{2K_2}{Lq}\right) = 16{,}949\left(1.33 - \frac{2(44.56)}{960(0.2624)}\right) = 16{,}602$$

$$Q_4''\left(\frac{16}{15} - \frac{2K_2}{Lq}\right) = 3{,}390\left(\frac{16}{15} - \frac{2(175.52)}{960(0.2624)}\right) = -1{,}108$$

For $\theta = 0°$: (see Table 7-5.)

$$M = \frac{Q_1''R}{\pi}\left(\frac{1}{2} - \frac{\cos\theta}{4} - \frac{\theta\sin\theta}{2}\right) - \left(\frac{Q_3''RG_2}{3}\right)\cos 2\theta - \left(\frac{Q_4''RG_4}{15}\right)\cos 4\theta$$

$$M = 2{,}564{,}883\ (0.5 - 1/4 - 0) - 303{,}305\ (1.00) - 8{,}691\ (1.00) = 329{,}225$$

$$T = -\frac{Q_1''}{4\pi}(\cos\theta + \theta\sin\theta) + \frac{Q_1''r}{\pi R}\left(1 - \frac{2K_1''}{qr}\right)\cos\theta + \frac{Q_o''}{L}\left[c + \frac{2}{q}(1-\mu^2)\right]$$
$$- Q_2''\left(\frac{4}{3} - \frac{2K_2}{Lq}\right)\cos 2\theta - Q_4''\left(\frac{16}{15} - \frac{2K_4}{Lq}\right)\cos 4\theta$$

$$T = -12{,}840(1.00 + 0) + 44{,}739(1.00) + 421 - 16{,}602(1.00) - (-1{,}108)1.00$$

$$T = -12{,}840 + 44{,}739 + 421 - 16{,}602 + 1{,}108 = +16{,}826$$

$$\sigma_5 = \frac{T}{A} - \frac{Mx}{I} = \frac{16{,}826}{17.32} - \frac{329{,}255(4.56)}{72.83} = 971 - 20{,}613 = -19{,}642$$

$$\sigma_5' = \frac{T}{A} - \frac{Mz}{I} = 971 + \frac{329{,}255(1.94)}{72.83} = 971 + 8{,}770 = +9{,}741$$

For $\theta = 45°$:

$$M = \frac{Q_1''R}{\pi}\left(\frac{1}{2} - \frac{\cos\theta}{4} - \frac{\theta\sin\theta}{2}\right) - \left(\frac{Q_2''RG_2}{3}\right)\cos 2\theta - \left(\frac{Q_4''RG_4}{15}\right)\cos 4\theta$$

$$= 2{,}564{,}883\left(0.5 - \frac{0.7071}{4} - \frac{0.7854(0.7071)}{2}\right) - 303{,}305(0) - 8{,}691(-1.00)$$

$$= 116{,}822 - 0 + 8{,}691 = +125{,}513$$

$$T = \frac{Q_1''}{4\pi}(\cos\theta + \theta\sin\theta) + \frac{Q_1''r}{\pi R}\left(1 - \frac{2K_1''}{qr}\right)\cos\theta + \frac{Q_0''}{L}\left[c + \frac{2}{q}(1-\mu^2)\right]$$

$$-Q_2''\left(\frac{4}{3} - 2\frac{K_2}{Lq}\right)\cos 2\theta - Q_4''\left(\frac{16}{15} - \frac{2K_4}{Lq}\right)\cos 4\theta = -1.0$$

$$T = -12{,}840[0.7071 + 1.5708(0.7071)] + 44{,}739(0.7071) + 421 - 16{,}602(0)$$

$$-(-1{,}108) - 1.0 = -23{,}341 + 31{,}635 + 421 - 0 - 1{,}108 = +7{,}607$$

$$\sigma_5 = \frac{T}{A} - \frac{Mx}{I} = \frac{7{,}607}{17.32} - \frac{(+125{,}513)(4.56)}{72.83} = +439 - 7{,}858 = -7{,}419$$

$$\sigma_5' = \frac{T}{A} - \frac{Mz}{I} = +540 + \frac{125{,}513(1.94)}{72.83} = +567 + 3{,}343 = +3{,}910$$

For $\theta = 90°(-)$:

$$M = \frac{Q_1''R}{\pi}\left(\frac{1}{2} - \frac{\cos\theta}{4} - \frac{\theta\sin\theta}{2}\right) - \left(\frac{Q_2''RG_2}{3}\right)\cos 2\theta - \left(\frac{Q_4''RG_4}{15}\right)\cos 4\theta$$

$$= 2{,}564{,}883\left[0.5 - 0 - \frac{1.5708(1.0)}{2}\right] - 303{,}305(-1.0) - 8{,}691(+1.0)$$

$$= -732{,}018 + 303{,}305 - 8{,}691 = -437{,}404$$

$$T = \frac{Q_1''}{4\pi}(\cos\theta + 2\theta\sin\theta) + \frac{Q_1''r}{\pi R}\left(1 - \frac{2K_1''}{qr}\right)\cos\theta + \frac{Q_0''}{L}\left[c + \frac{2}{q}(1-\mu^2)\right]$$

$$-Q_2''\left(\frac{4}{3} - \frac{2K_2}{Lq}\right)\cos 2\theta - Q_4''\left(\frac{16}{15} - \frac{2K_4}{Lq}\right)\cos 4\theta$$

$$T = -12{,}840[0 + 3.1416(1.0)] + 44{,}739(0) + 421 - 16{,}602(-1.0) + 1{,}108(1.0)$$

$$= -40{,}338 + 0 + 421 + 16{,}602 + 1{,}108 = -22{,}207$$

$$\sigma_5 = \frac{T}{A} - \frac{Mx}{I} = -\frac{22{,}207}{17.32} - \frac{(-437{,}404)4.56}{72.83} = -1{,}282 + 27{,}387 = +26{,}105$$

$$\sigma_5' = \frac{T}{A} + \frac{Mz}{I} = -3{,}211 - \frac{437{,}404(1.94)}{72.83} = -1{,}282 - 11{,}651 = -12{,}933$$

For $\theta = 90°(+)$:

Stresses are the same as for $\theta = 90°$ (–).

For $\theta = 135°$:

$$M = \frac{Q_1''R}{\pi}\left(\frac{1}{2} - \frac{\cos\theta}{4} - \frac{\theta\sin\theta}{2}\right) - \left(\frac{Q_2''RG_2}{3}\right)\cos 2\theta - \left(\frac{Q_4''RG_4}{15}\right)\cos 4\theta$$

$$= 2{,}564{,}883\left[0.5 + \frac{0.7071}{4} - \frac{2.3562(0.7071)}{2}\right] - 303{,}305(0) - 8{,}691(-1.00)$$

$$= -400{,}787 + 0 + 8{,}691 = -392{,}096$$

$$T = \frac{Q_1''}{4\pi}(\cos\theta + 2\theta\sin\theta) + \frac{Q_1''r}{\pi R}\left(1 - \frac{2K_1''}{qr}\right)\cos\theta + \frac{Q_0''}{L}\left[c + \frac{2}{q}(1-\mu^2)\right]$$

$$- Q_2''\left(\frac{4}{3} - \frac{2K_2}{Lq}\right)\cos 2\theta - Q_4''\left(\frac{16}{15} - \frac{2K_4}{Lq}\right)\cos 4\theta$$

$$= -12{,}840[-0.7071 + 2(2.3562)(0.7071)] + 44{,}739(-0.7071) + 421$$

$$- 16{,}602(-1.00) + 1{,}108(-1.00)$$

$$= -33{,}750 - 31{,}635 + 421 + 16{,}602 - 1{,}108$$

$$= -49{,}470$$

$$\sigma_5 = \frac{T}{A} - \frac{Mx}{I} = -\frac{49{,}470}{17.32} - \frac{(-392{,}096)4.56}{72.83} = -2{,}856 + 24{,}550 = +21.694$$

$$\sigma_5' = \frac{T}{A} + \frac{Mz}{I} = -2{,}856 + \frac{(-392{,}096)1.94}{72.83} = -2{,}856 - 10{,}444 = -13{,}300$$

For $\theta = 180°$:

$$M = \frac{Q_1''R}{\pi}\left(\frac{1}{2} - \frac{\cos\theta}{4} - \frac{\theta\sin\theta}{2}\right) - \left(\frac{Q_2''RG_2}{3}\right)\cos 2\theta - \left(\frac{Q_4''RG_4}{15}\right)\cos 4\theta$$

$$= 2{,}564{,}883(0.5 + 0.25 - 0) - 303{,}305(1.00) - 8{,}691(1.00)$$

$$= -1{,}923{,}662 - 303{,}305 - 8{,}691 = 1{,}611{,}666$$

$$T = \frac{Q_1''}{4\pi}(\cos\theta + 2\theta\sin\theta) + \frac{Q_1''r}{\pi R}\left(1 - \frac{2K_1''}{qr}\right)\cos\theta + \frac{Q_0''}{L}\left[c + \frac{2}{q}(1-\mu^2)\right]$$

$$- Q_2''\left(\frac{4}{3} - \frac{2K_2}{Lq}\right)\cos 2\theta - Q_4''\left(\frac{16}{15} - \frac{2K_4}{Lq}\right)\cos 4\theta$$

$$= -12{,}840(1.00 + 0) + 44{,}739(-1.00) + 223 - 16{,}602(1.00) - (1{,}108)1.00$$

$$= +12{,}840 - 44{,}739 + 421 - 16{,}602 + 1.108 = -46{,}972$$

$$\sigma_5 = \frac{T}{A} - \frac{Mx}{I} = -\frac{46{,}972}{17.32} - \frac{(+1{,}611{,}666)4.56}{72.83} = -2{,}712 - 100{,}909 = -103{,}621$$

$$\sigma_5' = \frac{T}{A} - \frac{Mz}{I} = +\frac{1{,}611{,}666(1.94)}{72.83} = -2{,}712 + 42{,}931 = +40{,}219$$

Step 4:

Summarize stresses and tabulate in Table 7-4. It will be noted that the stresses σ_6 and σ_6', at = 90° (+) are (–)10,018 and (+)9,075 psi, respectively. It will be noted that all values of σ_6 and σ_6' are below the allowable stress of 25,600 psi.

Concrete Piers

Concrete piers should be designed for the vertical reactions at the support, longitudinal forces resulting from frictional resistance as a result of longitudinal strain (Poisson's ratio) and temperature movements, and lateral forces caused by wind and earthquake forces. The resultant of all forces under the most unfavorable conditions should

Table 7-4 Summary of stresses for half-full condition

	Step 1		Step 2		Step 3		Summation	
θ°	σ_3 *psi*	σ'_3 *psi*	σ_4 *psi*	σ'_4 *psi*	σ_5 *psi*	σ'_5 *psi*	σ_6 *psi*	σ'_6 *psi*
0°	+1,261	–706	+32,980	–16,145	–19,642	+9,744	+14,599	–7,120
45°	–588	+130	+9,897	–5,704	–7,419	+3,910	+1,890	–1,664
90°(–)	–5,051	+2,149	–45,832	+19,499	+26,105	–12,933	–24,778	+8,715
90°(+)	+5,051	–2,149	–41,174	+24,157	+26,105	–12,933	–10,018	+9,075
135°	+588	–130	–24,383	+16,563	+21,694	–13,300	–2,101	+3,133
180°	–1,261	706	+127,611	–52,177	–103,621	+40,219	+22,729	–11,252

NOTES: $\sigma_6 = \sigma_3 + \sigma_4 + \sigma_5$

$\sigma'_6 = \sigma'_3 + \sigma'_4 + \sigma'_5$

Stresses are symmetrical about 0°–180° axis.

Table 7-5 Trigonometric data

θ°	Radians	sin θ	cos θ	sin 2θ	cos 2θ	sin 4θ	cos 4θ
0°	0	0	1.00	0	1.00	0	1.00
45°	π/4 = 0.7854	0.4041	0.7071	1.00	0	0	–1.00
90°(–)	π/2 = 1.5708	1.00	0	0	–1.00	0	1.00
90°(+)	π/2 = 1.5708	1.00	0	0	–1.00	0	1.00
135°	3π/4 = 2.3562	0.7071	–0.7071	–1.00	0	0	–1.00
180°	π = 3.1416	0	–1.00	0	1.00	0	1.00

intersect the base within the middle third to insure that the footing is in bearing (compression) throughout. The pier must be stable against sliding. The vertical component of the resultant of all forces should not be less than the horizontal component of all forces divided by the coefficient of sliding friction at the base of the pier. The friction coefficient may vary from 0.35 to 0.65, depending on the underlying material. The base of the pier should be placed below the frost line. Steel reinforcement of concrete piers is usually limited to that required for temperature and shrinkage crack control.

Concrete Anchors

Pipelines supported aboveground and having expansion joints or sleeve-type couplings require anchors at all points of changes in slope or alignment. Where expansion joints are used, a spacing of over 500 ft between anchors and expansion joints is not normally justified because of the accumulation of longitudinal forces and the desirability of more fixed points during installation. Buried pipelines with welded circumferential joints normally do not require anchors at points of changes in slope or alignment. Buried pipelines with sleeve-type coupled or similar field joints require anchors similar to those required for an aboveground installation. Anchor blocks and restraint of buried pipe are covered in chapter 13.

RING-GIRDER CONSTRUCTION FOR LOW-PRESSURE PIPE

General designs for two types of long-span pipe of the flow line variety are shown in Figure 7-16.

Type 1. Usually recommended for crossing canals and other low places where a single length of pipe for spans up to 60 ft can be used, type 1 pipe may be made and shipped from the factory in one length or in two lengths; in the latter case, a welded joint must be made in the field at the time of installation.

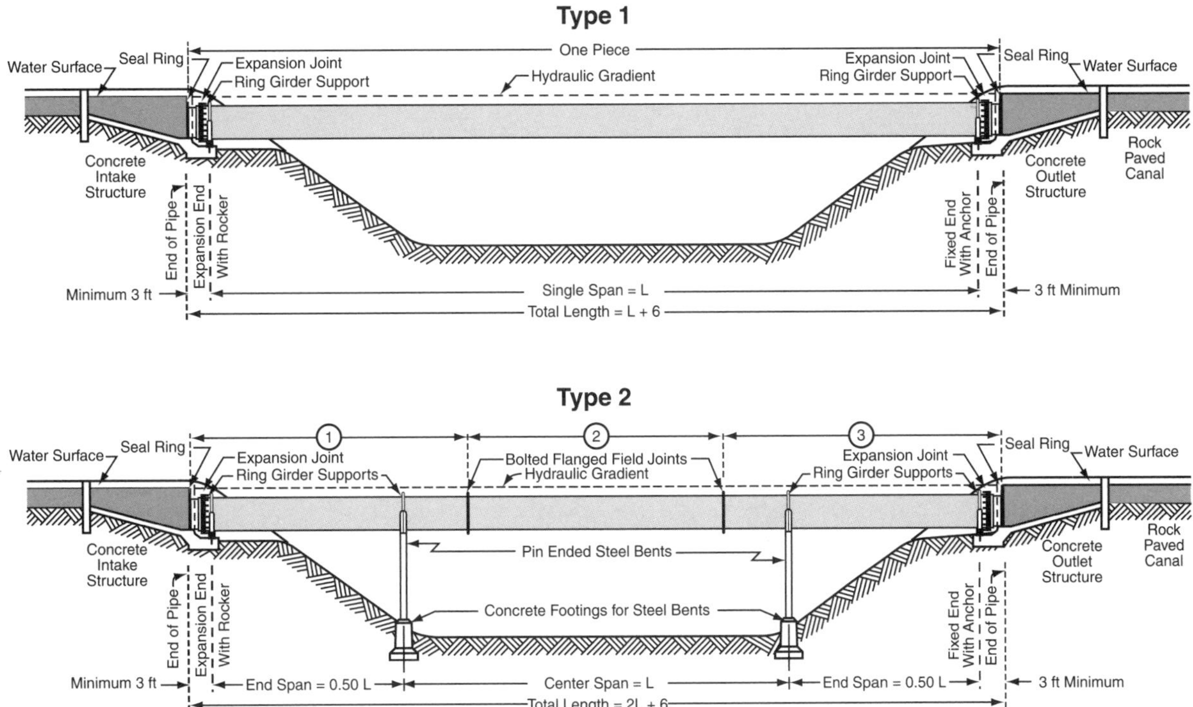

Figure 7-16 Long-span steel pipe for low pressures

Type 2. Used for crossing highways, canals, or rivers, where the length of the crossing makes it necessary to install two intermediate supporting columns, type 2 pipe is designed in three lengths with flanges welded to the ends of each length at points of contraflexure, together with expansion joints for both intake and outlet. This type is normally used for crossings 60 ft to 132 ft, with end spans half the length of the center span.

INSTALLATION OF RING GIRDER SPANS

In addition to proper design, long-span, ring-girder-supported steel pipelines require careful field erection, particularly in regard to alignment and camber, avoidance of movement caused by temperature differences on opposite sides of the pipe, and correct welding procedure. The following suggestions will be helpful, and more information has been published. (See Figure 7-17.)

Pipes such as these that may be exposed to low temperatures can affect the ability of the steel to resist brittle fracture. (See Effects of Cold Working on Strength and Ductility, chapter 1.) Steel should be properly selected, detailed, and welded to mitigate this effect.

Concrete Footings

Before assembling the pipe, concrete footings (but not the intake or outlet boxes) should be poured. If the pipe is supported on rollers, a pocket is left at the top of the footings as a base for the roller bed plates. If steel bents are used, anchor bolts are set in concrete footings to anchor the lower end of the pin-ended steel bents or the base plates. The concrete footings should be sized to allow for grouting these supporting members to their proper height.

Figure 7-17 111-in. pipe on ring girders

Expansion Joints

Expansion joints are installed in long-span steel pipe to allow for expansion or contraction caused by temperature changes. These joints are placed near the concrete headwalls and should be left entirely loose until the concrete has been allowed to set for at least two weeks. If expansion joints are tightened before concrete is poured, the pipe may pull loose from the green concrete. After concrete has set thoroughly, expansion joints are tightened and all danger of damage from pipe movement is eliminated.

To protect the expansion joint during shipment, the manufacturer may have to tack-weld steel ties to the inside of the pipe, tying the two pieces of pipe together across the joint. When this is done, the steel ties must by knocked loose from the pipe as soon as it is set in place and before the concrete is poured.

Assembling Pipe

Pipe being assembled should be supported by a temporary framework between piers. All bolts except expansion joint bolts should be tightened. When the pipe is in place, concrete intake and outlet boxes should be poured. Bed plates for the rollers or pin-ended steel bents can then be grouted in place to the proper height. Temporary supports and blocking should be removed before the pipe is filled with water, otherwise the structure will be subjected to undue stress.

Table 7-6 Values of moment of inertia and section modulus of steel pipe

Nominal Size*		Wall Thickness		Weight of Pipe & Water		Moment of Inertia		Section Modulus	
in.	*(mm)*	*in.*	*(mm)*	*lb/ft*	*(kg/m)*	*in.*^{4}	*mm*^{4}	*in.*^{3}	*mm*^{3}
6	150	0.105	2.67	19	28.28	8.45	3.5172×10^6	2.82	4.6212×10^4
		0.135	3.43	21	31.25	10.70	4.4537×10^6	3.57	5.8502×10^4
		0.188	4.78	28	41.67	14.47	6.0229×10^6	4.82	7.8986×10^4
		0.219	5.56	30	44.64	16.63	6.9219×10^6	5.54	9.0784×10^4
6⅝	166	0.105	2.67	20	29.76	11.43	4.7575×10^6	3.45	5.6535×10^4
		0.135	3.43	23	34.23	14.50	6.0353×10^6	4.38	7.1775×10^4
		0.188	4.78	28	41.67	19.66	8.1831×10^6	5.94	9.7339×10^4
		0.219	5.56	30	44.64	22.62	9.4151×10^6	6.83	1.1192×10^5
8	200	0.105	2.67	32	47.62	20.29	8.4453×10^6	5.07	8.3082×10^4
		0.135	3.43	33	49.11	25.80	1.0739×10^7	6.45	1.0570×10^5
		0.179	4.55	37	55.06	33.65	1.4006×10^7	8.41	1.3782×10^5
		0.188	4.78	38	56.55	35.13	1.4622×10^7	8.87	1.4535×10^5
		0.239	6.07	42	62.50	43.92	1.8281×10^7	10.98	1.7993×10^5
8⅝	216	0.105	2.67	35	52.09	25.51	1.0618×10^7	5.91	9.6848×10^4
		0.135	3.43	37	55.06	32.45	1.3507×10^7	7.52	1.2323×10^5
		0.179	4.55	41	61.01	42.37	1.7636×10^7	9.83	1.6108×10^5
		0.188	4.78	42	62.50	44.25	1.8418×10^7	10.26	1.6813×10^5
		0.239	6.07	44	65.48	55.40	2.3059×10^7	12.85	2.1057×10^5
10	250	0.135	3.43	49	72.92	50.91	2.1190×10^7	10.18	1.6682×10^5
		0.179	4.55	53	78.87	66.61	2.7725×10^7	13.32	2.1828×10^5
		0.188	4.78	54	80.36	69.59	2.8966×10^7	13.92	2.2811×10^5
		0.239	6.07	60	89.29	87.34	3.6354×10^7	17.47	2.8628×10^5
10¾	269	0.135	3.43	60	89.29	63.42	2.6397×10^7	11.80	1.9337×10^5
		0.179	4.55	65	96.73	83.06	3.4572×10^7	15.45	2.5318×10^5
		0.188	4.78	66	98.22	86.80	3.6129×10^7	16.15	2.6465×10^5
		0.239	6.07	70	104.17	109.05	4.5390×10^7	20.29	3.3249×10^5
12	300	0.135	3.43	66	98.22	88.56	3.6861×10^7	14.76	2.4187×10^5
		0.179	4.55	72	107.15	116.14	4.8341×10^7	19.36	3.1725×10^5
		0.188	4.78	74	110.12	121.39	5.0526×10^7	20.23	3.3151×10^5
		0.239	6.07	80	119.05	152.75	6.3579×10^7	25.46	4.1721×10^5
12¾	319	0.135	3.43	73	108.64	106.44	4.4304×10^7	16.70	2.7366×10^5
		0.179	4.55	79	117.56	139.67	5.8135×10^7	21.91	3.5904×10^5
		0.188	4.78	80	119.05	146.01	6.0774×10^7	22.90	3.7526×10^5
		0.239	6.07	88	130.96	183.86	7.6528×10^7	28.84	4.7260×10^5
14	350	0.135	3.43	87	129.47	141.32	5.8822×10^7	20.19	3.3085×10^5
		0.179	4.55	94	139.89	185.61	7.7257×10^7	26.52	4.3458×10^5
		0.188	4.78	95	141.38	194.07	8.0778×10^7	27.72	4.5425×10^5
		0.239	6.07	101	150.30	244.65	1.0183×10^8	34.95	5.7273×10^5

* Sizes under 45 in. are outside diameter sizes; those 45 in. and over are inside diameter sizes.

Table continued next page

Table 7-6 Values of moment of inertia and section modulus of steel pipe (continued)

Nominal Size*		Wall Thickness		Weight of Pipe & Water		Moment of Inertia		Section Modulus	
in.	*(mm)*	*in.*	*(mm)*	*lb/ft*	*(kg/m)*	*in.*^{4}	*mm*^{4}	*in.*^{3}	*mm*^{3}
16	400	0.135	3.43	113	168.16	211.71	8.8120×10^7	26.46	4.3360×10^5
		0.179	4.55	118	175.60	278.40	1.1588×10^8	34.80	5.7027×10^5
		0.188	4.78	119	177.09	291.16	1.2119×10^8	36.39	5.9633×10^5
		0.239	6.07	128	190.48	367.54	1.5298×10^8	45.94	7.5282×10^5
		0.250	6.35	129	191.97	383.66	1.5969×10^8	47.96	7.8592×10^5
		0.312	7.92	139	206.85	473.96	1.9728×10^8	59.25	9.7093×10^5
18	450	0.135	3.43	137	203.88	302.29	1.2582×10^8	33.59	5.5044×10^5
		0.179	4.55	145	215.78	397.88	1.6561×10^8	44.21	7.2447×10^5
		0.188	4.78	147	218.76	416.18	1.7323×10^8	46.24	7.5774×10^5
		0.239	6.07	158	235.13	525.94	2.1891×10^8	58.44	9.5766×10^5
		0.250	6.35	158	235.13	549.14	2.2857×10^8	61.02	9.9994×10^5
		0.312	7.92	170	252.99	679.28	2.8274×10^8	75.48	1.2369×10^6
20	500	0.135	3.43	165	245.55	415.60	1.7299×10^8	41.56	6.8105×10^5
		0.179	4.55	175	260.43	547.43	2.2786×10^8	54.74	8.9703×10^5
		0.188	4.78	176	261.92	572.69	2.3837×10^8	57.27	9.3849×10^5
		0.239	6.07	188	279.77	724.35	3.0150×10^8	72.44	1.1871×10^6
		0.250	6.35	189	281.26	756.44	3.1485×10^8	75.64	1.2395×10^6
		0.312	7.92	212	315.49	936.68	3.8988×10^8	93.67	1.5350×10^6
		0.375	9.53	225	334.84	1,113.47	4.6346×10^8	111.35	1.8247×10^6
22	550	0.179	4.55	208	309.54	730.41	3.0402×10^8	66.40	1.0881×10^6
		0.188	4.78	209	311.03	764.21	3.1809×10^8	69.47	1.1384×10^6
		0.239	6.07	222	330.37	967.27	4.0261×10^8	87.93	1.4409×10^6
		0.250	6.35	223	331.86	1,010.27	4.2051×10^8	91.84	1.5050×10^6
		0.312	7.92	237	352.69	1,252.07	5.2115×10^8	113.82	1.8652×10^6
		0.375	9.53	252	375.02	1,489.67	6.2005×10^8	135.42	2.2191×10^6
24	600	0.179	4.55	243	361.62	950.21	3.9551×10^8	79.18	1.2975×10^6
		0.188	4.78	244	363.11	994.27	4.1385×10^8	82.86	1.3578×10^6
		0.239	6.07	258	383.95	1,259.21	5.2412×10^8	104.93	1.7195×10^6
		0.250	6.35	259	385.43	1,315.35	5.4749×10^8	109.61	1.7962×10^6
		0.312	7.92	275	409.24	1,631.34	6.7901×10^8	135.94	2.2277×10^6
		0.375	9.53	291	433.05	1,942.30	8.0845×10^8	161.86	2.6524×10^6
		0.438	11.13	307	456.87	2,248.29	9.3581×10^8	187.36	3.0703×10^6
		0.500	12.70	322	479.19	2,549.36	1.0611×10^9	212.45	3.4814×10^6
26	650	0.179	4.55	280	416.68	1,210.19	5.0372×10^8	93.09	1.5255×10^6
		0.188	4.78	282	419.66	1,266.41	5.2712×10^8	97.42	1.5964×10^6
		0.239	6.07	296	440.50	1,604.67	6.6791×10^8	123.44	2.0228×10^6
		0.250	6.35	299	444.96	1,676.38	6.9776×10^8	128.95	2.1131×10^6

* Sizes under 45 in. are outside diameter sizes; those 45 in. and over are inside diameter sizes.

Table continued next page

Table 7-6 Values of moment of inertia and section modulus of steel pipe (continued)

Nominal Size*		Wall Thickness		Weight of Pipe & Water		Moment of Inertia		Section Modulus	
in.	*(mm)*	*in.*	*(mm)*	*lb/ft*	*(kg/m)*	*in.*^{4}	*mm*^{4}	*in.*^{3}	*mm*^{3}
		0.312	7.92	316	470.26	2,080.37	8.6591×10^8	160.03	2.6224×10^6
		0.375	9.53	333	495.56	2,478.43	1.0316×10^9	190.65	3.1242×10^6
		0.438	11.13	350	520.86	2,870.62	1.1948×10^9	220.82	3.6186×10^6
		0.500	12.70	366	544.67	3,257.00	1.3557×10^9	250.54	4.1056×10^6
28	700	0.179	4.55	320	476.21	1,513.74	6.3007×10^8	108.12	1.7718×10^6
		0.188	4.78	323	480.68	1,584.17	6.5938×10^8	113.16	1.8544×10^6
		0.239	6.07	338	503.00	2,008.15	8.3585×10^8	143.44	2.3506×10^6
		0.250	6.35	341	507.46	2,098.09	8.7329×10^8	149.86	2.4558×10^6
		0.312	7.92	359	534.25	2,605.06	1.0843×10^9	186.08	3.0493×10^6
		0.375	9.53	378	562.52	3,105.12	1.2924×10^9	221.79	3.6345×10^6
		0.438	11.13	396	589.31	3,598.36	1.4977×10^9	257.03	4.2120×10^6
		0.500	12.70	414	616.10	4,084.82	1.7002×10^9	291.77	4.7813×10^6
30	750	0.188	4.78	359	534.25	1,956.16	8.1421×10^8	130.41	2.1370×10^6
		0.250	6.35	376	559.55	2,585.16	1.0760×10^9	172.35	2.8243×10^6
		0.312	7.92	393	584.85	3,206.29	1.3346×10^9	213.75	3.5027×10^6
		0.375	9.53	410	610.15	3,829.41	1.5939×10^9	255.30	4.1836×10^6
		0.438	11.13	427	635.44	4,444.54	1.8503×10^8	296.31	4.8557×10^6
		0.500	12.70	444	660.74	5,042.17	2.0987×10^9	336.15	5.5085×10^6
32	800	0.188	4.78	404	601.22	2,376.85	9.8932×10^8	148.55	2.4343×10^6
		0.250	6.35	422	628.00	3,142.35	1.3079×10^9	196.40	3.2184×10^6
		0.312	7.92	441	656.28	3,898.86	1.6228×10^9	243.68	3.9932×10^6
		0.375	9.53	459	683.07	4,658.44	1.9390×10^9	291.16	4.7713×10^6
		0.438	11.13	477	709.85	5,408.88	2.2513×10^9	338.06	5.5398×10^6
		0.500	12.70	495	736.64	6,138.57	2.5551×10^9	383.66	6.2871×10^6
34	850	0.188	4.78	453	674.14	2,853.91	1.8790×10^9	167.88	2.7511×10^6
		0.250	6.35	472	702.41	3,774.34	1.5710×10^9	222.02	3.6383×10^6
		0.312	7.92	491	730.69	4,684.61	1.9499×10^9	275.57	4.5158×10^6
		0.375	9.53	511	760.45	5,599.23	2.3306×10^9	329.37	5.3974×10^6
		0.438	11.13	530	788.72	6,503.51	2.7070×10^9	382.56	6.2690×10^6
		0.500	12.70	550	818.49	7,383.42	3.0732×10^9	434.32	7.1172×10^6
36	900	0.188	4.78	504	750.03	3,390.87	1.4114×10^9	188.38	3.0870×10^6
		0.250	6.35	524	779.80	4,485.86	1.8672×10^9	249.22	4.0840×10^6
		0.312	7.92	545	811.05	5,569.44	2.3182×10^9	309.42	5.0705×10^6
		0.375	9.53	566	842.30	6,658.87	2.7716×10^9	369.94	6.0622×10^6
		0.438	11.13	586	872.06	7,736.68	3.2202×10^9	429.82	7.0435×10^6
		0.500	12.70	607	903.31	8,786.13	3.6571×10^9	488.12	7.9989×10^6
		0.625	15.88	647	962.84	10,868.29	4.5237×10^9	603.80	9.8945×10^6

* Sizes under 45 in. are outside diameter sizes; those 45 in. and over are inside diameter sizes.

Table continued next page

Table 7-6 Values of moment of inertia and section modulus of steel pipe (continued)

Nominal Size*		Wall Thickness		Weight of Pipe & Water		Moment of Inertia		Section Modulus	
in.	*(mm)*	*in.*	*(mm)*	*lb/ft*	*(kg/m)*	*in.*^{4}	*mm*^{4}	*in.*^{3}	*mm*^{3}
38	950	0.188	4.78	558	830.39	3,991.29	1.6613×10^9	210.07	3.4424×10^6
		0.250	6.35	579	861.64	5,281.61	2.1984×10^9	277.98	4.5553×10^6
		0.312	7.92	601	894.38	6,559.19	2.7301×10^9	345.22	5.6571×10^6
		0.375	9.53	623	927.12	7,884.41	3.2817×10^9	412.87	6.7657×10^6
		0.438	11.13	645	959.86	9,116.66	3.7946×10^9	479.83	7.8630×10^6
		0.500	12.70	666	991.11	10,356.13	4.3105×10^9	545.07	8.9321×10^6
		0.625	15.88	709	1,055.11	12,817.44	5.3350×10^9	674.61	1.1055×10^7
40	1,000	0.188	4.78	614	913.73	4,658.71	1.9391×10^9	232.94	3.8172×10^6
		0.250	6.35	637	947.96	6,166.31	2.5666×10^9	308.32	5.0525×10^6
		0.312	7.92	660	982.19	7,659.77	3.1882×10^9	382.99	6.2761×10^6
		0.375	9.53	683	1,016.41	9,162.94	3.8139×10^9	458.15	7.5077×10^6
		0.438	11.13	706	1,050.64	10,651.69	4.4336×10^9	532.59	8.7276×10^6
		0.500	12.70	729	1,084.87	12,102.84	5.0376×10^9	605.15	9.9166×10^6
		0.625	15.88	774	1,151.84	14,986.75	6.2379×10^9	749.35	1.2280×10^7
42	1,050	0.188	4.78	674	1,003.02	5,396.67	2.2463×10^9	256.99	4.2113×10^6
		0.250	6.35	698	1,038.74	7,144.66	2.9738×10^9	340.23	5.5754×10^6
		0.312	7.92	722	1,074.45	8,877.04	3.6949×10^9	422.72	6.9271×10^6
		0.375	9.53	746	1,110.17	10,621.49	4.4210×10^9	505.79	8.2884×10^6
		0.438	11.13	770	1,145.88	12,350.02	5.1405×10^9	588.10	9.6372×10^6
		0.500	12.70	794	1,181.60	14,035.69	5.8421×10^9	668.37	1.0953×10^7
		0.625	15.88	841	1,251.54	17,388.00	7.2374×10^9	828.01	1.3569×10^7
45	1,125	0.250	6.35	809	1,203.92	9,096.41	3.7862×10^9	399.84	6.5522×10^6
		0.312	7.92	840	1,250.05	11,417.88	4.7525×10^9	500.51	8.2019×10^6
		0.375	9.53	870	1,294.70	13,758.52	5.7267×10^9	601.47	9.8563×10^6
		0.438	11.13	901	1,340.83	16,118.42	6.7090×10^9	702.71	1.1515×10^7
		0.500	12.70	932	1,386.97	18,497.69	7.6993×10^9	804.25	1.3179×10^7
		0.625	15.88	993	1,477.74	23,324.77	9.7085×10^9	1,008.21	1.6522×10^7
48	1,200	0.250	6.35	912	1,357.20	11,028.20	4.5903×10^9	454.77	7.4523×10^6
		0.312	7.92	945	1,406.31	13,839.09	5.7603×10^9	569.22	9.3278×10^6
		0.375	9.53	977	1,453.93	16,671.75	6.9393×10^9	683.97	1.1208×10^7
		0.438	11.13	1,010	1,503.04	19,526.28	8.1274×10^9	799.03	1.3094×10^7
		0.500	12.70	1,043	1,552.15	22,402.80	9.3247×10^9	914.40	1.4984×10^7
		0.625	15.88	1,108	1,648.88	28,222.24	1.1747×10^{10}	1,146.08	1.8781×10^7
		0.750	19.05	1,174	1,747.10	34,130.98	1.4206×10^{10}	1,379.03	2.2598×10^7
51	1,275	0.250	6.35	1,021	1,519.41	13,215.78	5.5008×10^9	513.23	8.4103×10^6
		0.312	7.92	1,056	1,571.50	16,580.45	6.9013×10^9	642.34	1.0526×10^7
		0.375	9.53	1,090	1,622.09	19,969.67	8.3120×10^9	771.77	1.2647×10^7

* Sizes under 45 in. are outside diameter sizes; those 45 in. and over are inside diameter sizes.

Table continued next page

Table 7-6 Values of moment of inertia and section modulus of steel pipe (continued)

Nominal Size*		Wall Thickness		Weight of Pipe & Water		Moment of Inertia		Section Modulus	
in.	*(mm)*	*in.*	*(mm)*	*lb/ft*	*(kg/m)*	*in.*^{4}	*mm*^{4}	*in.*^{3}	*mm*^{3}
		0.438	11.13	1,125	1,674.18	23,383.52	9.7329×10^{9}	901.53	1.4773×10^{7}
		0.500	12.70	1,155	1,718.82	26,822.15	1.1164×10^{10}	1,031.62	1.6905×10^{7}
		0.625	15.88	1,229	1,828.95	33,774.18	1.4058×10^{10}	1,292.79	2.1185×10^{7}
		0.750	19.05	1,299	1,933.12	40,826.72	1.6993×10^{10}	1,555.30	2.5487×10^{7}
54	1,350	0.250	6.35	1,137	1,692.04	15,675.07	6.5245×10^{10}	575.23	9.4263×10^{6}
		0.312	7.92	1,174	1,747.10	19,661.86	8.1839×10^{10}	719.88	1.1797×10^{7}
		0.375	9.53	1,200	1,785.79	23,676.13	9.8547×10^{10}	864.88	1.4173×10^{7}
		0.438	11.13	1,247	1,855.74	27,717.97	1.1537×10^{10}	1,010.22	1.6555×10^{7}
		0.500	12.70	1,284	1,910.80	31,787.54	1.3231×10^{10}	1,155.91	1.8942×10^{7}
		0.625	15.88	1,357	2,019.43	40,010.33	1.6654×10^{10}	1,448.34	2.3734×10^{7}
		0.750	19.05	1,431	2,129.56	48,345.50	2.0123×10^{10}	1,742.18	2.8549×10^{7}
57	1,425	0.250	6.35	1,258	1,872.11	18,421.95	7.6678×10^{10}	640.76	1.0500×10^{7}
		0.312	7.92	1,296	1,928.66	23,103.18	9.6163×10^{10}	801.85	1.3140×10^{7}
		0.375	9.53	1,335	1,986.69	27,814.98	1.1577×10^{10}	963.29	1.5785×10^{7}
		0.438	11.13	1,373	2,043.24	32,557.47	1.3551×10^{10}	1,125.10	1.8437×10^{7}
		0.500	12.70	1,412	2,101.28	37,330.80	1.5538×10^{10}	1,287.27	2.1095×10^{7}
		0.625	15.88	1,490	2,217.36	46,970.46	1.9551×10^{10}	1,612.72	2.6428×10^{7}
		0.750	19.05	1,568	2,333.43	56,735.04	2.3615×10^{10}	1,939.66	3.1785×10^{7}
60	1,500	0.250	6.35	1,387	2,064.08	21,472.35	8.9375×10^{10}	709.83	1.1632×10^{7}
		0.312	7.92	1,427	2,123.60	26,924.30	1.1207×10^{10}	888.22	1.4555×10^{7}
		0.375	9.53	1,468	2,184.62	32,410.10	1.3490×10^{10}	1,067.00	1.7485×10^{7}
		0.438	11.13	1,508	2,244.15	37,929.85	1.5788×10^{10}	1,246.16	2.0420×10^{7}
		0.500	12.70	1,549	2,305.16	43,483.72	1.8099×10^{10}	1,425.70	2.3363×10^{7}
		0.625	15.88	1,631	2,427.19	54,694.33	2.2765×10^{10}	1,785.94	2.9266×10^{7}
		0.750	19.05	1,713	2,549.22	66,043.06	2.7489×10^{10}	2,147.74	3.5195×10^{7}
		0.875	22.23	1,795	2,671.25	77,531.04	3.2271×10^{10}	2,511.13	4.1150×10^{7}
		1.000	25.40	1,872	2,785.84	89,159.39	3.7111×10^{10}	2,876.11	4.7131×10^{7}
63	1,575	0.250	6.35	1,519	2,260.52	24,842.16	1.0340×10^{10}	782.43	1.2822×10^{7}
		0.312	7.92	1,569	2,334.92	31,145.10	1.2964×10^{10}	979.02	1.6043×10^{7}
		0.375	9.53	1,604	2,387.01	37,485.33	1.5603×10^{10}	1,176.01	1.9271×10^{7}
		0.438	11.13	1,646	2,449.51	43,862.94	1.8257×10^{10}	1,373.40	2.2506×10^{7}
		0.500	12.70	1,689	2,513.50	50,278.12	2.0927×10^{10}	1,571.19	2.5747×10^{7}
		0.625	15.88	1,775	2,641.48	63,221.71	2.6315×10^{10}	1,967.99	3.2250×10^{7}
		0.750	19.05	1,869	2,781.37	76,317.27	3.1766×10^{10}	2,366.43	3.8779×10^{7}
		0.875	22.23	1,945	2,894.47	89,566.00	3.7280×10^{10}	2,766.52	4.5335×10^{7}
		1.000	25.40	2,028	3,017.99	102,969.00	4.2859×10^{10}	3,168.28	5.1919×10^{7}

* Sizes under 45 in. are outside diameter sizes; those 45 in. and over are inside diameter sizes.

Table continued next page

Table 7-6 Values of moment of inertia and section modulus of steel pipe (continued)

Nominal Size*		Wall Thickness		Weight of Pipe & Water		Moment of Inertia		Section Modulus	
in.	*(mm)*	*in.*	*(mm)*	*lb/ft*	*(kg/m)*	*in.*^{4}	*mm*^{4}	*in.*^{3}	*mm*^{3}
66	1,650	0.250	6.35	1,660	2,470.35	28,547.28	1.1882×10^{10}	858.56	1.4069×10^{7}
		0.312	7.92	1,704	2,535.82	35,785.46	1.4895×10^{10}	1,074.24	1.7604×10^{7}
		0.375	9.53	1,749	2,602.79	43,064.52	1.7925×10^{10}	1,290.32	2.1145×10^{7}
		0.438	11.13	1,793	2,668.27	50,384.57	2.0972×10^{10}	1,506.83	2.4693×10^{7}
		0.500	12.70	1,838	2,735.24	57,745.80	2.4036×10^{10}	1,723.76	2.8247×10^{7}
		0.625	15.88	1,928	2,869.17	72,592.34	3.0215×10^{10}	2,158.88	3.5378×10^{7}
		0.750	19.05	2,018	3,003.11	87,605.39	3.6464×10^{10}	2,595.72	4.2536×10^{7}
72	1,800	0.375	9.53	2,055	3,058.17	55,833.09	2.3239×10^{10}	1,534.92.	2.5153×10^{7}
		0.500	12.70	2,152	3,202.52	74,832.06	3.1147×10^{10}	2,050.19	3.3597×10^{7}
		0.625	15.88	2,250	3,348.36	94,027.23	3.9137×10^{10}	2,567.30	4.2071×10^{7}
		0.750	19.05	2,348	3,494.20	113,419.94	4.7209×10^{10}	3,086.26	5.0575×10^{7}
78	1,950	0.375	9.53	2,384	3,547.77	70,901.74	2.9512×10^{10}	1,800.68	2.9508×10^{7}
		0.500	12.70	2,489	3,704.03	94,990.38	3.9538×10^{10}	2,404.82	3.9408×10^{7}
		0.625	15.88	2,595	3,861.78	119,308.79	4.9660×10^{10}	3,070.95	5.0324×10^{7}
		0.750	19.05	2,701	4,019.52	143,858.45	5.9878×10^{10}	3,619.08	5.9306×10^{7}
84	2,100	0.375	9.53	2,740	4,077.56	88,463.52	3.6821×10^{10}	2,087.63	3.4210×10^{7}
		0.500	12.70	2,853	4,245.72	118,478.18	4.9314×10^{10}	2,787.72	4.5683×10^{7}
		0.625	15.88	2,967	4,415.37	148,758.84	6.1918×10^{10}	3,489.94	5.7190×10^{7}
		0.750	19.05	3,081	4,585.02	179,307.07	7.4633×10^{10}	4,194.32	6.8733×10^{7}
90	2,250	0.375	9.53	3,119	4,641.57	108,709.30	4.5248×10^{10}	2,395.80	3.9260×10^{7}
		0.500	12.70	3,240	4,821.64	145,549.94	6.0582×10^{10}	3,198.90	5.2421×10^{7}
		0.625	15.88	3,362	5,003.19	182,695.46	7.6044×10^{10}	4,004.28	6.5618×10^{7}
		0.750	19.05	3,484	5,184.75	220,147.54	9.1632×10^{10}	4,811.97	7.8854×10^{7}
96	2,400	0.375	9.53	3,523	5,242.79	131,829.94	5.4872×10^{10}	2,725.17	4.4658×10^{7}
		0.500	12.70	3,652	5,434.76	176,460.14	7.3448×10^{10}	3,638.35	5.9622×10^{7}
		0.625	15.88	3,782	5,628.22	221,436.76	9.2169×10^{10}	4,553.97	7.4626×10^{7}
		0.750	19.05	3,912	5,821.68	266,761.59	1.1103×10^{11}	5,472.03	8.9671×10^{7}
102	2,550	0.375	9.53	3,649	5,430.30	158,016.29	6.5771×10^{10}	3,075.74	5.0402×10^{7}
		0.500	12.70	4,086	6,080.62	211,463.03	8.8017×10^{10}	4,106.08	6.7287×10^{7}
		0.625	15.88	4,224	6,285.99	265,300.52	1.1043×10^{11}	5,138.99	8.4213×10^{7}
		0.750	19.05	4,362	6,491.35	319,530.74	1.3300×10^{11}	6,174.51	1.0118×10^{8}
108	2,700	0.375	9.53	4,402	6,550.88	187,459.00	7.8026×10^{10}	3,447.52	5.6495×10^{7}
		0.500	12.70	4,569	6,799.40	250,813.57	1.0440×10^{11}	4,602.08	7.5415×10^{7}
		0.625	15.88	4,695	6,986.91	314,605.54	1.3095×10^{11}	5,759.37	6.4379×10^{7}
		0.750	19.05	4,841	7,204.18	378,837.35	1.5768×10^{11}	6,919.40	1.1339×10^{8}
114	2,850	0.375	9.53	4,881	7,263.71	220,349.30	8.3392×10^{10}	3,840.51	6.2935×10^{7}
		0.500	12.70	5,035	7,492.89	294,766.30	1.2269×10^{11}	5,126.37	8.4006×10^{7}
		0.625	15.88	5,188	7,720.57	369,669.79	1.5387×10^{11}	6,415.09	1.0512×10^{8}
		0.750	19.05	5,345	7,954.22	445,062.21	1.8525×10^{11}	7,706.71	1.2629×10^{8}

* Sizes under 45 in. are outside diameter sizes; those 45 in. and over are inside diameter sizes.

Table continued next page

Table 7-6 Values of moment of inertia and section modulus of steel pipe (continued)

Nominal Size*		Wall Thickness		Weight of Pipe & Water		Moment of Inertia		Section Modulus	
in.	*(mm)*	*in.*	*(mm)*	*lb/ft*	*(kg/m)*	*in.*^{4}	*mm*^{4}	*in.*^{3}	*mm*^{3}
120	3,000	0.500	12.70	5,545	8,251.85	343,575.02	1.4301×10^{11}	5,678.93	9.3061×10^{7}
		0.625	15.88	5,706	8,491.44	430,810.89	1.7932×10^{11}	7,106.16	1.1645×10^{8}
		0.750	19.05	5,869	8,734.01	518,588.23	2.1585×10^{11}	8,536.43	1.3989×10^{8}
126	3,150	0.500	12.70	6,079	9,046.52	397,494.49	1.6545×10^{11}	6,259.76	1.0258×10^{8}
		0.625	15.88	6,249	9,299.51	498,346.95	2.0743×10^{11}	7,832.57	1.2835×10^{8}
		0.750	19.05	6,419	9,552.50	559,795.37	2.3300×10^{11}	9,408.56	1.5418×10^{8}
132	3,300	0.500	12.70	6,638	9,878.41	456,779.50	1.9013×10^{11}	6,868.87	1.1256×10^{8}
		0.625	15.88	6,816	10,143.30	572,596.56	2.3833×10^{11}	8,594.32	1.4084×10^{8}
		0.750	19.05	6,994	10,408.19	689,067.49	2.8681×10^{11}	10,323.11	1.6917×10^{8}
138	3,450	0.500	12.70	7,221	10,746.00	521,684.34	2.1714×10^{11}	7,506.25	1.2301×10^{8}
		0.625	15.88	7,407	11,022.80	653,877.33	2.7216×10^{11}	9,391.42	1.5390×10^{8}
		0.750	19.05	7,593	11,299.60	786,784.58	3.2748×10^{11}	11,280.07	1.8485×10^{8}
144	3,600	0.500	12.70	7,829	11,650.80	592,463.77	2.4660×10^{11}	8,171.91	1.3391×10^{8}
		0.625	15.88	8,023	11,939.51	742,507.85	3.0905×10^{11}	10,223.86	1.6754×10^{8}
		0.750	19.05	8,217	12,228.21	893,329.02	3.7183×10^{11}	12,279.44	2.0122×10^{8}

* Sizes under 45 in. are outside diameter sizes; those 45 in. and over are inside diameter sizes.

REFERENCES

AISI. 1983. Welded Steel Pipe, *Steel Plate Engineering Data*, Vol. 3.

Barnard, R.E. 1948. Design Standards for Steel Water Pipe. *Jour. AWWA*, 40(1):24.

Bier, P.J. 1940. Siphon Self-Supporting in Long Spans. *Engineering News-Record*, 124:852.

———. 1949. *Welded Steel Penstocks—Design and Construction*. Engineering Monograph 3. Washington, D.C.: US Bureau of Reclamation.

Cates, W.H. 1950. Design Standards for Large-Diameter Steel Water Pipe, *Jour. AWWA*, 42:860.

Crocker, S., ed. 1954. *Piping Handbook*, 4th ed. New York: McGraw-Hill Book Co.

Foster, H.A. 1949a. Formulas Facilitate Design of Ring-Supported Pipes. *Civ. Engrg.*, 19:629.

———. 1949b. Formulas Indicate Earthquake Forces in Design of Ring Girder-Supported Pipes. *Civ. Engrg.*, 19:697.

Roark, R.J. 1954. *Formulas for Stress and Strain*. New York: McGraw-Hill Book Co.

Schorer, H. 1933. Design of Large Pipelines, *Trans. ASCE*, 98:101.

Timoshenko, S. 1936. *Theory of Elastic Stability*, 1st ed. Engrg. Soc. Monographs. New York: McGraw-Hill Book Co.

US Bureau of Reclamation. 1944. *Penstock Analysis and Stiffener Ring Design*. Bull. 5, Part V. Tech. Invest. Boulder Canyon Project, Final Design Reports. Denver, Colo.: US Bureau of Reclamation.

Wilson, W.M., and N.M. Newmark. 1933. *The Strength of Thin Cylindrical Shells as Columns*. Bull. 255. Urbana, Ill.: Univ. of Illinois, Engrg. Exp. Stn.

The following references are not cited in the text.

AISI. 1982. *Steel Penstocks and Tunnel Liners*. AISI. *Steel Plate Engineering Data*, Vol. 4.

Manuals and Reports on Engineering Practice #79. 1993. American Society of Civil Engineers.

Younger, J.E. 1935. *Structural Design of Metal Airplanes*. New York: McGraw-Hill Book Co.

Chapter 8

Pipe Joints

The pipe joint selected and the careful installation of the joints are important considerations for the design engineer and the inspector. Many kinds of joints are used with steel water pipe. Common types are bell-and-spigot rubber gasket joints, field-welded joints (both illustrated in Figure 8-1), flanges, couplings, and expansion joints. There are several types of couplings available: bolted, sleeve type; bolted, split-sleeve type; shouldered couplings; and single-point closure couplings. The types of expansion joints available include fabricated steel mechanical slip type, rubber arch type, and stainless steel bellows type. Bolted, sleeve-type couplings (AWWA C219), grooved and shouldered couplings (AWWA C606), and fabricated steel mechanical slip-type expansion joints (AWWA C221) are described in this chapter. Detailed information about all of these joints is available from the manufacturer.

BELL-AND-SPIGOT JOINT WITH RUBBER GASKET

Several types of rubber-gasket field joints (shown in Figures 8-1E, 8-1F, 8-1G, 8-1H, and 8-1I) have been developed for steel water-pipe service. Gasketed joints permit rapid installation in the field and, when properly manufactured and installed, they provide a watertight joint that will give long service without maintenance. The design of the joints allows flexibility in the line, permitting certain angular and longitudinal movement caused by settlement of the ground or other conditions while allowing the joints to remain watertight. The joints are easy to assemble, reducing the cost of laying the pipe. Any type of coating can be applied to the pipe in the shop and not be damaged at the joint during laying operations. The joint is self-centering and economical. Because of potential problems in maintaining joint integrity, gasketed joints should be manufactured carefully to maintain tight clearance between the bell and spigot.

The rubber gasket should conform to AWWA C200. The thrust at elbows, tees, laterals, wyes, reducers, valves, and dead ends should be considered. Joints should be restrained by welding (Figure 8-1), by harnessing, by anchors, or by thrust blocks (chapter 13). Calculations should consider the anchoring effect of soil friction (Ground Friction and Line Tension and Thrust Restraint).

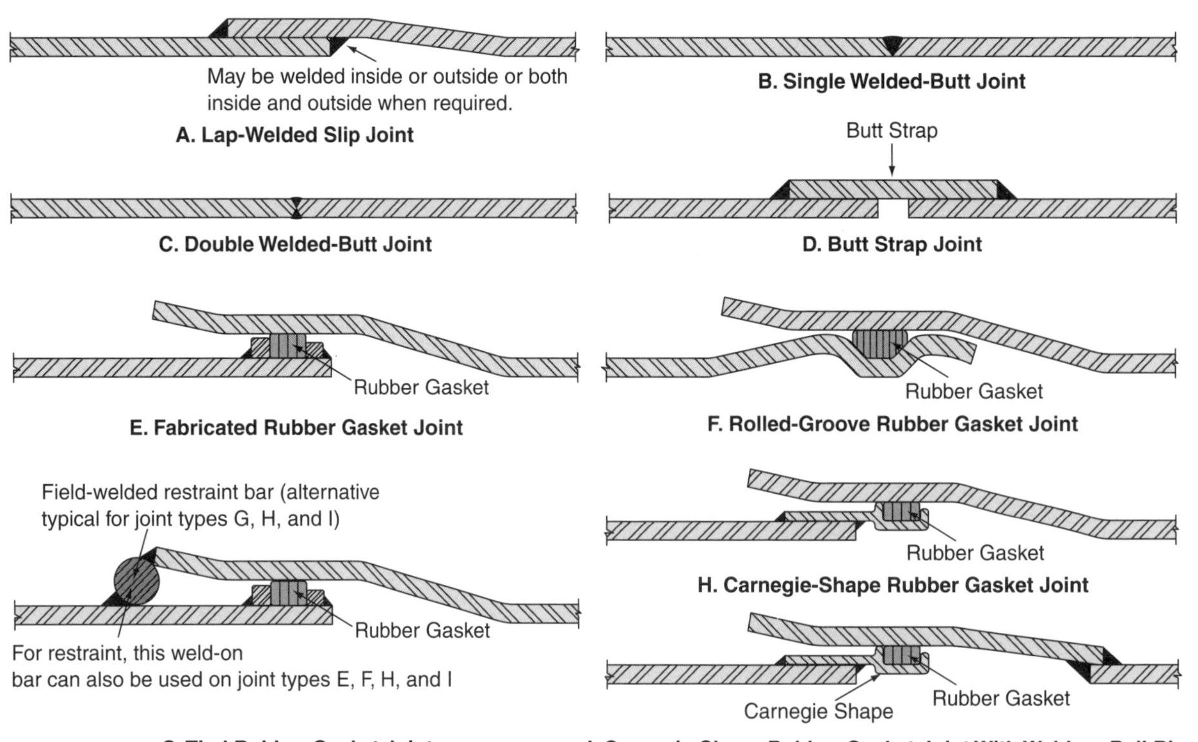

Figure 8-1 Welded and rubber-gasketed field joints

WELDED JOINTS

Field welding of joints in steel water pipe 24 in. (600 mm) in diameter and larger is a frequently used jointing method that results in strong, permanently tight joints. Welded lap joints having a single fillet weld (Figure 8-1A) have proved satisfactory for most installations. Single welded-butt joints (Figure 8-1B) and double welded-butt joints (Figure 8-1C) should withstand, to the limit of pipe-wall strength, longitudinal extension loading caused by settlement, washouts, and other disjointing forces. Where welded joints are used, the pipe should be left bare a sufficient distance back from the ends to avoid damaging the protective coatings by the heat produced during welding. These joints should be field-coated after welding. Field-welding in the interior of steel pipe with lining is ordinarily limited to 24-in. (600-mm) or larger pipe, because a worker must enter the pipe after welding to apply lining to the inside at the welded joints. Forced ventilation must be provided to ensure adequate air exchange for workers inside the pipe.

The slip joint is commonly used because of its flexibility, ease in forming and laying, watertight quality, and simplicity. Small angle changes can be made in this joint. It may be welded on the outside only, or if the diameter permits, on the inside only. In certain, special conditions, joints should be welded on the inside and outside, in which case a field testing method described in AWWA C206, *Standard for Field Welding of Steel Water Pipe* (latest edition) may be used.

AWWA C206 describes the requirements and techniques for satisfactory field welding. Where the pipe wall is thicker than ½ in. (12.7 mm) and the pipe is subject to temperatures below 40°F (4°C), the steel and welding procedures should be carefully selected to accommodate these adverse conditions.

BOLTED SLEEVE-TYPE COUPLINGS

Bolted sleeve-type couplings are used to join pipe sections of all diameters. Complete technical data is available in AWWA Standard C219. A typical bolted sleeve-type coupling is shown in Figure 8-2.

Bolted sleeve-type couplings provide tightness and strength with flexibility. They relieve expansion and contraction forces in a pipeline and provide sufficient flexibility so that the pipe may be laid on long radius curves and grades without using specials. The rubber gaskets are firmly held between the coupling parts and the pipe, and they join the lengths securely against high pressure, low pressure, or vacuum. The enclosed rubber gaskets are protected from normal damage and decay as a result of typical water service. These joints have been used successfully since 1891.

Acceptable axial movement in bolted sleeve-type couplings results from shear displacement of the rubber gaskets rather than from sliding of the gaskets on the mating surface of the pipe. When properly installed according to good pipeline practice, couplings 10 in. (250 mm) and larger should accommodate longitudinal pipe movement of 3/8 in. (9.5 mm) total per joint without leakage. If greater displacement is needed, true expansion joints should be provided rather than bolted sleeve-type couplings (see Expansion and Contraction—General). Bolted sleeve-type couplings transmit only minor tension or shear stresses across pipe joints, and they will not permit differential settlement at the joints when used alone. However, a degree of flexibility is possible when used in conjunction with another adjacent flexible joint. Bolted sleeve-type couplings are suitable for joining buried or exposed anchored pipes that are laid on curves established using deflections up to the maximum permitted at the coupling.

Restrained, harnessed, flanged, or welded joints may be needed to resist the unbalanced thrust at tees, elbows, valves, and fittings, or to resist the line pull in underwater crossings, if such forces are not resisted by external forces provided by thrust blocks or anchors. Thrust calculations should consider the anchoring effect of soil friction on buried pipe (discussed in Ground Friction and Line Tension and Thrust Restraint). Details of joint harness are given in chapter 13.

Pipe Layout When Using Bolted Sleeve-Type Couplings

When laying bolted sleeve-type coupled pipe on curves, separation measured on the pipe centerline should be determined using data supplied by the coupling manufacturer or by consulting AWWA C219.

Data for Pipe Layouts

The profile and alignment of pipelines is frequently staked on a curve. It is important to know what pipe lengths are needed to negotiate such curves and to know the offset necessary to locate properly the free end of the pipe section being laid. Tables showing radius of curves, pipe length, and offset deflections, as well as formulas and sketches showing dimensions, are available from coupling manufacturers.

FLANGES

Flanges commonly used for steel water pipe are of the slip-on type welded to the pipe. Flanges may be of two classes, as follows:

- Steel ring, hubless flange made from rolled plate, billet, or curved flat
- Forged steel, made with a low hub or with a welding neck by a rolling or forging process

The more costly welding-neck type of flange ordinarily is not justified for the comparatively low pressures usually found in waterworks service.

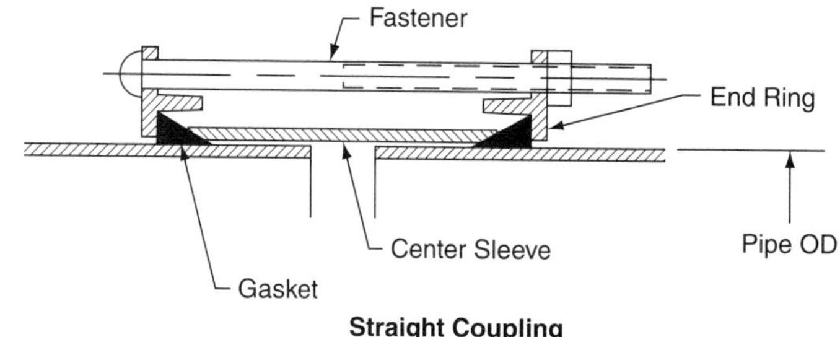

Straight Coupling

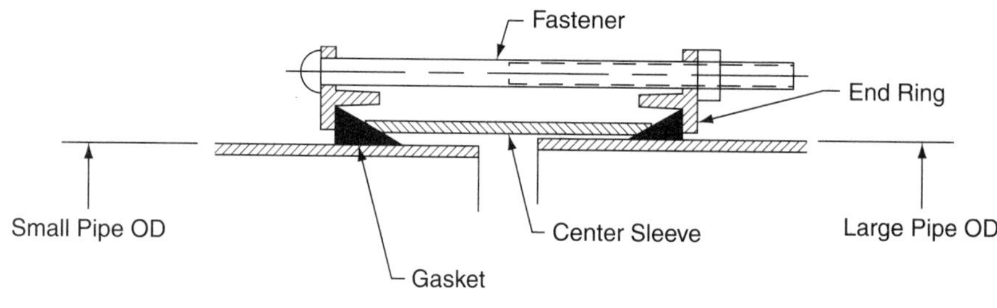

Transition Coupling

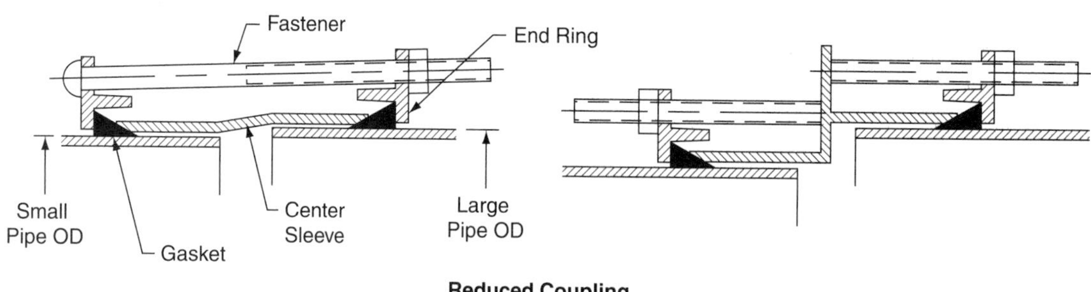

Reduced Coupling

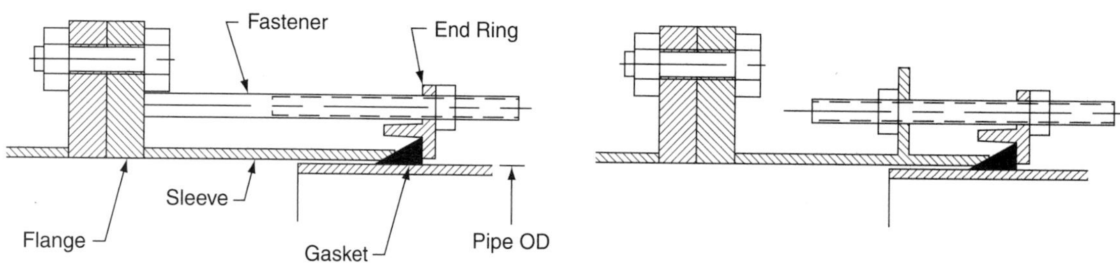

Flanged Coupling Adapter

Figure 8-2 Bolted sleeve-type couplings

Steel Ring Flanges

Field performance (Armco Steel Corp., Barnard 1950) conducted on full-size specimens demonstrated that satisfactorily tight joints can be obtained using steel ring flanges with 1⁄16-in. (1.6-mm) thick gaskets extending at least to the bolt holes. For dimensions and other details concerning the series of steel ring flanges and hub flanges as developed by an ASME-AWWA committee, refer to AWWA C207, *Standard for Steel Pipe Flanges for Waterworks Service—Sizes 4 in. Through 144 in. (100 mm Through 3,600 mm)*(latest edition).

Gaskets

Steel ring flanges conforming to AWWA C207 have been designed for use with rubber or synthetic ring gaskets that are either 1⁄16-in. (1.6-mm) or 1⁄8-in. (3.2-mm) thick. The gaskets should occupy the surface of the flange between the bolt holes and the inside diameter of the pipe or flange. Both the size of the gasket and the type of gasket material are integral and controlling factors in the design of a bolted joint. Recommendations by the gasket manufacturer should be observed.

Other Flanges

In waterworks service, flanges are frequently used for attaching pipe to pumps, valves, or other appurtenances having standard ASME drilling. Flanges of lesser thickness, but having the same drilling template, are obtainable from pipe and flange manufacturers, who will provide dimensional data.

Pressure Ratings

The flange pressure ratings in AWWA C207 are for temperatures normally encountered in waterworks service. The pressure rating for the flange should be based on the design of the maximum operating pressure plus the anticipated surge pressure. Test pressures should not exceed 125 percent of the rating.

When working pressures require flanges heavier than Class E (275 psi [1,896 kPa]) in AWWA C207, other flanges may be used (American National Standards Institute 1977). The cold-water (100°F [38°C]) rating of an ASME flange greatly exceeds the rating of an AWWA C207 class flange. For example, the 100°F (38°C) pressure rating for the 300-psi (2,068-kPa) ASME flange is 720 psi (4,964 kPa). When heavier flanges are selected, data on sizes above 24 in. (600 mm) must come from the flange manufacturer.

GROOVED-AND-SHOULDERED COUPLINGS

The grooved-and-shouldered coupling is a bolted, segmental, clamp-type, mechanical coupling having a housing that encloses a U-shaped rubber gasket. The housing locks the pipe ends together to prevent end movement, yet allows some degree of flexibility and alignment. The rubber gasket is tight under either pressure or vacuum service. The coupling is shown sectioned in Figures 8-3 and 8-4.

Ends of pipe must be specially prepared to accommodate grooved-and-shouldered couplings. This is done by grooving, banding, rolling, or welding adapters to pipe ends. Careful attention must be given pipe-end preparation so that the couplings will fit properly. Some typical grooved-and-shouldered joints are described in AWWA C606, *Standard for Grooved and Shouldered Type Joints* (latest edition).

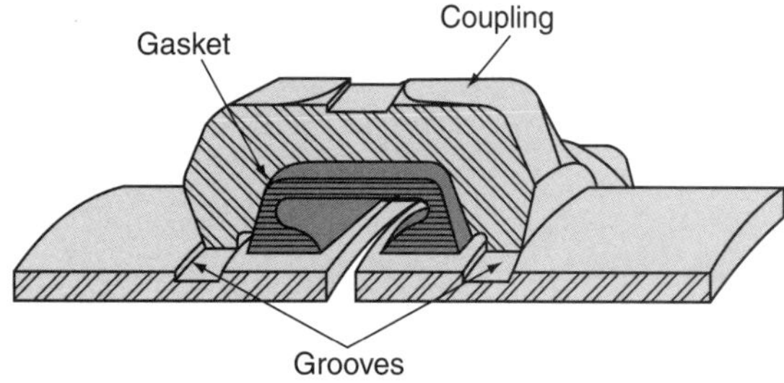

Figure 8-3 Grooved coupling

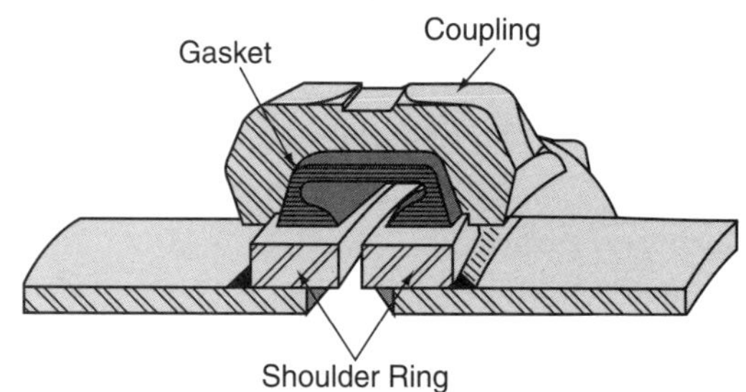

Figure 8-4 Shouldered coupling

EXPANSION AND CONTRACTION—GENERAL

The coefficient of expansion of steel is 6.5×10^{-6} in./in./°F (11.7×10^{-6} m/m/°C). The change in length of an unrestrained steel pipe can be determined using:

$$\Delta L = (6.5 \times 10^{-6}) L (\Delta t) \tag{8-1}$$

Where:

ΔL = change in length, in.
L = length, in.
Δt = change in temperature, °F

The expansion or contraction of an unrestrained steel pipe is about ¾ in. (19 mm) per 100 ft (30.5 m) of pipe for each 100°F (56°C) change in temperature.

Expansion and Contraction—Underground

Ordinarily, a buried pipeline under operating conditions will not experience significant changes in temperature, and thermal stresses will be minimal. However, during the construction and prior to the completion of backfilling, extreme changes in ambient temperatures may cause excessive expansion or contraction in the pipe. These extreme temperature changes and the resulting expansion and contraction may be avoided by backfilling the pipe as construction progresses.

For field-welded lines, AWWA Standard C206 describes a method that has been used satisfactorily to reduce the thermal stresses resulting from temperature variations. This method utilizes a special closure lap joint at 400- to 500-ft (122- to 152-m) intervals. The special closure is set so that the pipe is stabbed deeper than the normal closed position, all joints are welded except the closure, partial backfill is placed over all pipe except the closure joint to aid in cooling and contraction of the pipe, and the closure weld is made during the coolest part of the day. (See chapter 12.)

Forces caused by expansion and contraction should not be allowed to reach valves, pumps, or other appurtenances that might be damaged by these forces. Appurtenances can be protected by connecting the pipe and appurtenance with an expansion joint or bolted sleeve-type coupling, or by providing anchor rings and thrust blocks of sufficient size and weight to prevent the forces from reaching the appurtenance to be protected.

Expansion and Contraction—Aboveground

Expansion and contraction of exposed lines must be provided for where individual pipe sections are anchored and bolted-sleeve-type couplings are used for field joints.

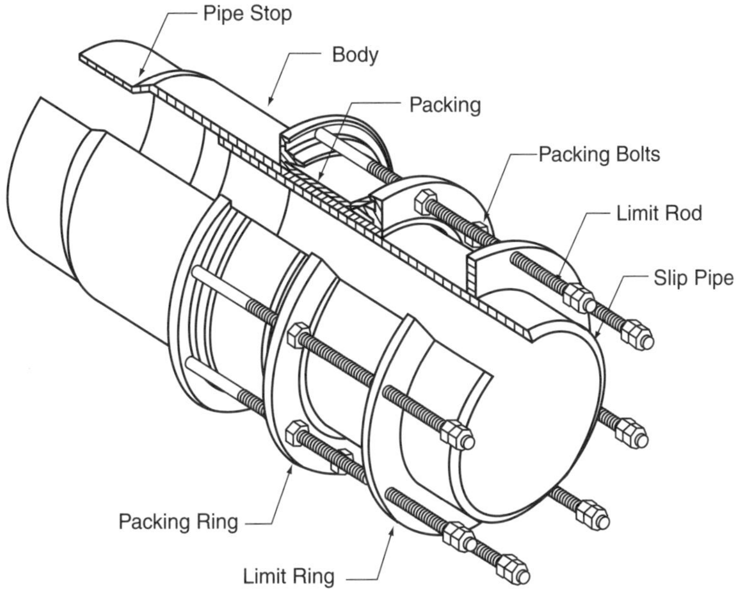

Figure 8-5 Typical expansion joint with limit rods

The joints will ordinarily allow enough movement so that expansion or contraction is not cumulative over several lengths.

On exposed field-welded lines, expansion joints may be located midway between the anchors if the pipeline is laid level. On slopes, the joint should be placed adjacent to or on the downhill side of the anchor point. Pipe ordinarily resists uphill movement; therefore, the pipe strength at the anchor block should be investigated to ensure that it is adequate to resist the downhill thrust. The coefficient of sliding friction for bare pipe bearing on supports should be determined. Spacing and positioning of expansion joints should be determined by site and profile requirements. Expansion joints in pipe on bridges should be at points where the bridge structure itself contains expansion joints.

Fabricated steel mechanical slip-type expansion joints (AWWA C221) are sometimes used. These joints permit linear movement of the slip pipe relative to the packing. Details of fabricated steel mechanical slip-type expansion joints are shown in Figures 8-5 and 8-6. The packing of expansion joints may consist of rubber rings only or alternate split rings of rubber compound and other approved materials.

The stuffing-box expansion joint is sometimes made double-ended. Limited-movement features are also added to both the single- and double-ended types. This is particularly important for double-ended expansion joints. Without restraint, such joints may creep along the pipe during repeated temperature-change cycles. When installing an expansion joint, the initial setting should be established with consideration of the current temperature and pipe length. Expansion joints must be installed in accordance with the manufacturer's instructions.

GROUND FRICTION AND LINE TENSION

When metal temperatures change, the grip of soil on buried pipe or the friction of sliding supports for exposed pipe offers resistance to longitudinal movement of the pipe. This resistance induces tensile or compressive stress in the pipe wall. The motivating force may also be caused by gravity on slopes.

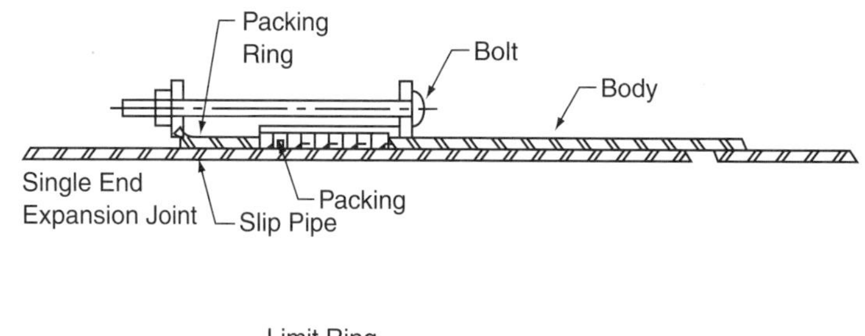

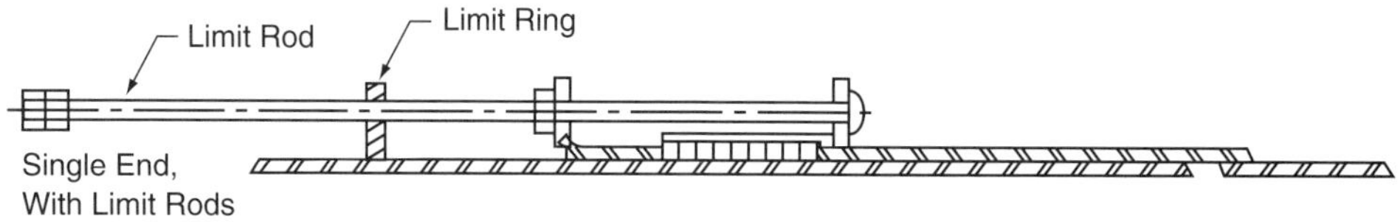

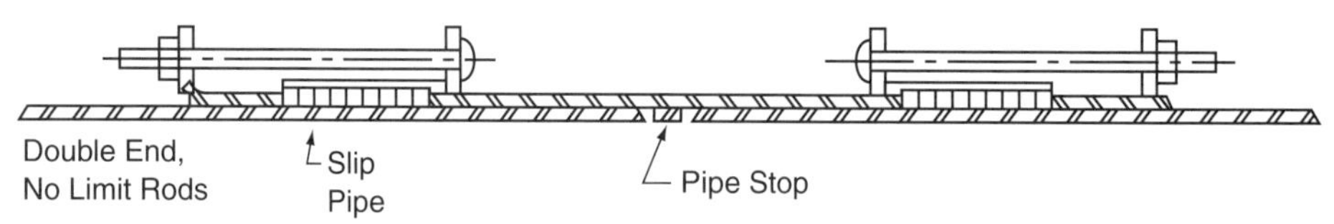

Figure 8-6 Typical expansion joint configurations

The change in longitudinal stress in a pipe with fixed ends due to a temperature change is determined by

$$\Delta S = E\varepsilon\Delta t \tag{8-2}$$

Where:

ΔS = change in stress, psi (MPa)
E = modulus of elasticity of steel, 30×10^6 psi (206,800 MPa)
ε = coefficient of expansion of steel, 6.5×10^{-6} in./in./°F (11.7×10^{-6} m/m/°C)
Δt = change in temperature, °F (°C)

A temperature change of 30°F (17°C) causes a theoretical stress of 5,850 psi (40.33 MPa) in the pipe wall if ends are restrained. The stress in all-welded buried steel pipe has been investigated (McClure and Jackson), with the result that some measured stresses were found to be higher and some lower than theoretical. It was found in all cases, however, that the soil restraint was sufficient to absorb the longitudinal stress in pipe approximately 100 ft (30.48 m) long. Chapter 13 includes discussion and design aid on frictional resistance between the pipe and ground.

GOOD PRACTICE

The requirements of installation and operation of a pipeline may dictate the use of more than one type of field joint. The type of internal lining and pipe diameter will also determine factors in joint selection. Bell-and-spigot rubber gasket joints cost the least on an installed-cost basis. Flanges are commonly used to join steel pipe to valves, meters, and other flanged accessories. Thermal stresses may be a consideration, and these can be accommodated by sleeve couplings, grooved-and-shouldered couplings, special welded joints, or expansion joints.

By agreement between the manufacturer and the purchaser, the ends of the pipe may be supplied with joint configurations and tolerances other than those described in this manual. In such cases, pipe ends shall conform to the dimensions, description, and tolerances of detailed drawings provided by the purchaser.

REFERENCES

American National Standards Institute. 1977. *Pipe Flanges and Flanged Fittings.* ANSI Standard B16.5. New York: American National Standards Institute.

American Water Works Association. ANSI/AWWA Standard C206, *Field Welding of Steel Water Pipe*. Latest edition. Denver, Colo.: American Water Works Association.

———. ANSI/AWWA Standard C207, *Steel Pipe Flanges for Waterworks Service—Sizes 4 In. Through 144 In. (100 mm Through 3,600 mm).* Latest edition. Denver, Colo.: American Water Works Association.

———. ANSI/AWWA Standard C606, *Grooved and Shouldered Type Joints.* Latest edition. Denver, Colo.: American Water Works Association.

Armco Steel Corp. *Steel Ring Flanges for Steel Pipe*. Bull. 47-A. Middletown, Ohio: Armco Steel Corp.

Barnard, R.E. 1950. Design of Steel Ring Flanges for Water Works Service, A Progress Report. *Jour. AWWA*, 42(10):931.

Killam, E.T. 1943. Mechanical Joints for Water Lines. *Jour. AWWA*, 35(11):1457.

McClure, G.M., and L.R. Jackson. *Slack in Buried Gas Pipelines*. Columbus, Ohio: Battelle Memorial Institute.

Chapter 9

Fittings and Appurtenances

Because of the wide range of design, welding and fabrication processes applicable to steel pipe provide a solution to most problems involving fittings and specials. The design of pipe layouts, especially intricate ones, is greatly facilitated by having standardized dimensions for the center-to-face distance or the center-to-end distance of fittings. AWWA C208, *Standard for Dimensions for Fabricated Steel Water Pipe Fittings* (latest edition), provides dimensions for welded steel pipe fittings in sizes 6 in. (150 mm) and larger. AWWA C200, *Standard for Steel Water Pipe 6 In. (150 mm) and Larger* (latest edition), specifies the manufacturing requirements for fittings and special joints.

The standard dimensions of fittings for screwed-joint pipe can be found in the catalogs of many manufacturers. Manufacturers can also provide the dimensions of compression fittings for use on standard plain-end pipe in the smaller sizes.

DESIGNATION OF FITTINGS

Fittings should be designated using standard methods to prevent misunderstandings. Figures 9-1 through 9-5 are diagrammatic. The desired deflection angle of the elbow or lateral should be indicated and refers to smooth as well as segmental fittings. On ordinary elbows, both ends are numbered the same because both are the same size. Therefore, only one diameter is needed for a standard or nonreducing elbow, together with the deflection angle. (Example: 54-in. (1,350-mm) OD, 90° elbow.)

Reducing crosses and elbows are always identified by first giving the size or outside diameter of the largest openings, then following with the sizes of the openings in the numerical sequence shown. (Examples: 48-in. (1,200-mm) OD × 36-in. (900-mm) OD 90° elbow; 24-in. (600-mm) OD × 22-in. (550-mm) OD × 8⅝-in. (216-mm) OD × 6⅝-in. (166-mm) OD cross.)

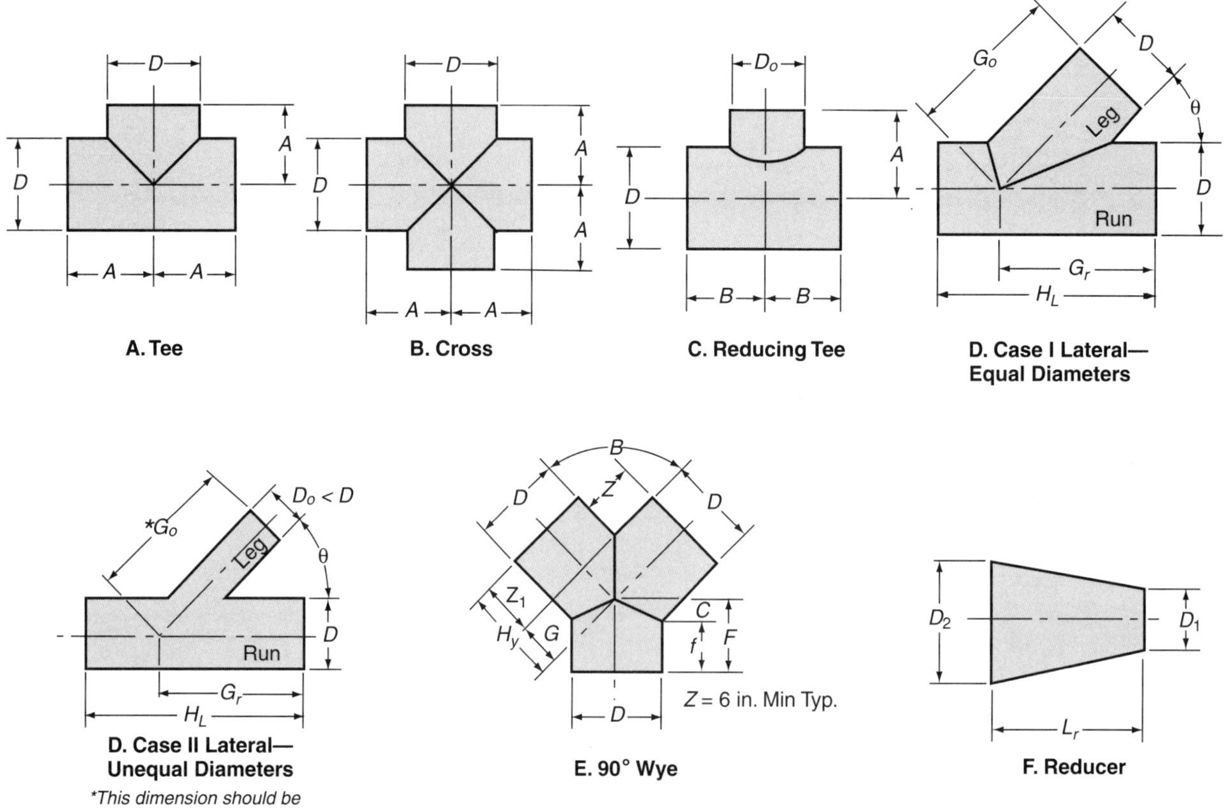

NOTE: See AWWA C208 for formulas to calculate dimensions.

Figure 9-1 Recommended dimensions for water pipe fittings (except elbows)

Tees and laterals are specified by giving the size of the largest opening of the run first, the opposite opening of the run second, and the size of the outlet or branch last. Example: 72-in. (1,800-mm) OD × 66-in. (1,650-mm) OD × 24-in. (600-mm) OD tee.

The size of side outlets on fittings should be specified last. When specifying side-outlet tees in reducing elbows, particular care should be exercised to show whether they are right- or left-hand. In addition to designating the size of the fitting, the purchaser should provide a complete specification for the types of ends or flanges desired.

ELBOWS AND MITER END CUTS

Small deflection angles can be taken at O-ring joints, mechanical couplings, and welded lap joints by pulling the joint asymmetrically. The pipe or coupling manufacturer can provide the allowable deflections.

Deflection angles up to 5° can be taken in welded butt joints by miter end cuts on one or both pipe ends (Figure 9-2A), provided the difference in the circumference of the true circle and the ellipse formed by the miter end cut does not result in a joint fit up that would exceed the allowable plate edge offset. Deflection angles up to 5° can be taken in welded lap joints using miter cut bell ends (Figure 9-2B). In this procedure, the pipe end is miter cut, and then the bell is expanded square with the face of the miter cut.

Deflection angles greater than those allowed above shall be taken using fabricated elbows as shown in AWWA C208 and Figure 9-2C, 9-2D, 9-2E, and 9-2F, except that the deflection per miter weld shall be limited to 15°. The radius of the elbow shall be at

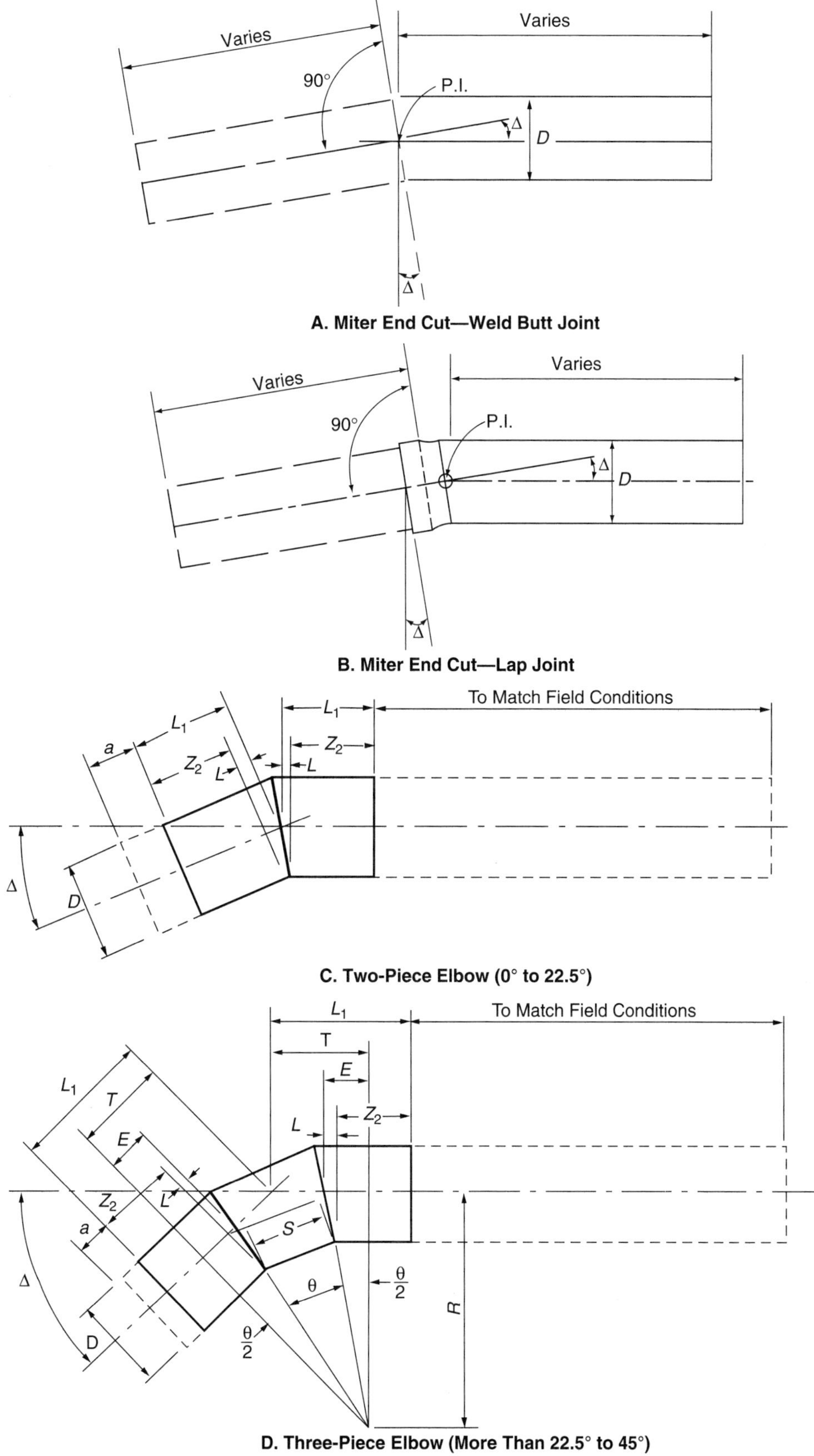

NOTE: See AWWA C208 for formulas to calculate dimensions.

Figure 9-2 Recommended dimensions for water pipe elbows

Figure continued next page.

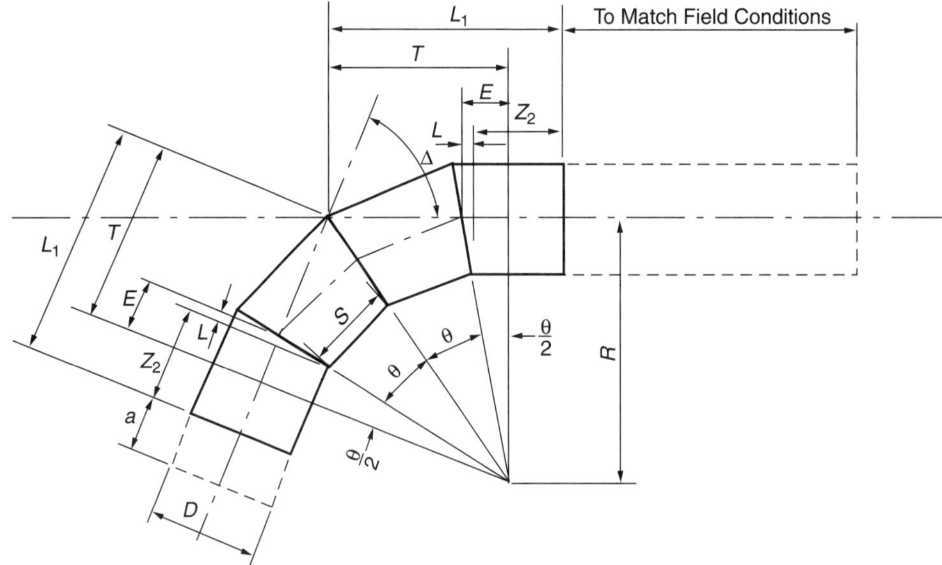

E. Four-Piece Elbow (More Than 45° to 67.5°)

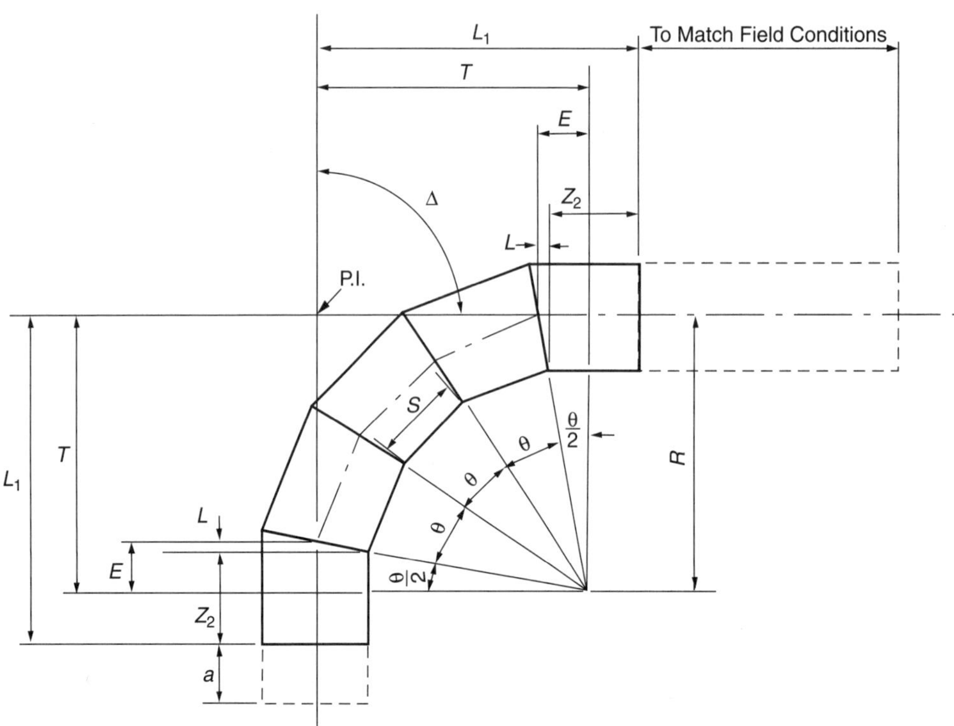

F. Five-Piece Elbow (More Than 67.5° to 90°)

NOTE: See AWWA C208 for formulas to calculate dimensions.

Figure 9-2 Recommended dimensions for water pipe elbows (continued)

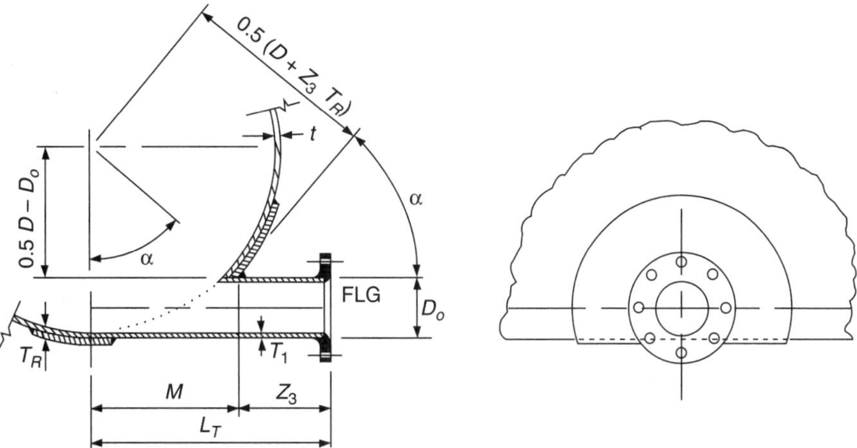

Figure 9-3 Tangent-type outlet (AWWA C208)

least 2.5 pipe diameters, or the wall thickness of the elbow section shall be calculated using the following formula:

$$t = \frac{PD}{Sf}\left[\frac{S}{2} + \frac{D}{3}\tan\frac{\theta}{2}\right] \qquad (9\text{-}1)$$

Where:

t = required elbow wall thickness (in.)
P = design internal pressure (psi)
f = allowable tensile stress at design pressure (psi)
D = outside diameter of pipe (in.)
S = segment length along inside of elbow as shown on Fig. 9-2D, E, and F
θ = segmented deflection angle as show on Fig. 9-2D, E, and F

The above formula may be simplified as follows:

$$S = 2\left(R - \frac{D}{2}\right)\tan\frac{\theta}{2} \qquad (9\text{-}2)$$

Where:

R = radius of elbow

Substituting for S in the above formula

$$t = \frac{PD}{2f}\left[1 + \frac{D}{3R - 1.5D}\right] \qquad (9\text{-}3)$$

Elbows may be designated with a constant diameter or with a different diameter on each end (Figure 9-5). Compound or combined elbows, in which the plane of the bend is neither horizontal nor vertical, require certain trigonometric computations. Usually, the plan angle and profile angles are known, and the true angle must be determined in the plane of the elbow and the elbow rotations. The computation method and applicable formulas for these elbow properties are shown in Figure 9-6.

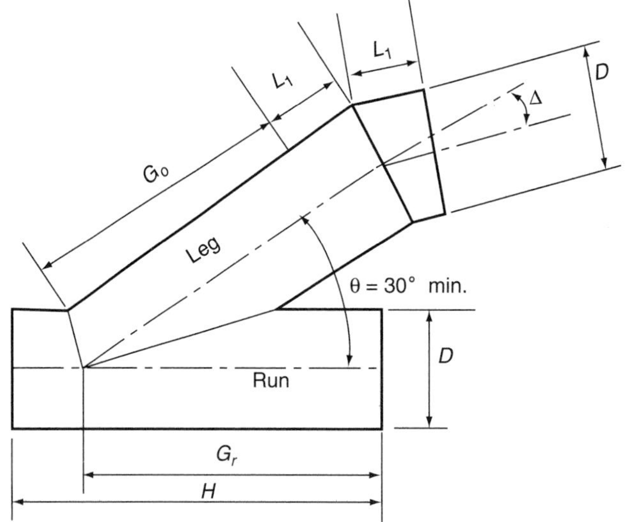

NOTE: See AWWA C208 for formulas to calculate dimensions.

Figure 9-4 Lateral less than 30 degrees

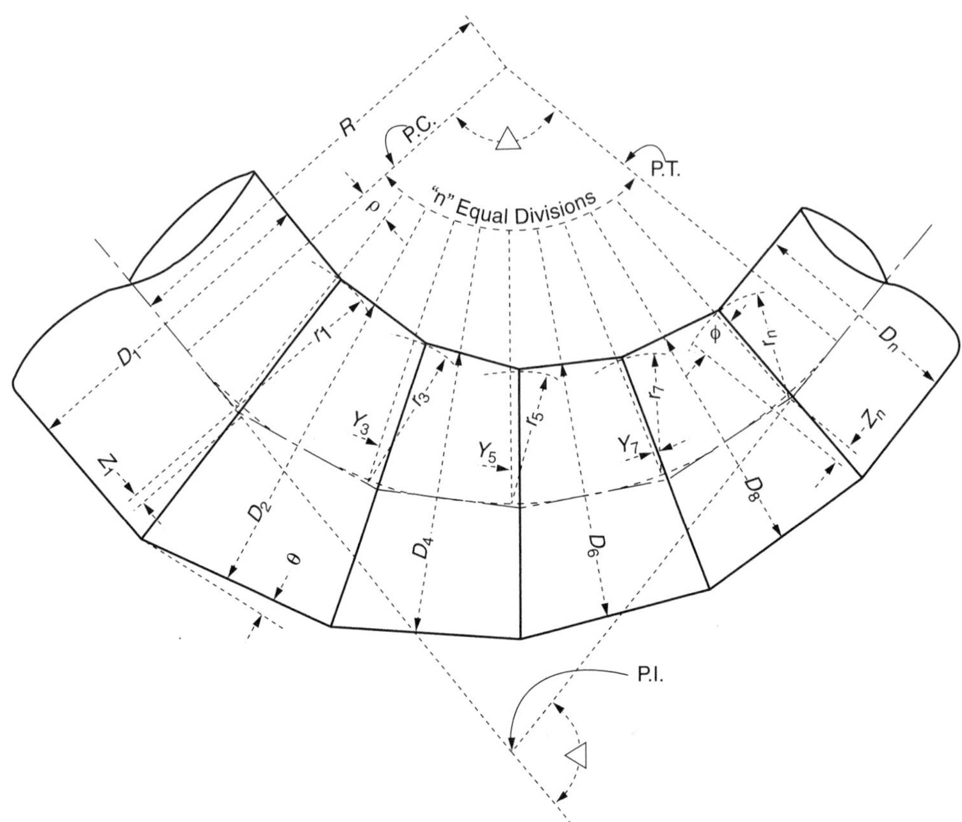

NOTE: See AWWA C208 for formulas to calculate dimensions.

Source: Reprinted with permission from Buried Steel Penstocks, *American Iron and Steel Institute, Washington, D.C., in cooperation with Steel Plate Fabricators Association, Des Plaines, Ill.*

Figure 9-5 Reducing elbow

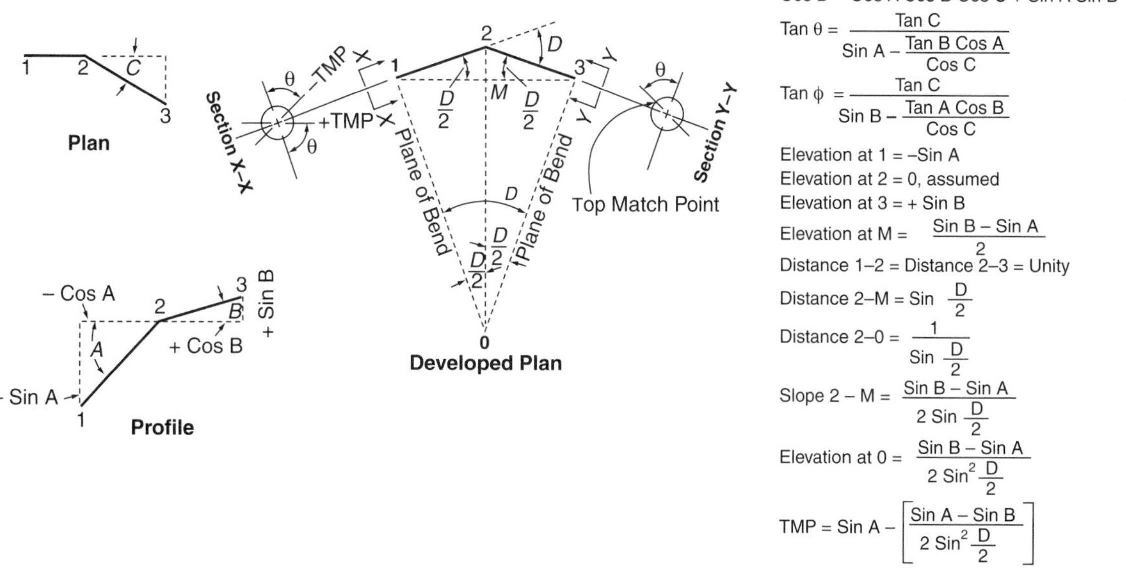

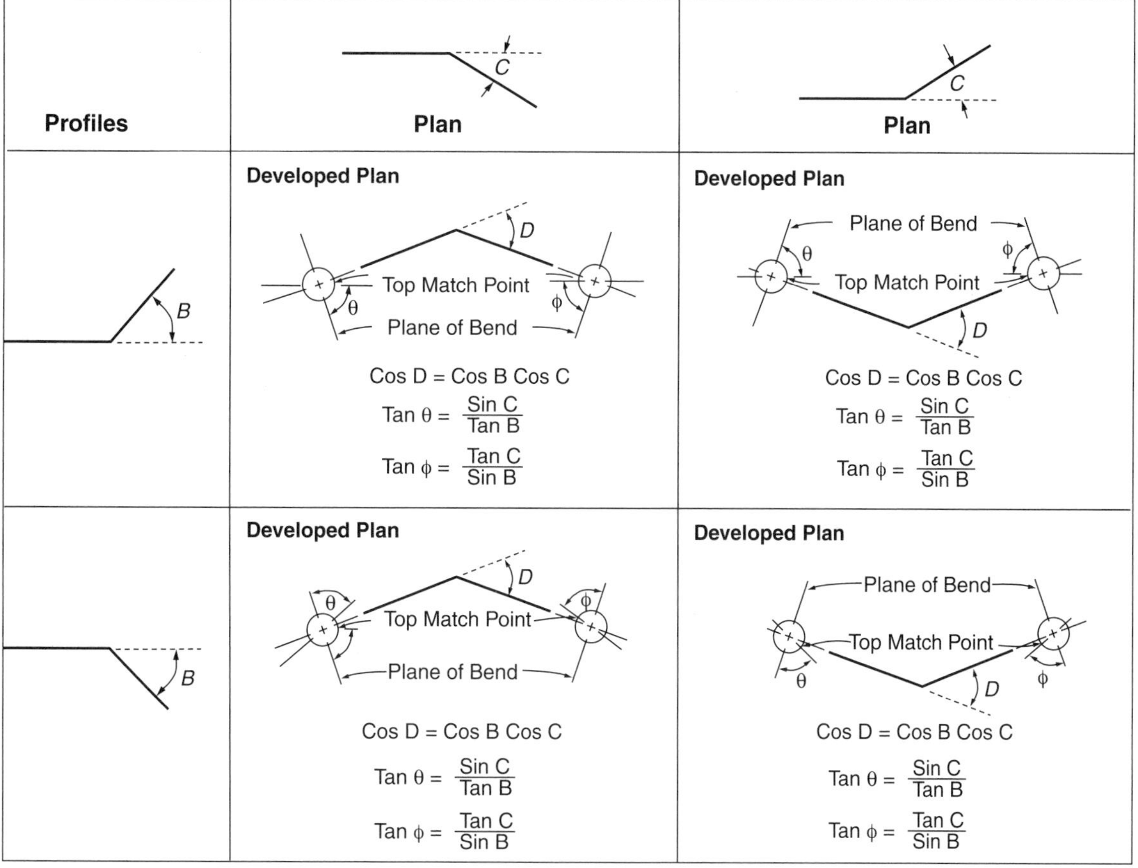

Figure 9-6 Computation method and formulas for compound pipe elbows

Figure continued next page

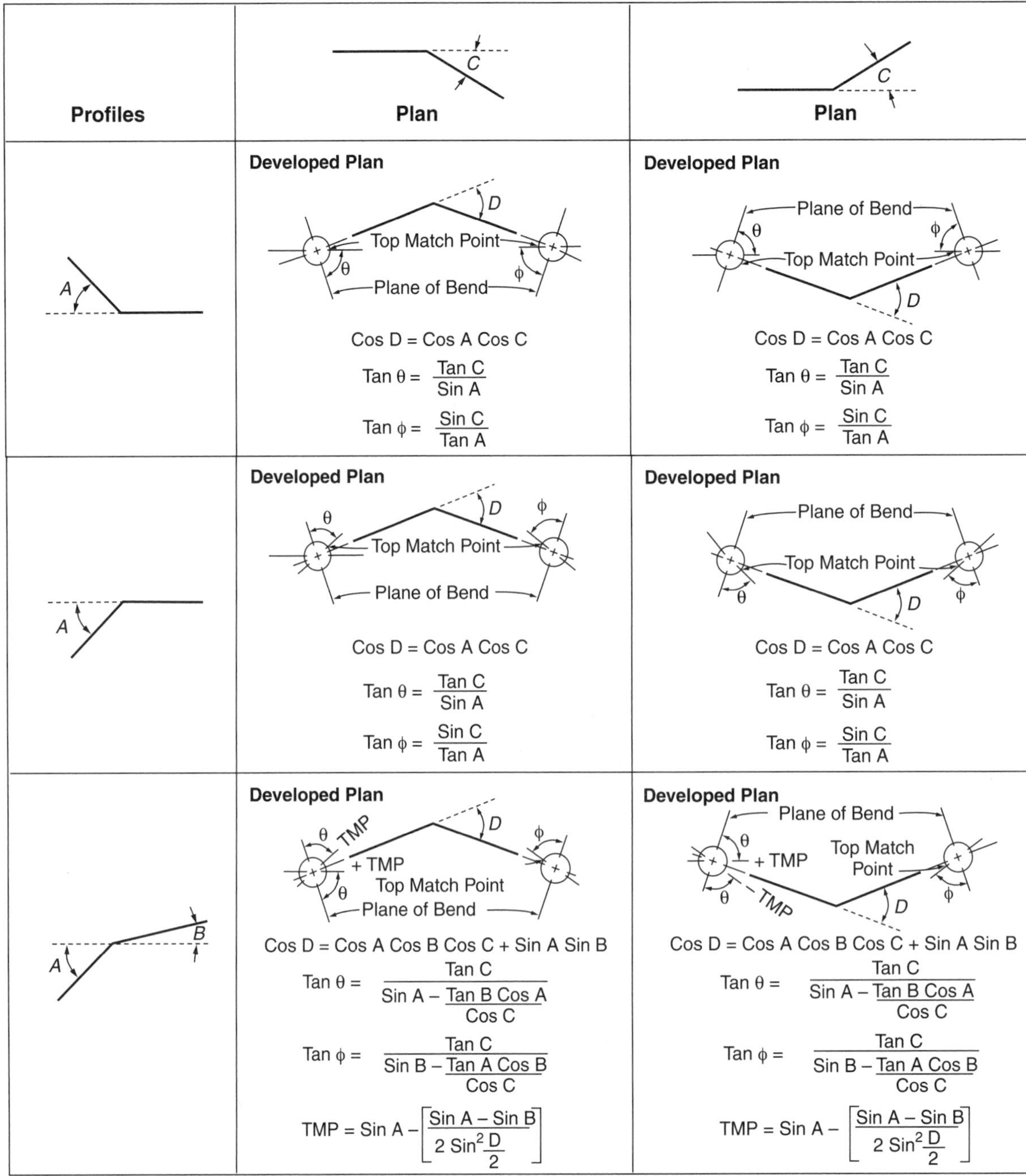

Figure 9-6 Computation method and formulas for compound pipe elbows (continued)

Figure continued next page

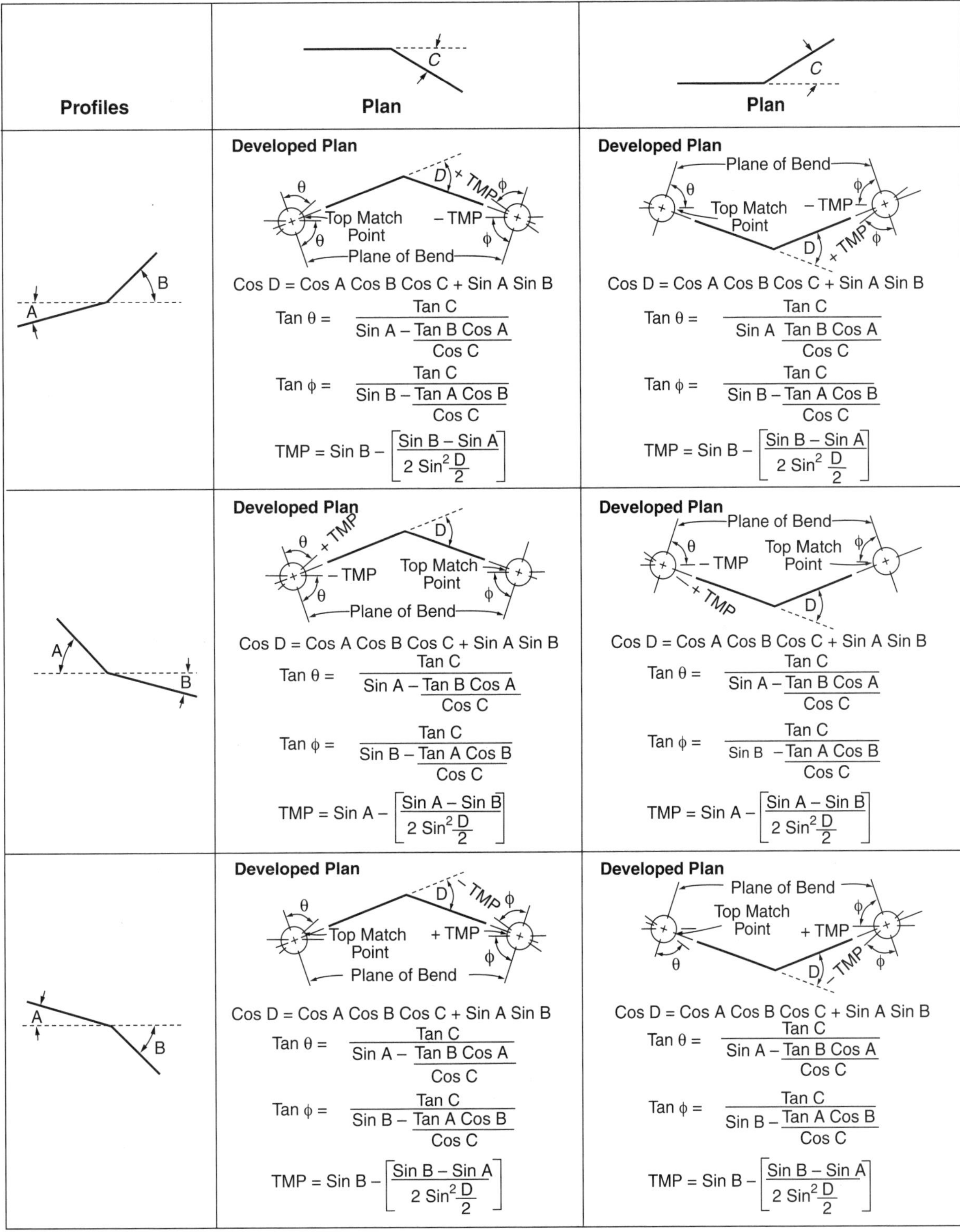

Figure 9-6 Computation method and formulas for compound pipe elbows (continued)

Figure continued next page

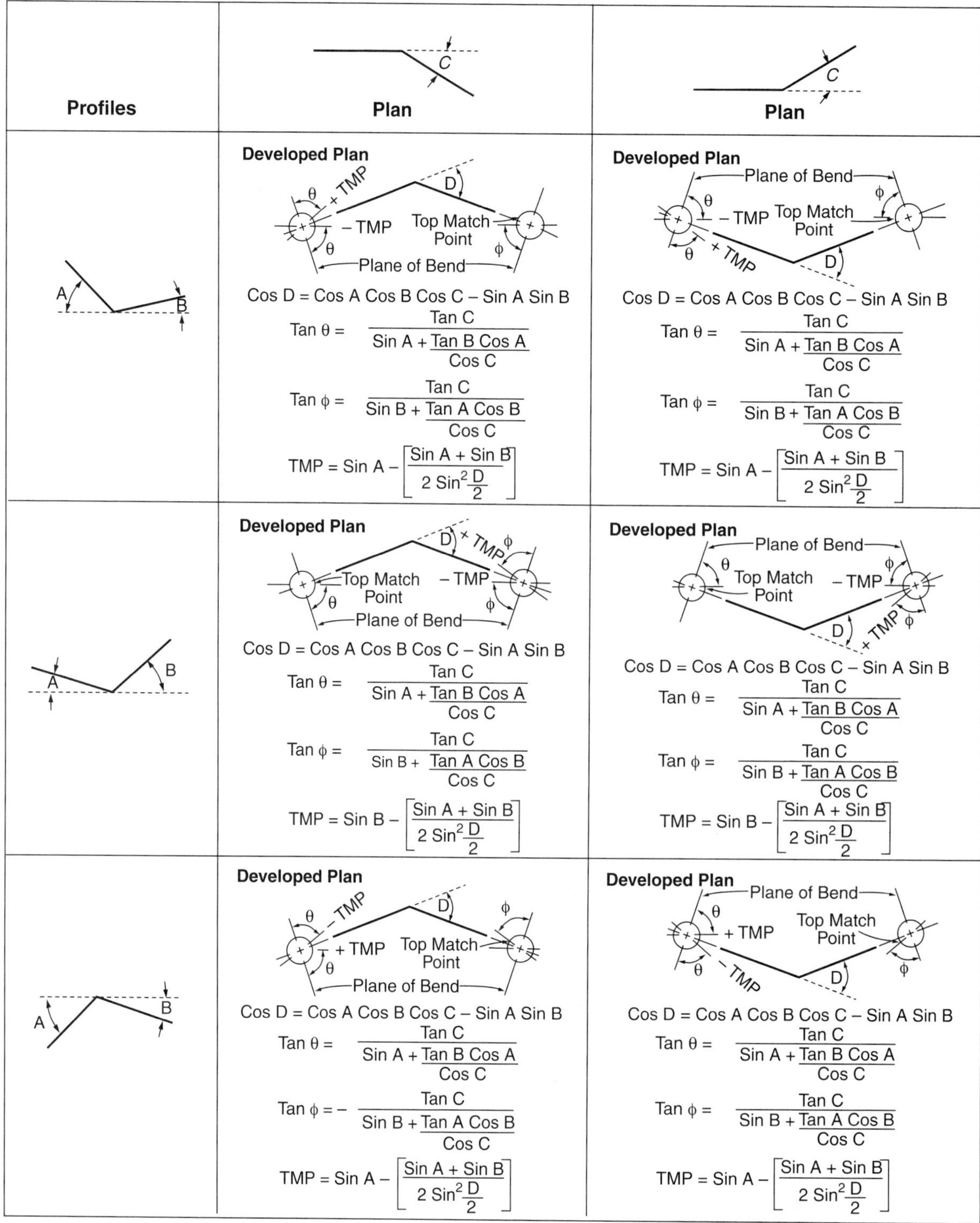

Figure 9-6 Computation method and formulas for compound pipe elbows (continued)

Figure continued next page

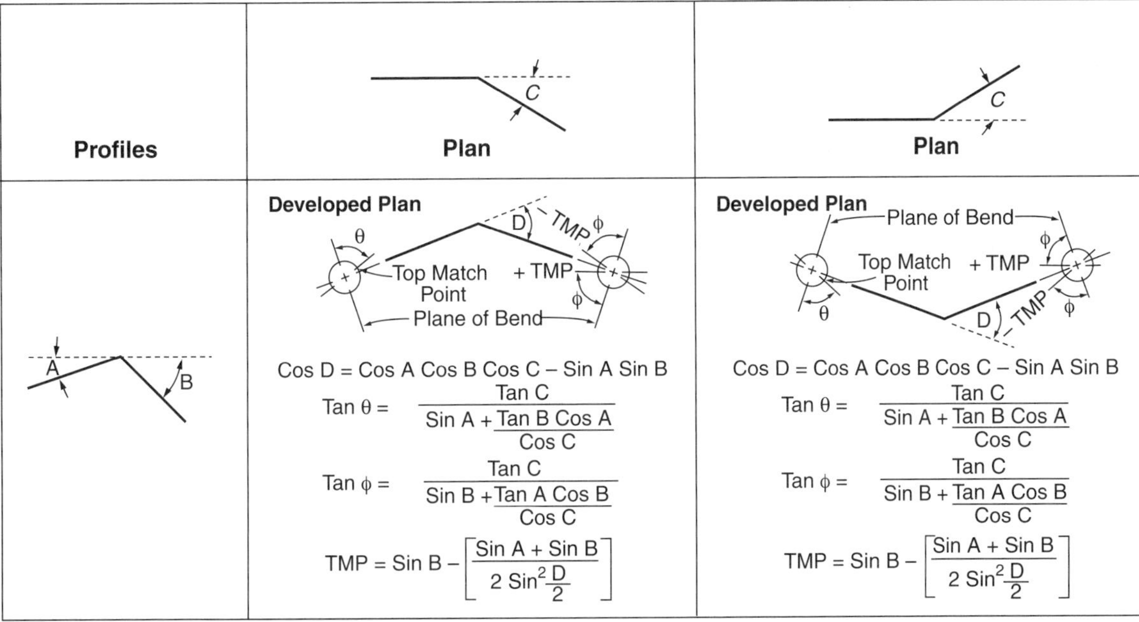

Figure 9-6 Computation method and formulas for compound pipe elbows (continued)

REDUCERS

Changes in diameter are accomplished using concentric or eccentric cones or reducers placed in the straight section of a line or combined with a mitered elbow, tee, or cross.

AWWA C208 shows the common length of a reducer to be $4(D_2 - D_1)$, which gives a half-apex angle of 7.0°–7.5°. For reducers of this half-apex angle or less, the wall thickness of the larger diameter pipe is adequate.

Reducers with half-apex angles over 8° should be designed in accordance with ASME Boiler and Pressure Vessel Code Section VIII, Division I, Par. UG-32 (g) and appendix 1-5.

BOLT HOLE POSITION

It is standard practice to attach flanges to pipe lengths so that the bolt holes straddle the vertical centerline. If flanged pipe is to be installed at various angles to the vertical centerline, this standard practice should be modified, and the proper data should be provided on drawings and in specifications so that the flange will be attached as needed.

DESIGN OF WYE BRANCHES, LATERALS, TEES, AND CROSSES

A full treatise on the design of wye branches, laterals, tees, and crosses for steel pipe, including a nomograph design method, is described in a paper prepared by the Department of Water and Power, City of Los Angeles (Swanson et al. 1955). The design using this nomograph method is presented in chapter 13. Examples are given for single-plate, two-plate, and three-plate design. Larger wye branches (up to 144-in. (3,600-mm) diameter and up to 150-psi (1,034-kPa) pressure, including safety factor) may be designed by extrapolation beyond the limits of the graphs and tables in the reference. Other data on the subject also have been published (Rudd).

TESTING OF FITTINGS

AWWA C200 (latest edition) describes nondestructive testing of weld seams on fittings and special sections. Special sections fabricated from previously hydrostatically tested straight pipe require testing of only those welded seams that were not previously tested in the straight pipe. Nondestructive testing methods include dye penetrant, magnetic particle, ultrasonic, x-ray, or other methods. AWWA C200 (latest edition) also permits hydrostatic testing of specials in lieu of other nondestructive testing. Maximum test pressure should not exceed 1.5 times the design pressure. This maximum should be observed because fitting tests are costly.

UNBALANCED THRUST FORCES

Piping systems are subjected to unbalanced thrust forces resulting from static and dynamic fluid action on the pipe. These forces must be absorbed or balanced if the piping systems are to maintain their integrity. Unbalanced thrust forces occur at changes in the flow direction, such as elbows, tees, laterals, wyes, and at reducers, valves, and dead ends. These thrust forces can be balanced by thrust blocks or by transmitting forces to the pipe wall by restrained, harnessed, flanged, or welded joints. Forces in the pipe shell are ultimately transferred to the soil. In many cases, it is desirable to combine blocking and transmitting forces to the pipe wall. Methods for handling these thrust forces and pertinent data are mentioned in chapter 13.

FRICTIONAL RESISTANCE BETWEEN SOIL AND PIPE

If an unblocked fitting is tied to buried pipe, preventing movement, and tension is placed on the pipe, it may be necessary to determine the length of the pipe on which the earth friction will overcome the disjointing force. chapter 13 includes discussion and design aids on this subject.

ANCHOR RINGS

Anchor rings for use in concrete anchor blocks or concrete walls may be simple ring flanges. Rings are proportioned to accept dead-end pull or thrust imposed by the internal pressure and any pipe thrust or pull as a result of temperature change, with approximately 500-psi (3,447-kPa) bearing on concrete. Care must be exercised to ensure that thrust rings are positioned so as to provide an adequate safety factor against punching shear of the concrete. See Wrapper-Plate Design, chapter 13, for sizing of fillet welds for attachment of rings.

NOZZLE OUTLETS

Outlets from steel mains can be easily arranged in any desired location according to size, shape, or position. Nozzles are welded to the main line with reinforcing collars. This work can be done in the shop during the fabrication of the pipe, or at trenchside, or after the pipe is installed. Shop lining and coating of nozzles and pipe is satisfactory and more economical than work done in the field. All outlets should be checked to determine whether reinforcement is required; however, outlets larger than about one third of the diameter of the line need special consideration for reinforcement, even for small pipe sizes.

If required for hydraulic efficiency, a reducer may be welded to the main pipe with the outlet welded to the reducer. The reinforcing of the shell must be computed on the larger diameter.

The end of the outlet nozzle should be prepared to receive the valve or fitting to be attached. This may call for a flange, a grooved or shouldered end for a mechanical coupling, a plain end for a flexible coupling joint, a grooved spigot end for a bell-and-spigot joint, or a threaded end.

CONNECTION TO OTHER PIPE MATERIAL

Care must be exercised when connecting dissimilar pipe materials, because of the possibility of galvanic corrosion. See chapter 10 for principles of this reaction. When connecting steel pipe to either gray or ductile-iron pipe, or to steel-reinforced concrete pipe, or to copper or galvanized pipe, an electrically insulating joint should be used. The insulating joint can be accomplished with an insulating gasket with sleeves and washers on a flanged connection or with an insulating sleeve-type flexible coupling. (See Insulating Joints.)

Any valves or other ferrous equipment connected to steel pipe should be encapsulated in polyethylene sheeting or coated with a coating compatible with the steel pipe coating. Similar precautions are not necessary when connecting to nonmetallic pipe, such as asbestos cement or plastic.

FLANGED CONNECTIONS

Flanged outlets can be assembled from a short piece of pipe using a steel ring flange, or a hub flange of the slip-on type can be used. Attachment of flanges should be in accordance with AWWA C207 (latest edition). The bolt holes in flanges straddle the vertical and horizontal centerlines. If the main line slopes, the flange should be rotated with reference to this slope to bring the attachments vertical.

Outlet nozzles should be as short as possible to reduce the leverage of any bending force applied to the outlet. In general, every outlet should have a valve firmly attached to the mainline and a flexible connection to the pipe downstream from this valve.

VALVE CONNECTIONS

Valves are self-contained devices that may not function properly or remain tight if subjected to external forces. If a valve is rigidly installed in a pipeline—for example, when flanged joints are used—the whole assembly of pipe and valves can be stressed by temperature changes, settlement, and exceptional surface loads. To prevent a valve from being strained there should be at least one flexible joint close to it.

It is good practice to provide for a flexible joint when fittings are flanged. This can be easily accomplished by installing a flexible coupling or a grooved-and-shouldered mechanical coupling immediately adjacent to one of the flanges. Such a coupling not only provides a satisfactory degree of flexibility but also makes installation and possible removal of the valve much easier. In such a situation, the center stop should be removed if a flexible coupling is used. The coupling, when loose, may be moved along the pipe to expose the joint and facilitate placement or removal.

BLOWOFF CONNECTIONS

Outlets for draining a pipeline should be provided at low points in the profile and upstream of line valves located on a slope. Short dips, such as those occurring in many pipelines in city streets when a line must pass under a large drain or other structure, can often be dewatered by pumping, when necessary.

The exact location of blowoff outlets is frequently determined by opportunities to dispose of the water. Where a pipeline crosses a stream or drainage structure, there usually will be a low point in the line; but if the pipeline goes under the stream or drain, it cannot be completely drained into the channel. In such a situation, a blowoff connection should be located at the lowest point that will drain by gravity and provide easy means for pumping out the section below the blowoff.

Blowoffs must, of course, be provided with a shutoff valve. If the pipeline is aboveground, the valve should be attached directly to the outlet nozzle on the bottom of the pipeline. A pipe attached to the valve will route the discharge to an appropriate location. The discharge pipe will usually require installation of an elbow at the blowoff valve, which must be securely blocked to avoid stresses on the attachment to the pipeline.

Usually the blowoff will be belowground. Because the operating nut of the valve must be accessible from the surface, the valve cannot be under the main but may be set with the stem vertical and just beyond the side of the pipeline. A typical detail of a blowoff is shown in AWWA C208 (latest edition) and in chapter 13.

MANHOLES

The most common type of manhole for access in waterworks is circular, having a short, flanged neck and a flat, bolted cover. Such manholes are commonly 24–36 in. (500–900 mm) in diameter.

Manholes should be placed in accessible locations. They provide access to the inside of the pipeline for many purposes besides inspection. In general, they will be most useful if located close to valves and sometimes close to the low points that might need to be pumped out for inspection or repair.

INSULATING JOINTS

Long steel pipelines frequently conduct electric currents originating from differences in ground potentials or stray currents. This phenomenon is explained in chapter 10. Where tests indicate the necessity, a long line is often separated into sections or insulated from other parts of a system by insulating joints. These joints can be provided at any flanged joint, but it is often necessary to make a joint at a particular place by installing a pair of flanges for this purpose.

Special insulating gaskets, sleeves, and washers are used to provide electrical insulation at the flanged joint. These insulating sleeves and washers are made of fabric-reinforced bakelite, micarta, Teflon®, or similar materials that have a long life and good mechanical strength.

The bolts of the insulated flanged joints must be carefully insulated by sleeves and washers. Insulating washers should be used at both ends of the bolts. Some purchasers specify flange holes 1⁄16-in. (1.6-mm) larger in diameter than normal flange holes.

It is important that insulating gaskets, sleeves, and washers be installed carefully so that the flanged joint will be properly insulating. After the installation of the insulated joint is complete, an electrical resistance test should be performed. The electrical resistance should be at least 10,000 ohms; if the resistance is less, the joint should be inspected for damage, the damage repaired, and the joint retested.

AIR-RELEASE VALVES AND AIR/VACUUM VALVES

Generally, air valves are sized at 1 in. per 1-ft (8.3 mm per 100-mm) diameter of pipe. Manufacturers' catalogs should be consulted for more accurate sizing information.

Air valves are installed to admit or vent air. The accumulation of air can become so large as to impair the pipe's flow capacity. There are basically three types as standardized in AWWA Standard C512 (latest edition): air-release valves, air/vacuum valves, and combination air valves, which combine the functions of an air-release valve and an air/vacuum valve.

Air-release valves are installed along the pipeline to provide for the continuous venting of accumulated air. An air-release valve consists of a chamber in which a float operates through levers to open a small orifice as air accumulates and to close as the water level rises. The float must operate against an air pressure equal to the water pressure and must be able to sustain the maximum pipeline pressure.

Air/vacuum valves are used to admit air into a pipe, preventing the creation of a vacuum that may be the result of a valve operation, the rapid draining or failure of a pipe, a column separation, or other causes. A vacuum can cause the pipe to collapse from atmospheric pressure.

Air/vacuum valves also serve to vent air from the pipeline while it is filling with water. An air/vacuum valve consists of a chamber with a float that is generally center-guided. The float opens and closes against a large orifice. As the water level recedes in the chamber, air re-enters the pipeline. The air/vacuum valve does not vent air under pressure. Air release or combination valves are needed to eliminate accumulated air under pressure.

Air valves, if not installed directly over the pipe, may be located adjacent to the pipeline. A horizontal run of pipe connects the air valve and the pipeline. The connecting pipe should rise gradually to the air valve to permit flow of the air to the valve for venting. The performance requirements of the valves are based on the venting capacity standard (cubic feet of free air per minute) and the pressure differential across the valves. The valves must be protected against freezing, and the vents from these valves must be located aboveground to prevent contamination when operating.

Manufacturers' catalogs should be consulted for sizing information. Figure 9-7 shows a typical pipeline and locations of air valves. Air valves should also be used on the draining side of mainline valves, upstream of Venturi meters, and on pump discharge locations. Valves on long ascents, descents, or horizontal runs should be spaced ¼ to ½ mile apart. For more detailed information, see AWWA C512, *Standard for Air Release, Air/Vacuum, and Combination Air Valves for Waterworks Service* (latest edition) and AWWA Manual of Water Supply Practice M51, *Air-Release, Air/Vacuum, and Combination Air Valves*.

CASING SPACERS

Casing spacers are used to ease installation and provide cathodic isolation of carrier pipelines within casing pipes. Spacers are mechanically attached around the outside of the pipeline at appropriate intervals to provide support and stiffness both during and after installation. Inert liner and runner materials provide dielectric insulation between the pipeline and the casing.

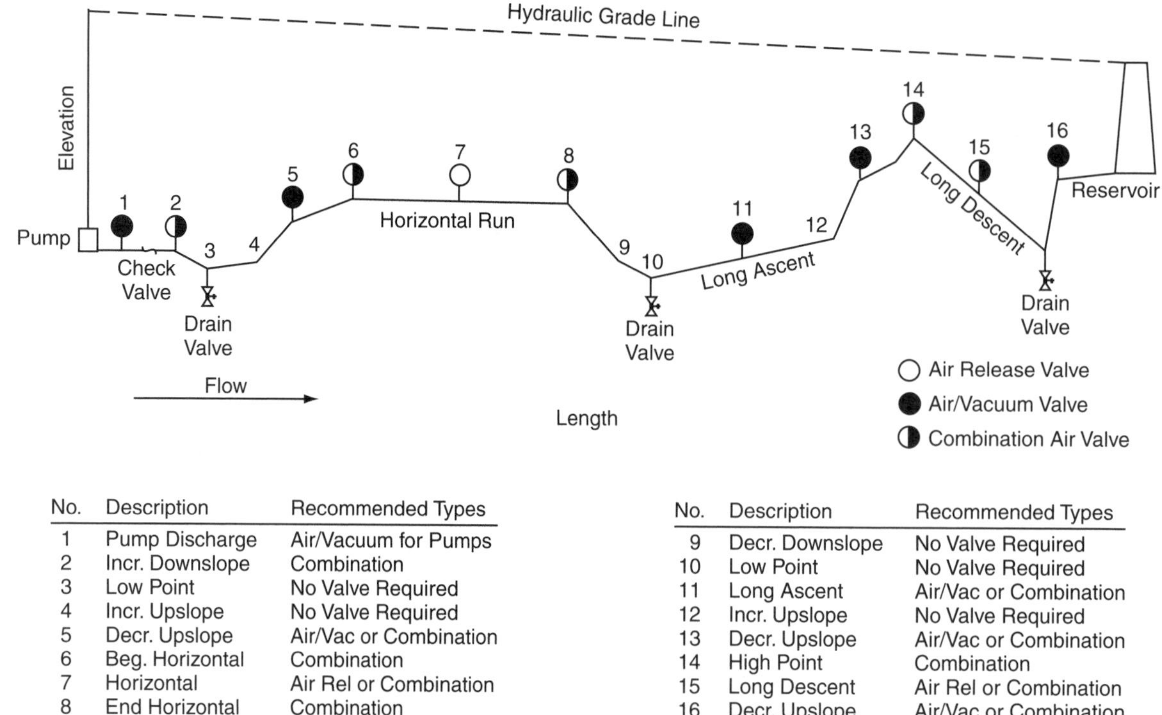

No.	Description	Recommended Types
1	Pump Discharge	Air/Vacuum for Pumps
2	Incr. Downslope	Combination
3	Low Point	No Valve Required
4	Incr. Upslope	No Valve Required
5	Decr. Upslope	Air/Vac or Combination
6	Beg. Horizontal	Combination
7	Horizontal	Air Rel or Combination
8	End Horizontal	Combination

No.	Description	Recommended Types
9	Decr. Downslope	No Valve Required
10	Low Point	No Valve Required
11	Long Ascent	Air/Vac or Combination
12	Incr. Upslope	No Valve Required
13	Decr. Upslope	Air/Vac or Combination
14	High Point	Combination
15	Long Descent	Air Rel or Combination
16	Decr. Upslope	Air/Vac or Combination

Figure 9-7 Sample pipeline profile illustrating air valve locations

Casing spacers can be fabricated to position the pipeline within the casing as necessary and to accommodate multiple pipelines within a single casing. Pipelines installed under highways, railroads, runways, and other structures often must be placed within casing pipes. Reasons to use casings include

- Compliance with requirements or regulations imposed by property owners where the pipeline is installed
- Prevention of damage to the pipeline from excessive loads of vehicles
- Prevention of damage to structures caused by soil erosion or settlement in the pipe zone
- Allowing free and easy access to pipeline for removal or replacement in the future
- Permitting boring or tunneling instead of open excavation where excavation would be impossible, prohibitively expensive, or would interfere with traffic or other utilities

GOOD PRACTICE

The standard-dimension fittings described in AWWA C208 (latest edition) should be used whenever possible. If drawings are not used in purchasing, the designation of fittings is always necessary. Design data should be used to determine if reinforcement is needed. When necessary, special welded steel-pipe fittings can be fabricated to meet unusual requirements and severe service conditions. When special steel-pipe fittings are designated, they should be accompanied with drawings to show their exact configuration.

REFERENCES

American Water Works Association. ANSI/AWWA Standard C200, *Steel Water Pipe—6 In. (150 mm) and Larger.* Latest edition. Denver, Colo.: American Water Works Association.

———. ANSI/AWWA Standard C207, *Steel Pipe Flanges for Waterworks Service—Sizes 4 In. Through 144 In. (*100 *mm Through* 3,600 *mm).* Latest edition. Denver, Colo.: American Water Works Association.

———. ANSI/AWWA Standard C208, *Dimensions for Fabricated Steel Water Pipe Fittings.* Latest edition. Denver, Colo.: American Water Works Association.

———. ANSI/AWWA Standard C512, *Air Release, Air/Vacuum, and Combination Air Valves for Water Works Service.* Latest edition. Denver, Colo.: American Water Works Association.

Rudd, F.O. *Stress Analysis of Wye Branches.* Engineering Monograph 32. Denver, Colo.: US Bureau of Reclamation.

Swanson, H.S., et al. 1955. Design of Wye Branches for Steel Pipe. *Jour. AWWA,* 47(6):581.

Timoshenko, S. 1940. *Strength of Materials.* Part II. New York: Van Nostrand Co.

This page intentionally blank.

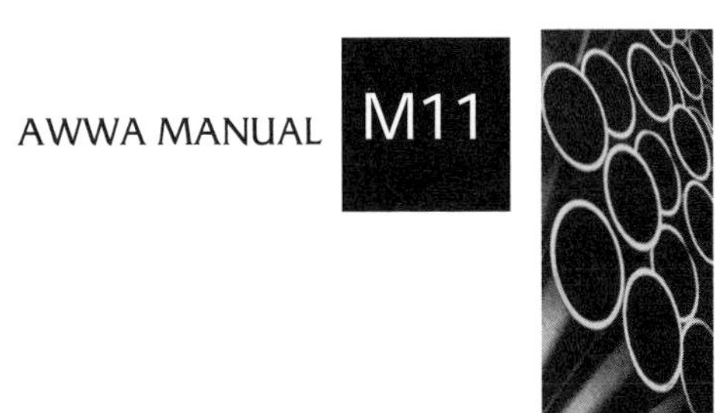

Chapter 10

Principles of Corrosion and Corrosion Control

Corrosion is the deterioration of a substance (usually a metal) or its properties because of a reaction with its environment (NACE 1975). Even though the process of corrosion is complex and the detailed explanations even more so, relatively nontechnical publications on the subject are available (*Manual on Underground Corrosion* 1952, Hertzberg 1956).

An understanding of the basic principles of corrosion leads to an understanding of the means and methods of corrosion control. Methods of corrosion control are discussed in this chapter and in chapter 11. Although many of these methods apply to all metals, both chapters specifically address corrosion and corrosion control of steel pipe.

GENERAL THEORY

All materials exposed to the elements eventually change to the state that is most stable under prevailing conditions. Most structural metals, having been converted from an ore, tend to revert to it. This reversion is an electrochemical process—that is, both a chemical reaction and the flow of a direct electric current occur. Such a combination is termed an *electrochemical cell*. Electrochemical cells fall into three general classes:

- Galvanic cells, with electrodes of dissimilar metals in a homogeneous electrolyte
- Concentration cells, with electrodes of similar material but with a nonhomogeneous electrolyte
- Electrolytic cells, which are similar to galvanic cells, but which have, in addition, a conductor plus an outside source of electrical energy

Three general types of corrosion are recognized: galvanic, electrolytic, and biochemical.

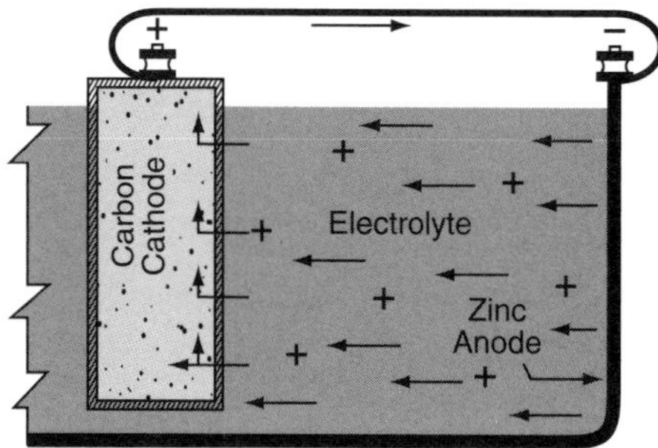

Figure 10-1 Galvanic cell—dissimilar metals

Galvanic Corrosion

Galvanic corrosion occurs when two electrodes of dissimilar materials are electrically connected and exposed in an electrolyte. An example is the common flashlight cell (Figure 10-1). When the cell is connected in a circuit, current flows from the zinc case (the anode) into the electrolyte, carrying ionized atoms of zinc with it. As soon as the zinc ions are dissolved in the electrolyte, they lose their ionic charge, passing it on by ionizing atoms of hydrogen. The ionic charge (the electric current) flows through the electrolyte to the carbon rod (the cathode). There, the hydrogen ions are reduced to atoms of hydrogen, which combine to form hydrogen gas. The current flow through the circuit, therefore, is from the zinc anode to the electrolyte, to the carbon rod cathode, and back to the zinc anode through the electrical conductor connecting the anode to the cathode. As the current flows, the zinc is destroyed but the carbon is unharmed. In other words, the anode is destroyed, but the cathode is protected.

If the hydrogen gas formed in the galvanic cell collects on the cathode, it will insulate the cathode from the electrolyte and stop the flow of current. As long as the hydrogen film is maintained, corrosion will be prevented. Removal or destruction of the hydrogen film will allow corrosion to start again at the original rate. Formation of the film is called *polarization*; its removal is called *depolarization*. Corrosion cells normally formed in highly corrosive soils or waters are such that the hydrogen formed on the cathode escapes as a gas and combines with dissolved oxygen in the electrolyte, therefore depolarizing the cathode and allowing corrosion to proceed.

In the flashlight battery, the zinc case is attacked, and the carbon is not. However, zinc or any other metal may be attacked when in circuit with one metal but not attacked when in circuit with another. A metal listed in Table 10-1 will be attacked if connected in a circuit with one listed beneath it in the table, if they are placed in a common electrolytic environment, such as water or moist soil.

The order in Table 10-1 is known as the *galvanic series*; it is true for neutral electrolytes. Changes in the composition or temperature of the electrolyte, however, may cause certain metals listed to shift positions or actually reverse positions in the table. For example, zinc is listed above iron in the table, and zinc will corrode when connected to iron in fresh water at normal temperature. However, when the temperature of the water is above about 150°F (66°C), the iron will corrode and protect the zinc. Therefore, the table cannot be used to predict the performance of all metal combinations under all conditions.

In the flashlight battery, dissimilar metals and a single electrolyte cause the electric current to flow. Similar metals in dissimilar electrolytes can also produce a current, as illustrated in Figure 10-2. In corrosion underground, differential oxygen concentration

Table 10-1 Galvanic series of metals and alloys*

Metal or alloy	
Magnesium and magnesium alloys	↑
Zinc	
Aluminum 2S	
Cadmium	
Aluminum 17ST†	
Steel or iron	
Cast iron	
Chromium-iron (active)	
Ni-Resist†	Anodic or Corroded End
18-8 Stainless steel (active)†	
18-8-3 Stainless steel (active)†	
Hastelloy C†	
Lead-tin solders	
Lead	
Tin	
Nickel (active)	
Inconel (active)†	
Hastelloy A†	Cathodic or Protected End
Hastelloy B†	
Brass	
Copper	
Bronzes	
Copper-nickel alloy	
Monel†	
Silver solder	
Nickel (passive)	
Inconel (passive)	
Chromium-iron (passive)	
18-8 Stainless steel (passive)	
18-8-3 Stainless steel (passive)	
Silver	
Graphite	
Gold	
Platinum	↓

Source: Hertzberg, L.B., Suggested Non-technical Manual on Corrosion for Water Works Operators. Jour. AWWA, *48:719 (June 1956).*

* A "passive" metal has a surface film of absorbed oxygen or hydrogen. A metal may be initially "active" and become "passive" to the other metal when the protective film is formed.

† Composition of items is as follows: Aluminum 17ST—95% Al, 4% Cu, 05% Mn, 0.5% Mg; Ni-Resist, International Nickel Co., New York, N.Y.—austenitic nickel and cast iron; 18-8 stainless steel—18% Cr, 8% Ni; 18-8-3 stainless steel—18% Cr, 8% Ni, 3% Mo; Hastelloy C, Union Carbide Carbon Co., Niagara Falls, N.Y.—59% Ni, 17% Mo, 14% Cr, 5% Fe, 5% W; Inconel International Nickel Co., New York, N.Y.—59–80% Ni, 10–20% Cr, 0–23% Fe; Hastelloy A—60% Ni, 20% Mo, 20% Fe; Hastelloy B—65% Ni, 30% Mo, 5% Fe; Monel—63–67% Ni, 29–30% Cu, 1–2% Fe, 0.4–1.1% Mn.

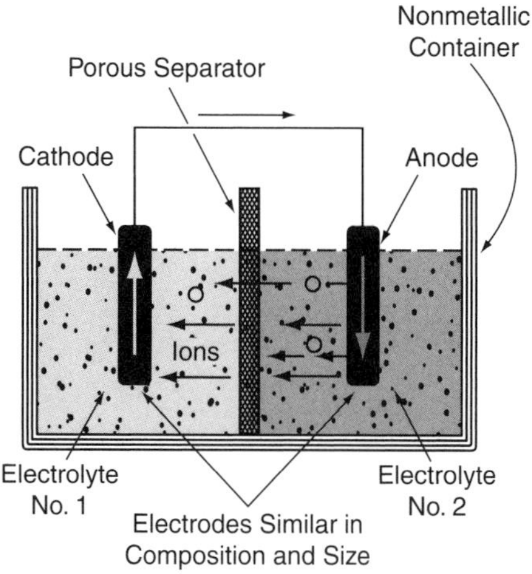

Figure 10-2 Galvanic cell—dissimilar electrolytes

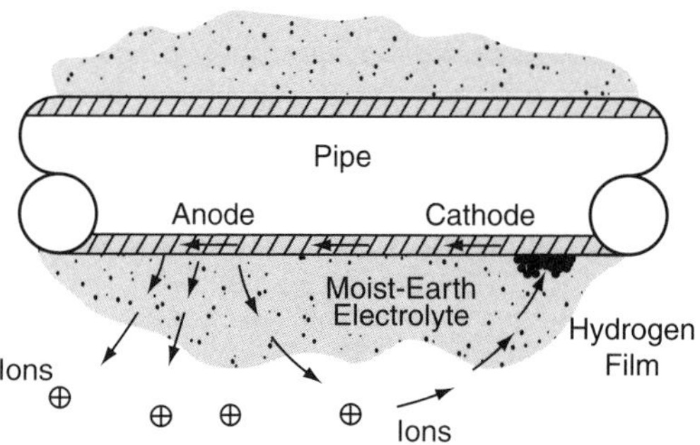

Moist earth is the electrolyte; two areas on the pipe are anode and cathode; pipe wall takes the place of wire in Figures 10-1 and 10-2. Pipe wall at anode will corrode like zinc battery case; pipe wall at cathode will not corrode but will tend to be coated with hydrogen gas, which if not removed, will tend to build resistance to current flow and thereby check corrosion of pipe wall at anode.

Figure 10-3 Galvanic cell on embedded pipe without protective coating

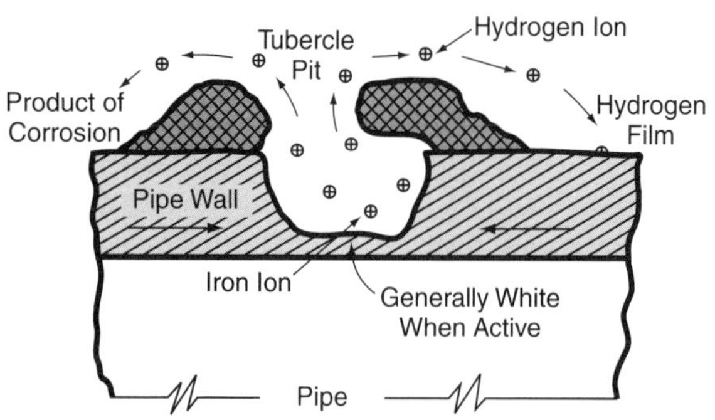

Detail of pipe wall at anode in Figure 10-3 is shown. As current leaves surface of anode, it carries with it small particles of metal (ions). These ions go into solution in soil (electrolyte) and are immediately exchanged for hydrogen ions, leaving metal behind as rusty scale or tubercle around pit area. In many soils, especially comparatively dry ones, this barnacle-like scab will "seal off" pit so that ions (electric current) cannot get through and cell becomes inactive as long as tubercle is not disturbed.

Figure 10-4 Galvanic cell—pitting action

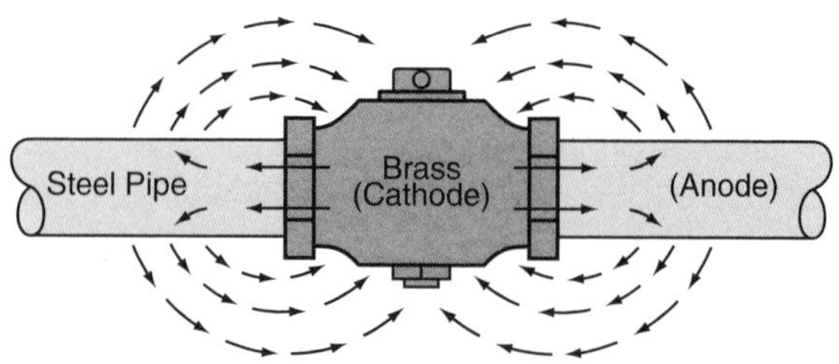

Brass valve is the cathode (protected area), steel pipe is anode (corroding area), and surrounding earth is the electrolyte. As long as cathode is small in area relative to anode, corrosion is not ordinarily severe or rapid. If these area proportions are reversed, corrosion may be much more rapid.

Figure 10-5 Corrosion caused by dissimilar metals in contact on buried pipe

in soils is one of the major reasons for dissimilarity in the electrolyte. Differential oxygen concentration (or differential aeration) may be caused by unequal compactness of backfill, unequal porosity of different soils or of one soil at different points, uneven distribution of moisture, or restriction of air and moisture movement in the soil caused by the presence of buildings, roadways, pavements, and vegetation.

The electrochemical cells described in the preceding paragraphs demonstrate the fundamental principles of the many kinds of electrochemical cells found in practice. The common forms of corrosion encountered on unprotected buried pipelines are shown in Figures 10-3 through 10-11.

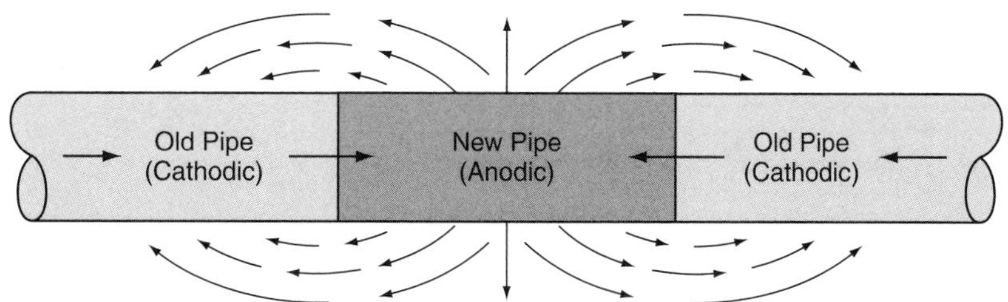

Although seldom considered, a galvanic cell is created by installing a piece of new pipe in an old line. New pipe always becomes the anode, and its rate of corrosion will depend on type of soil and relative areas of anode and cathode. Therefore, careful protective measures are essential.

Figure 10-6 Corrosion caused by dissimilar metals

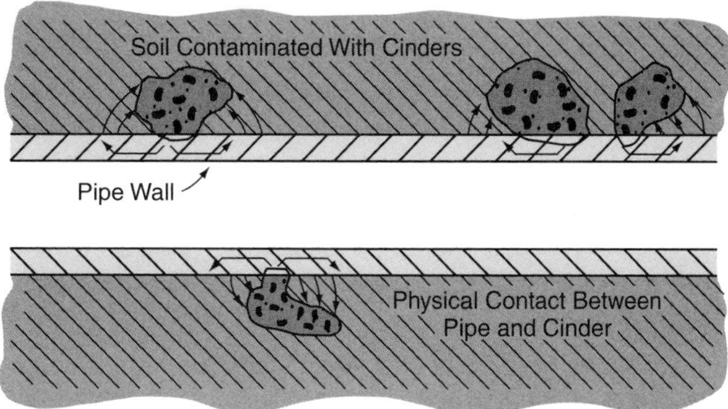

When metal pipe is laid in cinders, corrosive action is that of dissimilar metals. Cinder is one metal (cathode) and pipe the other (anode). Acid leached from cinders contaminates soil and increases its activity. No hydrogen collects on the cinder cathode, the cell remains active, and corrosion is rapid.

Figure 10-7 Corrosion caused by cinders

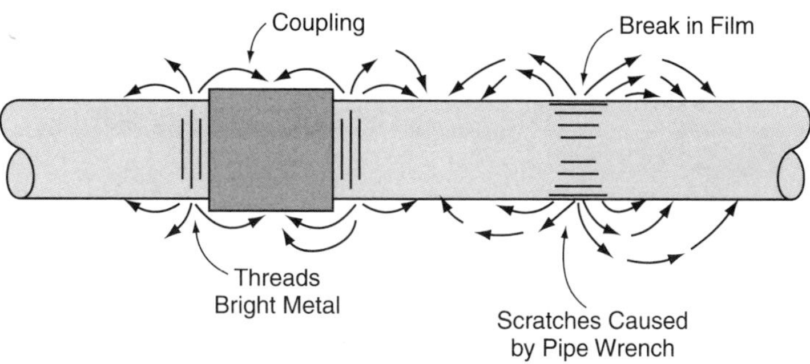

Bright scars or scratches of threads become anode areas in buried pipe, and the rest of the pipe is a cathodic area. In some soils, these bright areas can be very active and destructive because the small anode area and large cathode area produce the most unfavorable ratios possible.

Figure 10-8 Corrosion caused by dissimilarity of surface conditions

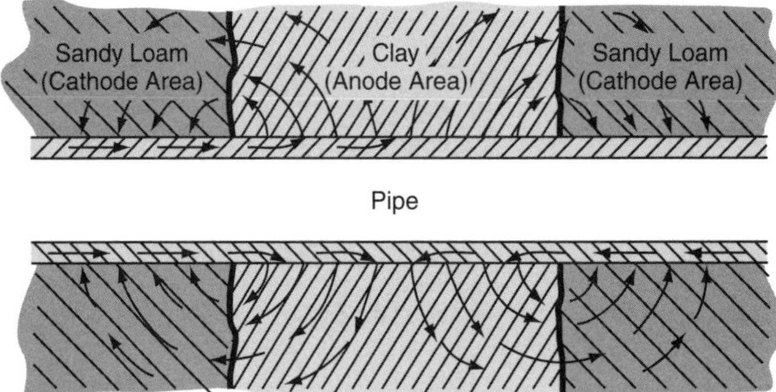

In this galvanic cell of dissimilar electrolytes (compare Figure 10-2), sections of pipe in sandy loam are cathodes (protected areas), sections in clay are anodes (corroding areas), and soil is electrolyte. If resistance to electric-current flow is high in the electrolyte, the corrosion rate will be slow. If resistance to current flow is low, corrosion rate will be high. Thus, knowledge of soil resistance to electric-current flow becomes important in corrosion control studies.

Figure 10-9 Corrosion caused by dissimilar soils

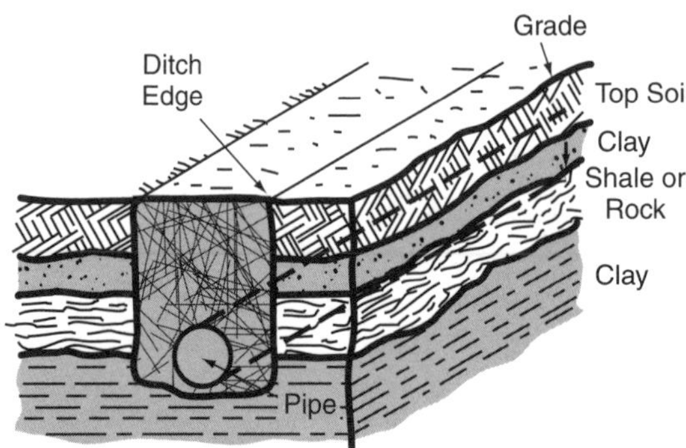

Dissimilarity of electrolytes, because of the mixture of soils, causes formation of galvanic cell. If large clods of dirt, originally from different depths in the excavation, rest directly against unprotected pipe wall, contact area tends to become anodic (corroding area), and adjacent pipe cathodic. Small, well-dispersed clods, such as result in trenching by machine, reduce cell-forming tendency. Galvanic cells having anode and cathode areas distributed around the circumference of pipe are often called short-path cells.

Figure 10-10 Corrosion caused by mixture of different soils

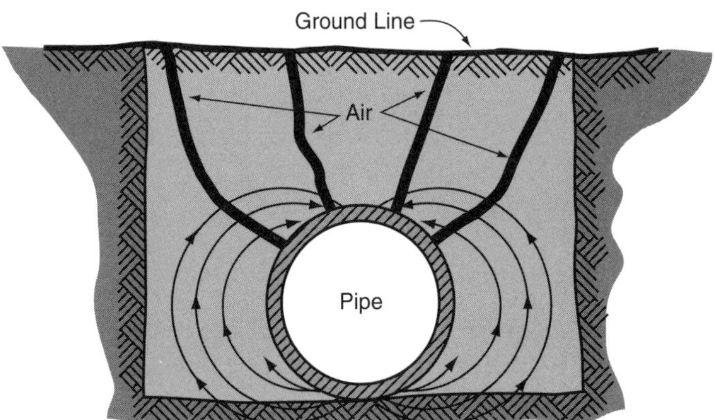

This is another galvanic cell of dissimilar-electrolyte type. Soil throughout the depth of the excavation is uniform, but a portion of the pipe rests on heavy, moist, undisturbed ground at the bottom of the excavation, while the remainder of the pipe is in contact with drier and more aerated soil backfill. Greatest dissimilarity—and the most dangerous condition—occurs along a narrow strip at the bottom of the pipe, which is the anode of the cell.

Figure 10-11 Corrosion caused by differential aeration of soil

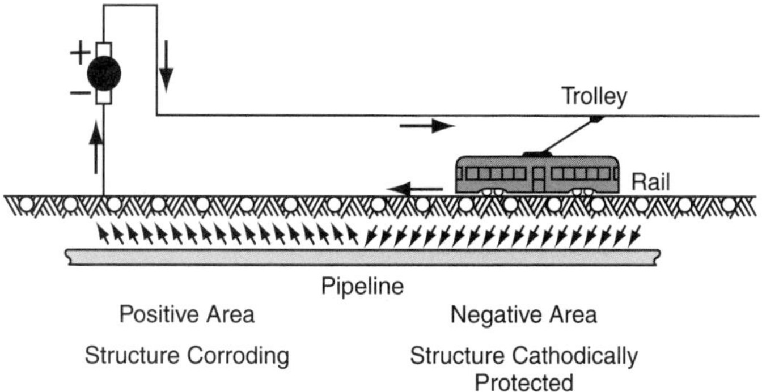

Figure 10-12 Stray-current corrosion caused by electrified railway systems

Electrolytic Corrosion

The transportation industry and other industries use direct current (DC) electricity for various purposes in their operations. It is common practice with DC circuits to use the ground as a return path for the current. In such cases, the path of the current may stray some distance from a straight line between two points in a system in order to follow the path of least resistance. Even where metallic circuits are provided for handling the direct currents, some of the return current may stray from the intended path and return to the generator either through parallel circuits in the ground or through some metallic structure. Because these currents stray from the desired path, they are commonly referred to as stray earth currents or stray currents.

The diagrammatic sketch of an electric street-railway system shown in Figure 10-12 is an example of a system that can create stray DC currents. Many modern subway systems operate on the same principle. In Figure 10-12, the direct current flows from the generator into the trolley wire, along this wire to the streetcar, and through the trolley of the car to the motors driving it. To complete the circuit, the return path of the current is intended to be from the motors to the wheels of the car, then through the rails to the generator at the substation. However, because of the many mechanical joints along these tracks, all of which offer resistance to the flow of the electricity, a portion of the current, seeking an easier path to the substation, leaves the rails, passes into the ground, and returns to the substation through the moist earth. If, in its journey through the ground, the current passes near buried metal pipe—which offers an easier path for return than does the ground around it—the current will flow along the metal walls of the pipe to some point near the substation. There it will leave the pipe to flow through the ground back to the rail, and finally return to the substation generator.

Areas of the pipe where the current is entering are not corroded. Where the current leaves the pipe, however, steel is destroyed at the rate of about 20 lb per ampere-year of current discharged. To combat electrolysis, an insulated metal conductor must be attached to the pipe where it will remove and return the current to the source, rather than allow the current to escape from the pipe wall. Figure 10-13 diagrammatically shows this method.

Biochemical Corrosion

Certain soil bacteria create chemicals that may result in corrosion. Bacterial corrosion, or anaerobic-bacterial corrosion, is not so much a distinct type of corrosion as it is another cause of electrochemical corrosion. The bacteria cause changes in the physical and chemical properties of the soil to produce active pseudogalvanic cells. The

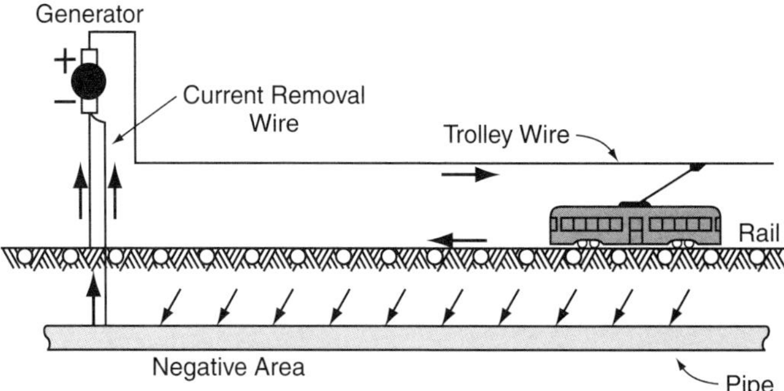

Figure 10-13 Control of stray-current corrosion

bacterial action may remove the protective hydrogen film. Differential aeration plays a major role in this activity.

The only certain way of determining the presence of anaerobic bacteria, the particular kind of microorganism responsible for this type of corrosion, is to secure a sample of the soil in the immediate vicinity of the pipe and develop a bacterial culture from that sample. Inspection under a microscope will determine definitely whether harmful bacteria are present.

Stress and Fatigue Corrosion

Stress corrosion is caused by tensile stresses that slowly build up in a corrosive atmosphere. With a static loading, tensile stresses are developed at the metal surfaces. At highly stressed points, accelerated corrosion occurs, causing increased tensile stress and failure when the metal's safe yield is exceeded.

Corrosion fatigue occurs from cyclic loading. In a corrosive atmosphere, alternate loadings cause corrosion fatigue substantially below the metal's failure in noncorrosive conditions.

Crevice Corrosion

Crevice corrosion in a steel pipeline is caused by a concentration cell formed where the dissolved oxygen of the water varies from one segment of the pipe metal to another. In a crevice area, the dissolved oxygen is hindered from diffusion, creating an anodic condition that causes metal to go into solution.

Severity of Corrosion

Severity of corrosion in any given case will depend on many different factors, some of which may be more important than others. The factors most likely to affect the rate of corrosion are

- relative positions of metals in the galvanic series
- size of anode area with respect to cathode area
- location of anode area with respect to cathode
- resistance of metallic circuit
- type and composition of electrolyte
- conductivity or resistivity of electrolyte
- uniformity of electrolyte
- depolarizing conditions

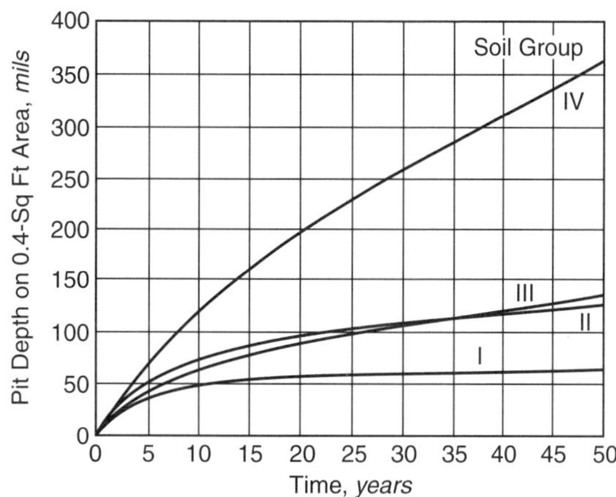

Source: Barnard, R.E., A Method of Determining Wall Thickness of Steel Pipe for Underground Service. Jour. AWWA. *29:6:791 (June 1937).*

NOTE: Soil groups are defined in Table 10-2.

Figure 10-14 Corrosion rate in various soils

Soil-Corrosion Investigations

The first organized soil-corrosion investigation was begun by the National Bureau of Standards (NBS) (now the National Institute for Science and Technology) in 1911. The purpose at that time was to study the effect of stray currents from street-railway lines on buried metallic structures. In its initial investigation, the bureau found that in many instances where rather severe corrosion was anticipated, little damage was observed, whereas in others, more corrosion was found than seemed to be indicated by the electrical data associated with the corroded structure. These observations led to a second investigation, undertaken in 1921. Originally about 14,000 specimens were buried at 47 test sites, but the number was subsequently increased to 36,500 specimens at 128 test sites. The American Petroleum Institute and the American Gas Association collaborated in analyzing the results of the latter tests.

Burial sites were selected in typical soils representing a sampling of areas in which pipe was or might be buried. The purpose of the investigation was to determine whether corrosion would occur in pipelines in the absence of stray currents under conditions representative of those encountered by working pipelines.

The NBS soil corrosion tests are probably the most extensive, well coordinated, and best analyzed of any test made for the same purpose. A final report on the studies made between 1910 and 1955, including over 400 references, has been published (National Bureau of Standards 1957). An important finding was that in most soils, the corrosion rate decreased with time. This is largely because of the fact that corrosion products, unless removed, tend to protect the metal.

Figure 10-14, taken from the NBS reports, clearly shows the decrease in corrosion rate with time in all but the worst soil group. Only a very small percentage of pipe is ever buried in soil belonging to that group. Modern methods of corrosion prevention requiring coatings and cathodic protection generally make it unnecessary to allow extra wall thickness as a safeguard against corrosion. Tables 10-2 and 10-3 give summary data on the corrosivity of soils and the relationship of soil corrosion to soil resistivity.

Table 10-2 Soils grouped in order of corrosive action on steel

Group I—Lightly Corrosive
Aeration and drainage good. Characterized by uniform color and no mottling anywhere in soil profile and by very low water table. Includes
1. Sands or sandy loams
2. Light, textured silt loams
3. Porous loams or clay loams thoroughly oxidized to great depths

Group II—Moderately Corrosive
Aeration and drainage fair. Characterized by slight mottling (yellowish brown and yellowish gray) in lower part of profile (depth 18–24 in.) and by low water table. Soils would be considered well drained in an agricultural sense, as no artificial drainage is necessary for crop raising. Includes
1. Sandy loams
2. Silt loams
3. Clay loams

Group III—Badly Corrosive
Aeration and drainage poor. Characterized by heavy texture and moderate mottling close to surface (depth 6–8 in.) and with water table 2–3 ft below surface. Soils usually occupy flat areas and would require artificial drainage for crop raising. Includes
1. Clay loams
2. Clays

Group IV—Unusually Corrosive
Aeration and drainage very poor. Characterized by bluish-gray mottling at depths of 6–8 in. with water table at surface or by extreme impermeability because of colloidal material contained. Includes
1. Muck
2. Peat
3. Tidal marsh
4. Clays and organic soils
5. Adobe clay

Table 10-3 Relationship of soil corrosion to soil resistivity

Soil Class	Description	Resistivity, *ohm-cm*
1	excellent	10,000–6,000
2	good	6,000–4,500
3	fair	4,500–2,000
4	bad	2,000–0

INTERNAL CORROSION OF STEEL PIPE

Corrosion of the internal surfaces of a pipe is principally caused by galvanic cells (Eliassen and Lamb 1953). The extent of corrosion of the interior of an unlined pipe depends on the corrosivity of the water carried. Langelier (1936) developed a method for determining the corrosive effect of different kinds of water on bare pipe interiors, and Weir (1940) extensively investigated and reported the effect of water contact on various kinds of pipe linings. Although some unlined pipes have been pitted through by some waters, the principal result of interior corrosion is a reduction in flow capacity. This reduction is caused by a formation of tubercles of ferric hydroxide, a condition known as *tuberculation* (Linsey and Franzini 1979). It is primarily to maintain flow capacity that pipe linings have been developed. Where internal corrosion is allowed to persist, quality of water deteriorates, pumping and transmission capacity decreases, efficiency diminishes, and costly replacement becomes inevitable. Serious accidents and loss of revenues from system shutdowns are also possible. The occurrence of these problems can be reduced by using quality protective linings.

ATMOSPHERIC CORROSION

Atmospheric corrosion of exposed pipelines is usually insignificant, except in industrial and seacoast areas. Where such corrosion is significant, the maintenance problem incurred is similar to that for bridges or other exposed steel structures.

METHODS OF CORROSION CONTROL

The electrochemical nature of corrosion suggests three basic methods of controlling it on underground and underwater pipelines. First, pipe and appurtenances can be isolated and electrically insulated from the surrounding soil and water using a protective coating. Second, electric currents can be imposed to counteract the currents associated with corrosion. Third, an inhibitive environment can be created to prevent or reduce corrosion.

To implement the first method, satisfactory and effective protective coatings have been developed. Cathodic protection, implementing the second method, is being used more widely in corrosion control. Inhibitive coatings implement the third method by providing an environment in which oxidation or corrosion of steel is inhibited. By judicious use of all of these methods, any required degree of corrosion control can be economically achieved (Peabody 1967). The combination of coating with supplemental cathodic protection is the most effective method of corrosion control for buried and submerged steel water pipelines.

Coatings and linings are described in chapter 11.

CATHODIC PROTECTION

Cathodic protection systems reverse the electrochemical corrosive force by creating an external circuit between the pipeline to be protected and an auxiliary anode (sacrificial metal) immersed in water or buried in the ground at a predetermined distance from the pipe. Direct current applied to the circuit is discharged from the anode surface and travels through the surrounding electrolyte to the pipe (cathode) surface.

Two methods are available for generating a current of sufficient magnitude to guarantee protection. In the first method, sacrificial-anode material such as magnesium or zinc is used to create a galvanic cell. The electrical potential generated by the cell causes current to flow from the anode to the pipe, returning to the anode through a simple connecting wire (Figure 10-15). This system is generally used where small amounts of current are applied at a number of locations, most often on coated pipelines in lightly or moderately corrosive soils.

The second method of current generation is to energize the circuit with an external DC power supply, such as a rectifier. This technique, commonly referred to as the impressed current method, uses relatively inert anodes (usually graphite or silicon cast iron) connected to the positive terminal of a DC power supply, with the pipe connected to the negative terminal (Figure 10-16). This system is generally used where large amounts of current are required at relatively few locations, and in many cases it is more economical than sacrificial anodes.

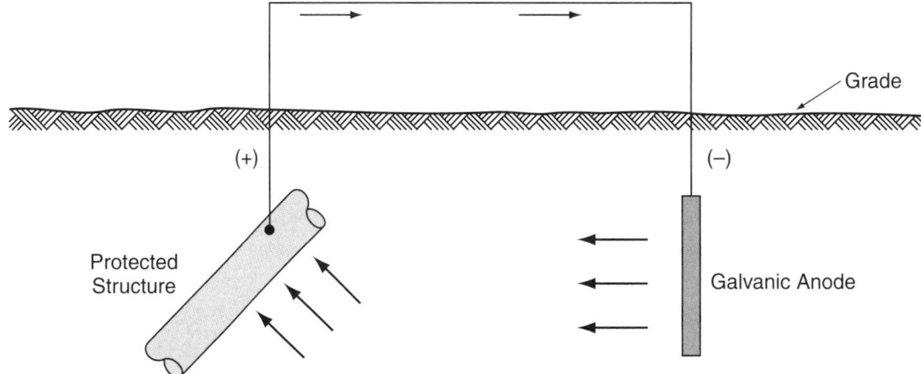

Figure 10-15 Cathodic protection—galvanic anode type

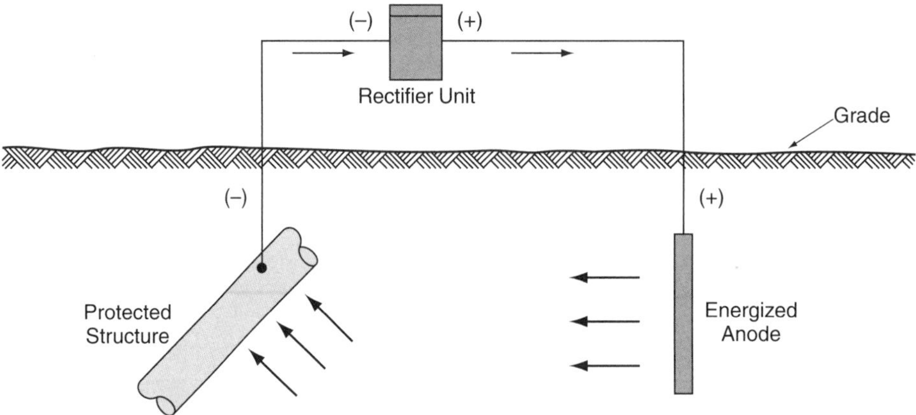

Figure 10-16 Cathodic protection—rectifier type

Bonding of Joints

When a pipeline is cathodically protected or when a pipeline is to be installed with the possibility of future cathodic protection, joint bonding is required to make the line electrically continuous (Figures 10-17 and 10-18). It is usually desirable to bond all joints at the time of installation, because the cost later will be many times greater. In addition to bonding, the pipeline should have test leads connected to it at appropriate intervals to permit monitoring of the pipeline, whether under cathodic protection or not. Field-welded lines require no additional bonding.

Current Required

For any cathodic protection system to be effective, sufficient current must flow from the soil to the pipe to ensure that no part of the structure acts as an anode. This is normally achieved when the potential between the pipe and a copper–copper sulphate electrode in contact with the soil and adjacent to it is at least –0.85 volts.

There are, however, other criteria for protection that may be more appropriate to specific sites, and in some cases, a combination of tests may be required to ensure that protection is achieved.

Design of Cathodic Protection Systems

In many situations, cathodic protection for steel pipelines will not be installed until proven necessary. However, all joints in steel pipe should be electrically bonded and electrical test stations provided along the pipeline as necessary.

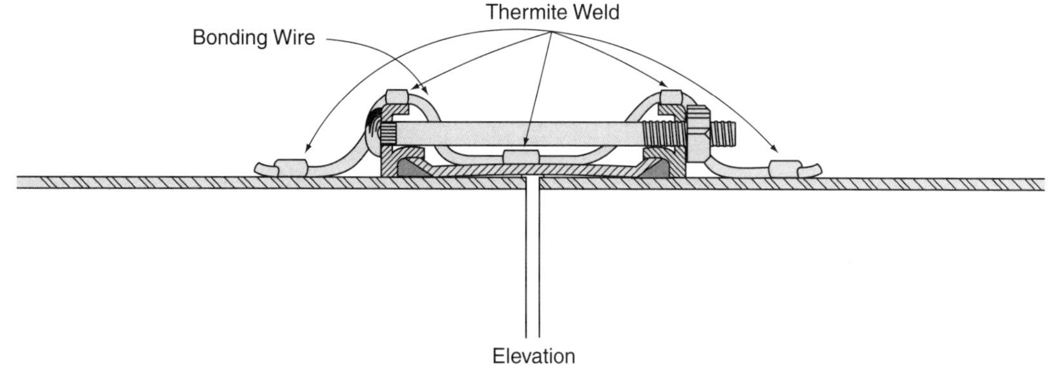

Figure 10-17 Bonding jumpers installed on sleeve-type coupling

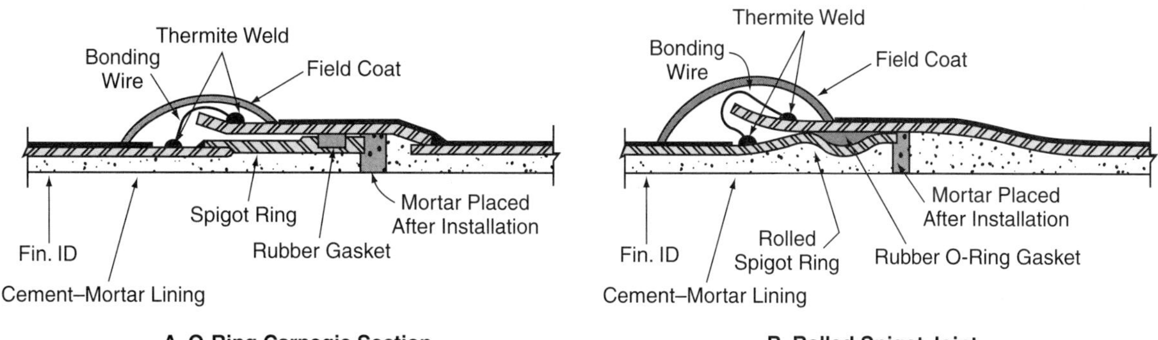

Figure 10-18 Bonding wire for bell-and-spigot rubber-gasketed joint

Corrosion Survey

A corrosion survey, including chemical–physical analyses of the soil, must be performed along the pipeline right-of-way. Some of the measurements taken include soil resistivity, soil pH, and tests for stray currents.

REFERENCES

Barnard, R.E. 1937. A Method of Determining Wall Thickness of Steel Pipe for Underground Service. *Jour. AWWA*, 29(6):791.

Eliassen, R., and J.C. Lamb III. 1953. Mechanism of Internal Corrosion of Water Pipe. *Jour. AWWA*, 45(12):1281.

Hertzberg, L.B. 1956. Suggested Nontechnical Manual on Corrosion for Water Works Operators. *Jour. AWWA*, 48:719.

Langelier, W.F. 1936. The Analytical Control of Anticorrosion Water Treatment. *Jour. AWWA*, 28:1500.

Linsey, R.K., and J.B. Franzini. 1979. *Water Resources Engineering*. New York: McGraw-Hill Book Co.

Manual on Underground Corrosion. 1952. New York: Columbia Gas System Service Corp.

NACE. 1975. *NACE Basic Corrosion Course*. Houston, Texas: NACE.

National Bureau of Standards. 1957. *Underground Corrosion*. NBS Circ. No. 579.

Peabody, A.W. 1967. *Control of Pipeline Corrosion*. Katy, Texas: National Association of Corrosion Engineers.

Weir, P. 1940. The Effect of Internal Pipe Lining on Water Quality. *Jour. AWWA*, 32:1547.

The following references are not cited in the text.

API Coating Tests. 1934. Prod. Bull. 214. New York: American Petroleum Institute.

Corrosion Control in Water Utilities. 1980. California–Nevada American Water Works Association Section: Corrosion Control Committee.

Davis, C.V., ed. 1969. *Handbook of Applied Hydraulics.* New York: McGraw-Hill Book Co.

Denison, I.A. 1936. *Electrolytic Measurement of the Corrosiveness of Soils.* NBS Res. Paper. RP 918.

Logan, K.H. 1937. *ASTM Symposium on Corrosion Testing Procedures.* Chicago Meetings.

McComb, G.B. 1965. *Pipeline Protection Using Coal Tar Enamels.* St. Louis, Mo.

A Preliminary Study of the Rate of Pitting of Iron Pipe in Soils. 1933. Prod. Bull. 212. New York: American Petroleum Institute.

Scott, G.N. 1933. *Adjustment of Soil Corrosion Pit Depth Measurements for Size and Sample.* Prod. Bull. 212. New York: American Petroleum Institute.

Steel Plate Engineering Data. Vol. 3. 1980. Washington, D.C.: American Iron & Steel Institute and Steel Plate Fabricators Assoc., Inc.

Chapter 11

Protective Coatings and Linings

Coatings for corrosion control are extremely effective when properly used. They are considered to be the primary line of defense against corrosion of steel pipeline systems. Coating costs are only a fraction of pipeline costs, yet coating is the major means of ensuring long-term operation by preventing pipeline deterioration and corrosion leaks.

REQUIREMENTS FOR GOOD PIPELINE COATINGS AND LININGS

The requirements for a coating vary with the type of construction, the aggressiveness of the environment, and the system operating conditions. The effectiveness of a good protective pipeline coating depends on its permanence and its physical resistance to the hazards of transportation, installation, temperature change, soil stress, and pressure; resistance to water penetration or absorption; effective electrical insulative properties; and chemical inertness to soil, air, water, organic acids, alkalies, and bacterial action. Coating effectiveness also depends on such general characteristics as the ease of application, high adhesion, compatibility of use with cathodic protection, and reasonable cost (NACE 1983).

The requirements of a lining also vary with the system and the environment. In addition to the factors considered for coatings, linings must be evaluated on their smoothness (low flow resistance), and they must meet toxicological requirements for potable water.

SELECTION OF THE PROPER COATING AND LINING

Selection and recommendation of the lining and coating materials for use on underground, aboveground, and underwater steel pipelines is one of the most important responsibilities of the engineer. Selection of materials involves assessing the magnitude of the corrosion, installation, and service hazards. Testing procedures have been developed to aid the engineer in evaluating and selecting the coating system that best meets a system's needs (American Society for Testing and Materials 1977b, g, h, 1979a). Requirements for external coating and internal lining are different, therefore each should be considered separately according to the anticipated corrosion severity.

Coating Selection

The corrosion potential for the exterior of steel pipe is difficult to evaluate because of the variety of environments encountered. Resistivity of the soil (see Table 10-3, chapter 10) is the most important parameter for evaluating soil corrosivity. Soil chemical and physical analyses, pH, moisture content, and existence of stray electrical currents are also important factors in the decision-making process. If the pipe is subjected to atmospheric conditions, the climate and environmental conditions, as well as governmental or their agencies' regulations, must be evaluated to determine the proper coating system.

After the level of soil corrosivity is assessed, the other conditions that affect the long-term performance of protective coatings should be considered (NACE 1983). Among these are

- distorting stresses exerted on the coating during compaction and settling of the backfill
- mechanical stresses created by certain soils having very high expansion and shrinkage during wet and dry cycles
- penetration by growing roots
- action of bacteria and fungus in soil surrounding the pipeline
- penetration by rocks, clods, or debris in the backfill
- attack by soil chemicals or industrial wastes, chemicals, and solvents that may be present along the pipeline route

Coating performance depends on putting the pipeline into service with the least amount of coating damage. The coating system selected must not only meet the corrosion-control needs but must also allow for economical transportation, handling, storage, and pipeline construction with minimal coating damage or repair. To ensure precise control of the coating application and quality, many types of coatings are applied in a plant or shop. The coating manufacturer can provide a guide to the proper protection during transportation, handling, and storage of pipe that has been coated in this manner. General guidelines are given in a later section of this chapter. There are several recognized testing procedures for evaluating coating system characteristics related to transportation, storage, and construction (American Society for Testing and Materials 1977a, c, d, e, f, 1979b). Among the characteristics to be considered are

- resistance of the coating to cold flow or penetration during mechanical loading
- resistance of the coating to ultraviolet exposure and temperature cycling during outdoor storage
- resistance of the coating to abrasion and impact

Lining Selection

The function of an internal lining is to prevent internal corrosion and to produce and maintain a smooth surface to enhance flow capacity. Cement–mortar linings and coatings for steel waterlines are durable and have provided many years of excellent service. Pipe surfaces covered with cement–mortar are protected by the alkaline cement environment, which passivates the steel and prevents iron corrosion in most natural environments. The passivation occurs quickly in newly coated surfaces and is not destroyed by moisture and oxygen absorbed through the mortar coating. Cement–mortar linings provide low hydraulic frictional resistance, and any leached products from mortar lining carrying soft water can be certified as nontoxic and anticorrosive.

Coal-tar enamel, coal-tar epoxy, two-component liquid epoxy, and fusion-bonded epoxy exhibit excellent corrosion-resistance properties and provide the required smoothness to maintain flow capacity. They protect steel water lines by electrically insulating the coated pipe surfaces from the environment. When reinforced, the coatings provide additional resistance to physical damage.

Regardless of the lining material selected, the effects of cavitation and silts on the lining should be considered.

RECOMMENDED COATINGS AND LININGS

Current AWWA standards list coatings and linings for steel water pipe that are believed to be the most reliable, as proved in practice. However, the AWWA Steel Pipe Committee is alert to the possibilities of new developments, and additions to and modifications of existing standards will be made as deemed advisable. The current list of AWWA coating and lining standards for pipe protection is as follows:

AWWA C203, Standard for Coal-Tar Protective Coatings and Linings for Steel Water Pipelines—Enamel and Tape—Hot-Applied. AWWA C203 (latest edition) describes the material and application requirements for shop-applied, coal-tar protective coatings and linings for steel water pipelines intended for use under normal conditions. The standard describes coal-tar enamel applied to the interior and exterior of pipe, special sections, connections, and fittings. It also covers hot-applied coal-tar tape applied to the exterior of special sections, connections, and fittings.

Coal-tar enamel is applied over a coal-tar or synthetic primer. External coal-tar enamel coatings use bonded nonasbestos-felt and fibrous-glass mat to reinforce and shield the coal-tar enamel. The applied external coating is usually finished with either a coat of whitewash or a single wrap of kraft paper.

Internally, the coal-tar enamel is used without reinforcement or shielding. The hot enamel is spun into the pipe and provides a smooth internal lining having low hydraulic frictional resistance.

The standard provides a rigid yet reasonable manufacturer's guide for the production of the coating; requires tests of material and its behavior to ensure the purchaser that the product has the desired qualities; and provides directions for the effective application of the coating.

AWWA C205, Standard for Cement–Mortar Protective Lining and Coating for Steel Water Pipe—4 In. (100 mm) and Larger—Shop Applied. AWWA C205 (latest edition) describes the material and application requirements to provide protective linings and coatings for steel water pipe by shop application of cement mortar.

Cement mortar is composed of Portland cement, sand, and water, well mixed and of the proper consistency to obtain a dense, homogeneous lining or coating. Internally, the cement mortar is centrifugally compacted to remove excess water and produce a smooth, uniform surface. Externally, the coating is a reinforced cement mortar, pneumatically or mechanically applied to the pipe surface. Reinforcement consists of spiral wire, wire fabric, or ribbon mesh. The standard provides a complete guide for the application and curing of the mortar lining and mortar coating. Cement mortar can also be used as a rock shield for flexible coated pipe.

AWWA C209, Standard for Cold-Applied Tape Coatings for the Exterior of Special Sections, Connections, and Fittings for Steel Water Pipelines. AWWA C209 (latest edition) describes the use of a cold primer and cold-applied tape on the exterior of special sections, connections, and fittings for steel water pipelines installed underground in any soil under normal or average conditions. Tapes with both polyvinyl chloride and polyethylene backing are listed. The thicknesses of the tapes vary; however, all tapes may be sufficiently overlapped to meet changing performance requirements. Cold-applied tapes provide ease of application without the use of special equipment and can be applied over a broad application temperature range. If severe construction or soil conditions exist where mechanical damage may occur, a suitable overwrap of an extra thickness of tape or other wrapping may be required.

AWWA C210, Standard for Liquid Epoxy Coating Systems for the Interior and Exterior of Steel Water Pipelines. AWWA C210 (latest edition) describes a liquid epoxy coating system, suitable for potable water service, which will provide corrosion protection to the interior and exterior of steel water pipe, fittings, and special sections installed underground or underwater. The coating system consists of one coat of a two-part chemically cured, inhibitive epoxy primer, and one or more coats of a two-part chemically cured epoxy finish coat. The finish coat may be a coal-tar epoxy coating, or it may be an epoxy coating containing no coal tar. The coating system may alternately consist of one or more coats of the same epoxy coating without the use of a separate primer, provided the coating system meets the performance requirements of AWWA C210.

These coatings are suitable when used for corrosion prevention in water service systems. The products are applied by spray application, preferably airless.

AWWA C213, Standard for Fusion-Bonded Epoxy Coating for the Interior and Exterior of Steel Water Pipelines. AWWA C213 (latest edition) describes the material and application requirements for fusion-bonded epoxy protective coatings for the interior and exterior of steel water pipe, special sections, welded joints, connections, and fittings of steel water pipelines installed underground or underwater under normal construction conditions. The epoxy coatings provide corrosion prevention in potable water systems.

Fusion-bonded epoxy coatings are heat activated, chemically cured coating systems. The epoxy coatings are supplied in powder form. Except for welded field joints, they are plant- or shop-applied to preheated pipe, special sections, connections, and fittings using fluid bed, air, or electrostatic spray.

AWWA C214, Standard for Tape Coating Systems for the Exterior of Steel Water Pipelines. AWWA C214 (latest edition) describes the materials, the systems, and the application requirements for prefabricated cold-applied tapes for the exterior of all diameters of steel water pipe mechanically installed. For normal construction conditions, prefabricated cold-applied tapes are applied as a three-layer system consisting of (1) primer, (2) corrosion preventive tape (inner layer), and (3) mechanical protective tape (outer layer). The primer is supplied in liquid form consisting of solid ingredients carried in a solvent. The corrosion preventive tape and the mechanical protective tape are supplied in suitable thicknesses and in roll form. The standard describes the application at coating plants.

AWWA C215, Standard for Extruded Polyolefin Coatings for the Exterior of Steel Water Pipelines. AWWA C215 (latest edition) describes the materials, systems, and application requirements for shop-applied extruded polyolefin coatings for the exterior of steel water pipe up to 120-in. (3,048-mm) diameter. The standard describes two types of extrusion, crosshead and side, and three types of applications, as follows: Type A—crosshead-die extrusion, consisting of an adhesive and an extruded polyolefin sheath for pipe diameters from ½ through 36 in. (13–914 mm);

Type B—side extrusion, consisting of an extruded adhesive and an extruded polyolefin sheath for pipe diameters from 2 through 120 in. (51 through 3,048 mm); and Type C—tape and side extrusion, consisting of a primer/inner-layer tape and an extruded polyolefin sheath for spiral and longitudinal welded pipe with pipe diameters from 4 through 120 in. (100 through 3,050 mm).

AWWA C216, Standard for Heat-Shrinkable Cross-Linked Polyolefin Coatings for the Exterior of Special Sections, Connections, and Fittings for Steel Water Pipelines. AWWA C216 (latest edition) sets minimum requirements for protective exterior coatings consisting of heat-shrinkable cross-linked polyolefin coatings and their application to special sections, connections, and fittings to be used on underground and underwater steel water pipelines. These coatings may be field or shop applied in accordance with the provisions of the standard.

AWWA C217, Standard for Cold-Applied Petrolatum Tape and Petroleum Wax Tape Coatings for the Exterior of Special Sections, Connections, and Fittings for Buried Steel Water Pipelines. AWWA C217 (latest edition) describes exterior tape coatings that consist of cold-applied petrolatum or petroleum wax primer and saturant petrolatum or petroleum wax tape coatings and their applications to special sections, connections, and fittings to be used with buried or submerged steel water pipelines. Such primers and tapes are not intended for use with exposed steel joints or sections of steel pipe where coatings of cement mortar or concrete have been applied directly onto bare steel pipe.

AWWA C218, Standard for Coating the Exterior of Aboveground Steel Water Pipelines and Fittings. AWWA C218 (latest edition) describes several alternative coating systems for the protection of exterior surfaces of steel pipelines and associated fittings used by the water supply industry in aboveground locations. The coating systems described are not necessarily equivalent in terms of cost or performance but are presented so that the purchaser can select the coating system that best meets the site-specific project requirements. Coating systems included are 1-91, a three-coat alkyd system; 2-91, a four-coat alkyd system, 3-91, a three-coat alkyd/silicone-alkyd system; 4-91, a three-coat epoxy/urethane system; 5-91, a three-coat inorganic zinc/epoxy/urethane system; 6-91, a two- or three-coat epoxy/coal-tar epoxy system; 7-91, a two- or three-coat water reducible epoxy-polyamide system; 8-91, a three-coat water-reducible acrylic or alkyd-modified acrylic emulsion system; and 9-95, a two- or three-coat epoxy/high-build alaphatic polyurethane over existing coating substrates.

AWWA C222, Standard for Polyurethane Coatings for the Interior and Exterior of Steel Water Pipe and Fittings. AWWA C222 (latest edition) describes the materials and application processes for shop- and field-applied polyurethane interior linings and exterior coatings for steel water pipe, special sections, welded joints, connections, and fittings installed underground or underwater.

AWWA C224, Standard for Two-Layer Nylon-11-Based Polyamid Coating System for the Interior and Exterior of Steel Water Pipe, Connections, Fittings, and Special Sections. AWWA C224 (latest edition) describes two-layer polyamide (Nylon-11-based) coating systems used in potable water-handling equipment installed aboveground, belowground, or underwater. Polyamid coatings systems are thermoplastic and are ordinarily applied in a shop or manufacturing facility.

AWWA C225, Standard for Fused Polyolefin Coating Systems for the Exterior of Steel Water Pipelines. AWWA C225 (latest edition) describes the materials and application of fused polyolefin coating systems for buried service. This system is applied in pipe-coating plants, both portable and fixed. Normally, these prefabricated, polyolefin coatings are applied as a three-layer system consisting of a liquid adhesive, a corrosion-protection inner layer, and a mechanical-protection outer layer.

AWWA C602, Standard for Cement–Mortar Lining of Water Pipelines—4 In. (100 mm) and Larger In Place. AWWA C602 (latest edition) describes the materials and application processes for the cement–mortar lining of pipelines in place, describing both newly installed pipes and older pipelines. Detailed procedures are included for surface preparation and application, surface finishing, and curing of the cement mortar.

EPOXY-BASED POLYMER CONCRETE COATINGS

Coatings of epoxy-based polymer concrete may be applied as a corrosion and mechanical protection for bare steel or over dielectric coatings. Some applications may include "slick-bore" or directional-drill operations, and bottom-tow applications.

The original epoxy-based polymer concrete coatings are designed to provide mechanical- and abrasion-resistant protection to FBE-coated pipe. Epoxy polymer concrete can be applied directly over the FBE as a secondary coating and provides a smooth surface that allows the pipeline to be pulled under a crossing in a "slick bore" or directional-drill operation with much less drag resistance, while providing optimum protection to the FBE corrosion coating. Other applications for epoxy polymer concrete over FBE-coated pipe include severe handling conditions, bottom tow applications on submarine pipelines, and as a rock shield.

COATING APPLICATION

This manual does not provide details on methods of coating and paint application, but the importance of obtaining proper application cannot be overemphasized. Effective results cannot be secured with any coating material unless adequate care is taken in preparing the surfaces for coating, in applying the coating, and in handling the pipe after coating. AWWA standards provide the requirements for obtaining good coating work. The coating manufacturer, the applicator, and the engineer should cooperate to ensure that the work is performed properly. Many excellent sources of information have been published describing the protection of steel pipe, the pitfalls of coating work, and the means of avoiding these problems (AWWA C217, C218 (latest editions).

Coating of Special Sections, Connections, and Fittings

The coating and lining of special sections, connections, and fittings are described in AWWA Standards C203, C205, C209, C210, C213, C214, C215, C216, C217, C218, C222, C224, C225, and C602 (latest editions). The materials used are the same as those specified for use with steel water pipe. The methods of application may differ from those prescribed for pipe because of the variety of physical configurations encountered.

Pipe joints are normally coated in the field with materials similar to those used on the main body of the pipe. These are described in the appropriate AWWA coating standards.

GOOD PRACTICE

The AWWA standards for protective coatings have been carefully prepared by experienced individuals and are based on the best current practice. They should be referred to in the job specification. Modification should be made only by experienced coating specialists.

For AWWA Standards to be complete for bidding purposes, the purchaser's job specifications must provide the supplementary details required in each standard.

REFERENCES

American Society for Testing and Materials. 1977a. ASTM Standard G6-77, *Test for Abrasion Resistance of Pipeline Coatings.* Philadelphia, Pa.: American Society for Testing and Materials.

———. 1977b. ASTM Standard G9-77, *Test for Water Penetration into Pipeline Coatings.* Philadelphia, Pa.: American Society for Testing and Materials.

———. 1977c. ASTM Standard G10-77, *Test for Bendability of Pipeline Coatings.* Philadelphia, Pa.: American Society for Testing and Materials.

———. 1977d. ASTM Standard G13-77, *Test for Impact Resistance of Pipeline Coatings (Limestone Drop Test).* Philadelphia, Pa.: American Society for Testing and Materials.

———. 1977e. ASTM Standard G14-77, *Test for Impact Resistance of Pipeline Coatings (Falling Weight Test).* Philadelphia, Pa.: American Society for Testing and Materials.

———. 1977f. ASTM Standard G17-77, *Test for Penetration Resistance of Pipeline Coatings (Blunt Rod).* Philadelphia, Pa.: American Society for Testing and Materials.

———. 1977g. ASTM Standard G19-77, *Test for Disbonding Characteristics of Pipeline Coatings by Direct Soil Burial.* Philadelphia, Pa.: American Society for Testing and Materials.

———. 1977h. ASTM Standard G20-77, *Test for Chemical Resistance of Pipeline Coatings.* Philadelphia, Pa.: American Society for Testing and Materials.

———. 1979a. ASTM Standard G8-79, *Test for Cathodic Disbonding of Pipeline Coatings.* Philadelphia, Pa.: American Society for Testing and Materials.

———. 1979b. ASTM Standard G11-79, *Test for Effects of Outdoor Weathering on Pipeline Coatings.* Philadelphia, Pa.: American Society for Testing and Materials.

American Water Works Association. ANSI/AWWA Standard C203, *Coal-Tar Protective Coatings and Linings for Steel Water Pipelines—Enamel and Tape—Hot Applied.* Latest edition. Denver, Colo.: American Water Works Association.

———. ANSI/AWWA Standard C205, *Cement–Mortar Protective Lining and Coating for Steel Water Pipe—4 In. (100 mm) and Larger—Shop Applied.* Latest edition. Denver, Colo.: American Water Works Association.

———. ANSI/AWWA Standard C209, *Cold-Applied Tape Coatings for the Exterior of Special Sections, Connections, and Fittings for Steel Water Pipelines.* Latest edition. Denver, Colo.: American Water Works Association.

———. ANSI/AWWA Standard C210, *Liquid-Epoxy Coating Systems for the Interior and Exterior of Steel Water Pipelines.* Latest edition. Denver, Colo.: American Water Works Association.

———. ANSI/AWWA Standard C213, *Fusion-Bonded Epoxy Coating for the Interior and Exterior of Steel Water Pipelines.* Latest edition. Denver, Colo.: American Water Works Association.

———. ANSI/AWWA Standard C214, *Tape Coating Systems for the Exterior of Steel Water Pipelines.* Latest edition. Denver, Colo.: American Water Works Association.

———. ANSI/AWWA Standard C215, *Extruded Polyolefin Coatings for the Exterior of Steel Water Pipelines.* Latest edition. Denver, Colo.: American Water Works Association.

———. ANSI/AWWA Standard C216, *Heat-Shrinkable Cross-Linked Polyolefin Coatings for the Exterior of Special Sections, Connections, and Fittings for Steel Water Pipelines.* Latest edition. Denver, Colo.: American Water Works Association.

———. ANSI/AWWA Standard C217, *Cold-Applied Petrolatum Tape and Petroleum Wax Tape Coatings for the Exterior of Special Sections, Connections, and Fittings for Buried or Submerged Steel Water Pipelines.* Latest edition. Denver, Colo.: American Water Works Association.

———. ANSI/AWWA Standard C218, *Coating the Exterior of Aboveground Steel Water Pipelines and Fittings.* Latest edition. Denver, Colo.: American Water Works Association.

———. ANSI/AWWA Standard C222, *Polyurethane Coatings for the Interior and Exterior of Steel Water Pipe and Fittings.* Latest edition. Denver, Colo.: American Water Works Association.

———. ANSI/AWWA Standard C602, *Cement–Mortar Lining of Water Pipelines 4 In. (100 mm) and Larger in Place.* Latest edition. Denver, Colo.: American Water Works Association.

NACE 1983. *Control of External Corrosion on Underground or Submerged Metallic Piping Systems.* NACE Standard RP0169. Houston, Texas: NACE.

Paint Manual. 1951. Denver, Colo.: US Bureau of Reclamation. (available from US Government Printing Office, Washington, D.C.).

Steel Structures Painting Manual. 1954. Good Painting Practice—Vol. 1. Systems and Specifications—Vol. 2. Pittsburgh, Pa.: Steel Structures Painting Council.

Chapter 12

Transportation, Installation, and Testing

The detailed procedures for transporting, trenching, laying, backfilling, and testing any steel pipeline depend on many controlling factors, including the character and purpose of the line; its size, operating pressure, and operating conditions; its location—urban, suburban, or rural—and the terrain over which it is laid—flat, rolling, or mountainous. Procedures also are affected by trench depth, character of the soil, and backfill. This chapter briefly discusses several of the more common requirements of installation, omitting precise details that vary in individual installations. Throughout the chapter, the importance of the engineering properties of the excavated soil and the backfill soil should be considered. The principles of soil mechanics properly applied to excavation and backfill practices lead to safer working conditions and to better and more economical pipeline installations (Sowers 1956, Reitz 1956).

TRANSPORTATION AND HANDLING OF COATED STEEL PIPE

Lined and coated steel pipe is readily transported by truck, rail, or ship and has been successfully transported to all parts of the United States and to other parts of the world. Regardless of which mode of transportation is used, lined and coated steel pipe is valuable cargo and should be handled as such.

Modes of Transportation

Requirements for packaging, stowing, and restraining pipe during transit depend on the mode of transportation.

Rail. Flat railroad cars can be loaded to approximately 17 ft (5.2 m) above the top of the rail and to widths of 10 ft (3 m). Cars are normally available for shipping 40-, 60-, or 80-ft (12.2-, 18.3-, or 24.4-m) lengths of pipe. Pipe can be restrained on the cars using stake pockets or made into floating loads in accordance with current Association of American Railroads rules. An inspector from the railroad will check each car for proper loading before accepting it for shipment.

Water. Constant pitching and rolling motions should be anticipated for pipe stowed aboard ships. Small pipe must be packaged, and large pipe must be stowed in such a manner to ride with or offset the pitching and rolling motion. Adequate padded timbers or similar barriers must be used to keep pipe from rubbing together. In many cases, flat racks or containers can be used. Air bags can help prevent pipe shifting inside the container.

The surveyor who is commonly responsible for checking loading arrangements should verify that all dock and ship handling equipment is approved for use on coated pipe. Pipe is normally shipped on a cubic-foot freight basis. The feasibility of nesting smaller diameter pipe inside larger pipe to reduce freight costs should be investigated; however, such nesting must be padded to ensure that lining and coating integrity is maintained.

Truck. Most coated pipe is carried on flat-bed trucks and trailers directly to the job site. This one-time handling between shipper and customer avoids damage sometimes caused by multiple loading and unloading sessions. The shipper should caution the trucking firms against use of tie-down equipment that could injure the coating.

Air. Delivery of the pipe to distant sites can be expedited by airplane, and delivery into otherwise inaccessible locations may require cargo helicopters. The air carrier should be contacted to obtain maximum length, width, height, and weight limitations for the route involved. Generally, the carriers will require pipe to be strapped directly to pallets suitable for handling.

Loading and unloading. Loads should be prepared in a manner that will protect the lined and coated pipe. Sufficient stringers should be used to layer the pipe without placing too much load on a single bearing point. When plain-end pipe is shipped, a pyramid load with the full length of pipe resting on adjacent pipe should be considered. Interior stulls should be used where the pipe wall is too light to maintain roundness during shipment. Contoured blocks may be necessary to give proper support to some loads. Pipe should not be allowed to roll or fall from the conveyance to the ground.

Handling equipment. Both loading and unloading of coated pipe should be performed with equipment that will not damage the pipe coating. Approved equipment for handling coated pipe includes nylon straps, wide canvas or padded slings, wide padded forks, and skids designed to prevent damage to the coating. Unpadded chains, sharp edges on buckets, wire ropes, narrow forks, hooks, and metal bars are unacceptable.

Stringing. If the pipe is to be distributed along the right-of-way in rock or gravelly terrain, both ends (at about one-quarter length from the ends) should be laid on padded wood blocks, sandbags, mounds of sand, or other suitable supports to protect the pipe coating.

TRENCHING

Depth

Trenches should be dug to grade as shown in the profile. Where no profile is provided, the minimum cover should be generally selected to protect the pipe from transient loads where the climate is mild; and it should be determined by the depth of the frost line in freezing climates. The profile should be selected to minimize high points where air may be trapped. The depth of trench in city streets may be governed by existing utilities or other conditions.

Width

Where the sides of the trench will allow reasonable side support, the trench width that must be maintained at the top of the pipe, regardless of the depth of excavation, is the narrowest practical width that will allow proper densification of pipe-zone bedding and backfill materials. If the pipe-zone bedding and backfill require densification by compaction, the width of the trench at the bottom of the pipe should be determined by the space required for the proper and effective use of tamping equipment, but it should never be less than the pipe OD plus 20 in. (508 mm).

When mechanical joints are assembled on pipe in the trench, bell holes must be provided at each joint and holes excavated to permit removal of the slings without damage to the pipe coating. To avoid imposing excessive external loads on the pipe, the trench width should be kept to the minimum width consistent with the backfill-compaction equipment and the type of joint used.

Bottom Preparation

Flat-bottom trenches should be excavated to a minimum depth of 2 in. (51 mm) below the established grade line of the outside bottom of the pipe. The excess excavation should then be filled with loose material from which all stones and hard lumps have been removed. The loose subgrade material should be graded uniformly to the established grade line for the full length of the pipe. Steel pipe should not be set on rigid blocks on the trench bottom that would cause concentration of the load on small areas of pipe coating or cause deformation of the pipe wall.

Where the bottom of the trench is covered with solid, hard objects that might penetrate the protective coating, a bedding of crushed rock or sand, 3–6 in. (75–152 mm) thick, should be placed under the barrel of the pipe. Screened earth also has been used successfully for such a bedding, where it will remain dry during pipe installation and backfill. It may be advantageous to shape the trench bottom under large steel pipe for full arc contact.

Overexcavation and Special Subgrade Densification

When required, the trench should be overexcavated to a depth of at least 6 in. (152 mm) below the bottom of the pipe (Figures 12-1 and 12-2) where the trench bottom is unstable, or where it includes organic materials, or where the subgrade is composed of rock or other hard and unyielding materials. The overexcavation should be replaced with well-densified material to a depth of approximately 2 in. (76 mm) below the bottom of the pipe, and the remaining subgrade should be completed with loose material, as shown in Figures 12-1 and 12-2. Voids formed by the removal of boulders and other large interfering objects extending below normal excavation limits should be refilled with material as described above.

Regulations

All applicable local, state, and federal laws and regulations should be carefully observed including those relating to the protection of excavations, the safety of persons working therein, and provision of the required barriers, signs, and lights.

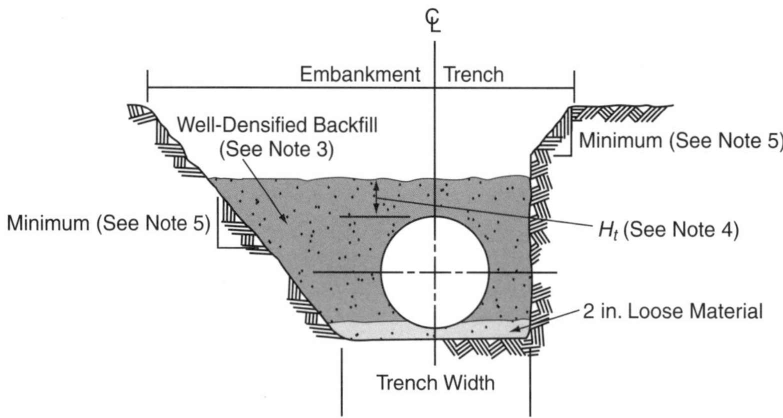

Figure 12-1 Densified pipe zone bedding and backfill

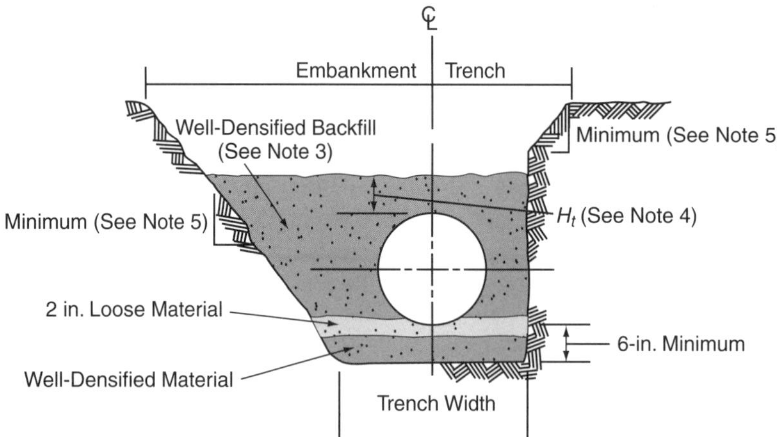

Figure 12-2 Special subgrade densification

Notes to Figures 12-1 and 12-2

1. Soil densities are expressed as a percentage of maximum dry soil density as determined by AASHTO T99 (American Association of State Highway and Transportation Officials 1981; Standard Proctor) or ASTM D698 (1978a).
2. Class C1, C2, and C3 backfills require that the contractor prepare a firm but yielding subgrade.
3. Well-densified material shall conform to the following relative dry densities as a percentage of the laboratory standard maximum dry soil density as determined by AASHTO T99 (1981) for compacted, cohesive soils:

Specified Bedding Class	Dry Density
C1	95%
C2	90%
C3	85%

For free-draining soils, the relative density shall be at least 70 percent as determined by ASTM D2049-69 (1969; withdrawn, replaced by ASTM D4253-83 [1983a] and ASTM 4254-83 [1983b]). Comparative soil density tests are shown in Table 12-1.

4. Pipe zone backfill height over top of pipe (H_t) shall be 12 in. (305 mm) minimum for pipe diameter larger than 24 in. (600 mm) and 6 in. (152 mm) minimum for pipe diameter 24 in. (600 mm) or less.
5. Side slopes shall be a minimum of $^3/_4$:1 or as required by OSHA, other safety orders, or by the soils engineer.
6. Figures 12-1 and 12-2 represent Class C bedding as shown in ASCE Manual No. 37 (1969; WPCF Manual of Practice 9).

Table 12-1 Comparison of standard density tests

Test	Compactive Energy per Volume	
	ft–lbf/cu ft	*(kJ/m³)*
Standard AASHTO (Standard Proctor) AASHTO T99-74, ASTM D698-78 (1978a)	12,400	593.7
Modified AASHTO (Modified Proctor) AASHTO T180-74 (1974), ASTM D1557-78 (1978b)	56,250	2,693.3

* Natural in-place deposits of soils have densities from 60 percent to 100 percent of maximum obtained by the standard AASHTO compaction method. The designer should be sure that the *E* value used in design is consistent with this specified degree of compaction and the method of testing that will be used during construction.

INSTALLATION OF PIPE

Handling and Laying

Care similar to that exercised during loading, transporting, unloading, and stringing should be observed during installation of the pipe in the trench. Dielectrically coated pipe may require additional special care when handled at temperatures above or below that recommended by the manufacturer

Coated pipe should not be strung on rough ground when stored at the trench site, nor should it be rolled on such a surface. Rolling of coated pipe should be permitted only when joint ends are bare and rails are provided on which to roll the exposed steel.

While handling and placing pipe in the trench, fabric slings should be used. The pipe should not be dragged along the bottom of the trench or bumped. It should be supported by the sling while preparing to make the joint. The coating on the underside of the pipe should be inspected while it is suspended from the sling, and any visible damage to the coating should be repaired before lowering the pipe into the trench.

Pipe should be laid to lines and grades shown on the contract drawings and specifications, except where modified by the manufacturer's detailed layout drawings or laying schedule, as reviewed by the purchaser. All fittings and appurtenances should be at the required locations, and all valve stems and hydrant barrels should be plumb. The pipe trench should be kept free from water that could impair the integrity of bedding and joining operations. On grades exceeding 10 percent, the pipe should be laid uphill or otherwise held in place by approved methods.

Special means of supporting the pipe may be provided, but under no conditions should pipe sections be installed permanently on timbers, earth mounds, pile bents, or other similar supports unless specific pipe designs for these special conditions have been provided.

Slight deflections for horizontal and vertical angle points, long radius curves, or alignment corrections may be made by unsymmetrical closure of joints. The manufacturer can provide data to the purchaser and the contractor indicating maximum joint offsets and deflections for each type of joint supplied.

Assembly of Pipe

Pipe larger than 24 in. (600 mm) in diameter is normally assembled in the trench except under the most unusual conditions. Smaller-diameter pipe joined by welding or couplings may be assembled aboveground in practicable lengths for handling and then lowered into the trench by suitable means allowing for progressive lowering of the assembled run of pipe. If the method of assembling pipe aboveground prior to lowering it into the trench, care must be taken to limit the degree of curvature of the pipe during the lowering operation so as to not exceed the yield strength of the pipe material or damage the lining or coating materials on the pipe. Pipe deflection at any joint should be limited to the manufacturer's recommendation during the lowering operation. Pipe that has O-ring rubber gaskets as seals must be assembled section by section in the trench.

Trestle and ring-girder construction is often used for highway, river, and similar crossings. Generally, such installation presents no unusual problems, providing three principal requirements are met for the field-welded spans (Garrett 1948).

- The centerline of the pipe sections is maintained in proper alignment when the sections are tacked for welding.
- Correct welding procedures and competent welding operators are employed to ensure that the welded joint will be as strong as the steel in the pipe (AWWA 1982).
- Bows or bends in the pipe caused by direct rays from the sun are prevented. (This can be achieved by providing a sun shield over the pipe.)

When pipe is installed on the decks of highway bridges, saddles are generally used to support the pipe at proper intervals, and hold-down clamps are provided as required. Usually the only expansion joints needed for welded pipe are those that occur where the bridge contains an expansion joint in its construction. Steel pipe is also often suspended from or attached to the underside of highway bridges, with appropriate attention given to the flexibility of the bridge's structure. Exposed pipelines in any location should be protected against freezing in areas where such a possibility exists.

Field-Welded Joints

Technical requirements for good field welding are contained in AWWA C206, *Standard for Field Welding of Steel Water Pipe* (latest edition). Practical data for field use have been published (Price and Garrett 1943). If pipe that has been lined and coated is to be field welded, a short length of the pipe barrel at either end must be left bare so that the heat of the welding operation will not adversely affect the protective coating. The length of the unprotected section may vary depending on the kind of protective coating and the pipe wall thickness. Care must be exercised when cutting and welding on pipes with combustible linings and coatings to avoid the risks of fire.

Following the completion of the weld, the gaps in the lining and coating must be filled, normally with the same material as that used for the pipe. For pipe 24 in. (600 mm) in diameter and larger, the joints should be repaired from the inside. Where workers must enter the pipe to complete the lining, proper ventilation must be provided. Joints in pipe smaller than 24 in. (600 mm) should be repaired from the outside, using handholes. With mortar-lined pipe, the joints may be repaired by placing mortar on the lining at the bell end of the unassembled pipe, stabbing the pipe, and then pulling an inflated ball through the joint. Outside coatings of pipe joints can be easily applied after welding for any diameter pipe or any coating system used.

The use of welded joints results in a rigid pipeline. This stiffness provides a considerable advantage where long, unsupported spans are required. It is also advantageous in restraining elbows in soils of low bearing capacity.

Welded joints are capable of resisting thrusts caused by closed valves or by changes in the direction of a pipeline. Welded joints may be provided to transmit such thrusts over a sufficient distance to absorb the force through skin friction provided by the backfill material against the pipe. In such cases, accurate computation of the thrust and strength of the weld must be made, particularly for larger pipe under high pressures, to determine if the weld is sufficiently strong to transmit the force from one pipe section to the next.

Except during the construction period when an open trench exists, pipe with welded joints will usually have no problems with excessive thermal expansion and contraction. Where immediate shading or backfill of welded-joint steel pipe is impractical, it is advisable to weld the pipe in sections of approximately 400 to 500 ft (122 to 152 m) and leave the end joint unwelded, as described in chapter 8. If the final open joints are then welded in the early morning hours when the pipe is typically coolest, a minimum of temperature stress will occur in the pipeline.

Pipe laid on piers above the ground can be continuously welded; however, it is necessary to provide for thermal expansion and contraction.

Bell-and-Spigot Rubber-Gasket Joints

Under normal laying conditions, work should proceed with the bell end of the pipe facing the direction of laying. Before setting the spigot in place, the bell should be thoroughly cleaned and then lubricated.

After the O-ring rubber gasket has been placed in the spigot groove, it should be adjusted so the tension on the rubber is uniform around the circumference of the joint. Following assembly, the pipe joint should be checked with a thin metal feeler gauge to ensure that proper gasket placement exists in the spigot groove and that the proper amount of joint lap has been achieved.

Installation Practice for Flanged Joints

The method of flange assembly and sequence of tightening bolts and controlling torque are very important and become more critical with larger diameter flanges and bolts.

Gasket Installation Procedure

Regardless of the type of gasket used, basic procedures must be followed if the joint is to be assembled, tested, and put into operation with minimum difficulty. The following procedures should be followed whether bolt stress will be achieved with ordinary stud wrench, preheating studs, using tensioning devices, using torque wrenches, or using hydraulic wrenches.

1. Inspect the gasket seating surfaces. Look for tool marks, cracks, scratches, or pitting by corrosion and make sure that the gasket seating surface is appropriate for the type of gasket being used. Radial tool marks on a gasket seating surface are virtually impossible to seal, regardless of the type gasket used; reject this type of defect.
2. Inspect the gasket. Confirm that the material is as specified, and look for any possible defects or damage to the gasket.
3. Inspect and clean each stud or bolt, nut, washer, and the facing on the flanges against which the nuts will rotate. Look for severe galling, pitting, etc. If any of the above-mentioned items are damaged beyond repair, replace the item.
4. Lubricate all thread contact areas and nut facings. Do not construct the joint without applying the proper lubrication to the threaded surfaces and to the nut facings. The lubricant is necessary to provide a low coefficient of friction for installation and to achieve proper torque values.
5. With ring gasket installation, loosely install the stud bolts on the lower half of the flange. Insert the gasket between the flange facing to allow the bolts to center the gasket on the assembly. Install the remaining bolts and nuts and bring all to a hand-tight or snug condition.

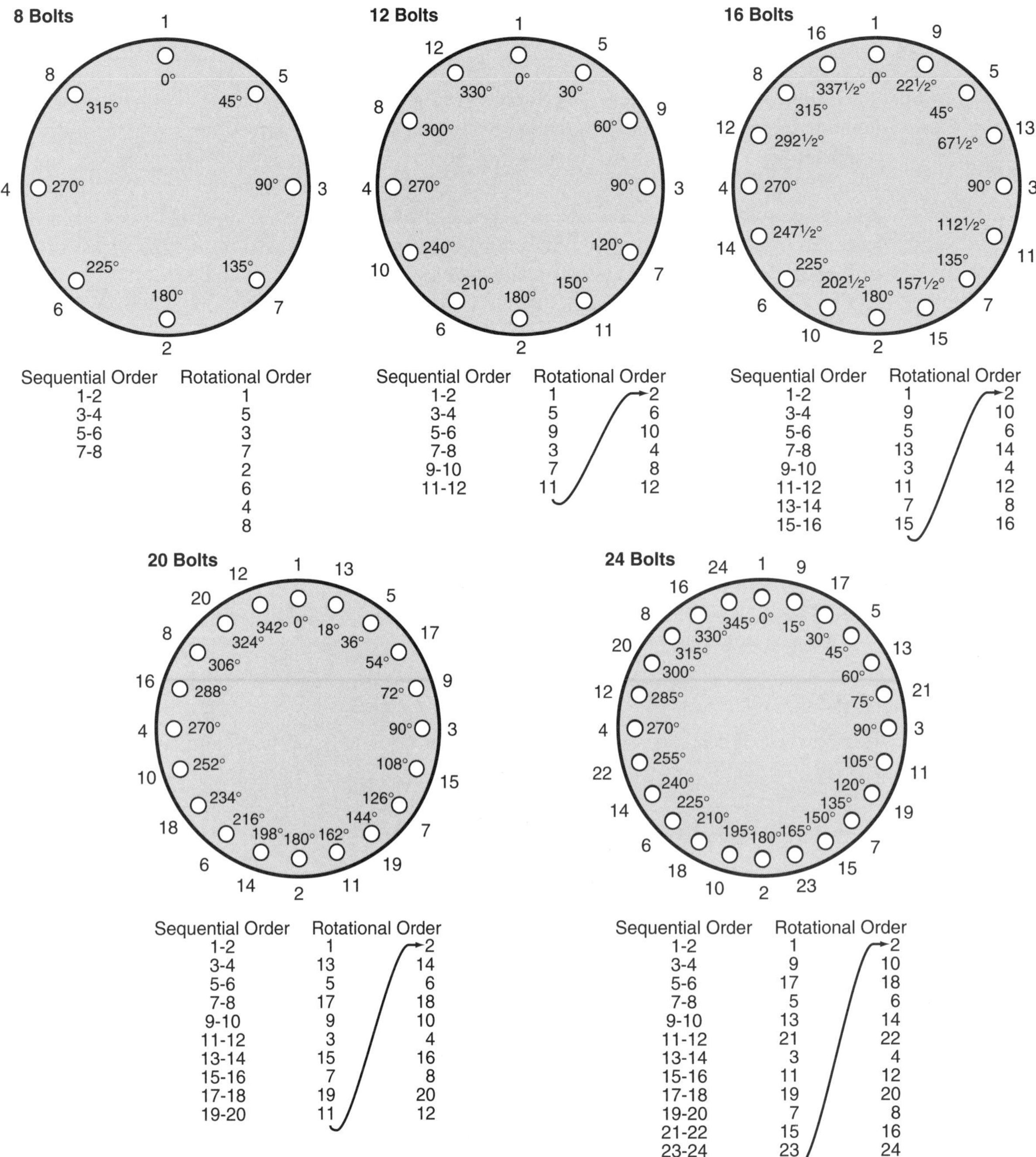

NOTE: Patterns similar to the above are followed on larger flanges.

Figure 12-3 Bolt torque sequence

With full-face gaskets, at least two bolts or studs must be placed in the top half of one flange, then the gasket placed over these bolts prior to mating the flanges. The remaining bolts and nuts can be installed to a hand-tight condition through the flange and gasket.

6. Torque the bolts to a maximum of 30 percent of the final torque value required following the recommended sequence. (See Figure 12-3 for bolting sequence). Number the bolts so that the torquing requirements can be followed. With any gasket material, it is extremely important to follow a proper bolting sequence.

7. Repeat Step 6, increasing the torque to approximately 50 to 60 percent of the final torque required.
8. Continue with a rotational pattern of retorquing all studs or bolts to the desired amount until no further rotation of the nuts can be achieved. This may require several retorquings because as one bolt is torqued, it will relieve the stress on the adjacent bolt until equilibrium is achieved.

Torque Values

Probably the only true measurement of bolt stress is by bolt or stud elongation. In practice, however, this would be a costly and impractical approach to determine the true measure of bolt stress. As a consequence, the industry trend today is the use of torque wrenches, tensioning devices, hydraulic wrenches, or drilling the studs and inserting heaters to preheat the stud to a specific temperature that will ultimately create the proper tension on the bolt.

Bolts

AWWA C207 specifies ASTM A307, grade B for Class B and D flanges, and ASTM A193, grade B7 for Class E and F flanges. The allowable design stress on A307 grade B bolts is 7,000 psi (48.3 MPa) in ASME B&PV Section VIII Table UCS-23. The minimum tensile stress is 60,000 psi (413.7 MPa). Bolts can and often must be torqued above the design stress to achieve gasket seal. ASTM A193, grade B7 studs with ASTM A194 nuts can be used and are an economic alternative in larger diameters. For studs 2½ in. (64 mm) and under, tensile strength is 125,000 psi (861.8 MPa) minimum, yield strength is 105,000 psi (172.4 MPa) minimum, proof load is 87,000 psi (600 MPa) and allowable design strength is 25,000 psi (172.4 MPa). For studs over 2½ in. to 4 in. (64 mm to 102 mm) tensile strength is 115,000 psi (793 MPa) minimum, yield strength is 95,000 psi (655 MPa) minimum, proof load is 80,000 psi (551.6 MPa) and allowable design stress is 23,000 psi (158.6 MPa).

Table 12-2 shows the theoretical bolt loads required for various gasket materials at safe working pressure of flange, maximum bolt load for ASTM A307, grade B bolts and ASTM A193, grade B7 studs, and theoretical torque required to obtain this bolt load with various lubricants. Note: from 54 to 63 percent of friction loss is in flange-to-nut surface contact. It is, therefore, very important to lubricate this surface and to use steel washers if the surface is rough. It may be necessary to torque bolts beyond the values shown to get a watertight joint. However, the maximum bolt loads given in Table 12-3 for ASTM A307 grade B bolts or ASTM A193 grade B7 studs, should not be exceeded.

$$T = \frac{KDW}{12} \tag{12-1}$$

Where:

T = torque, in lbf · ft (N · m)
D = nominal bolt diameter, in in.
W = bolt load, in lbf (N)
K = torque constant

$$\text{Based on } K = f_n\frac{R_n}{D} + f_t\frac{R_t}{D}\sec B + \frac{R_t}{D}\tan C$$

Where:

f_n = coefficient of friction at bearing face of nut
f_t = coefficient of friction at thread contact surface
R_n = effective radius of friction at bearing face of nut
R_t = effective radius of friction at bearing face of thread
D = bolt diameter
B = thread half-angle (30° for 60° basic thread form), Sec B = 1.15
C = helix angle of thread
K = 0.45–0.52 for nonlubricated threads, f = 0.4
K = 0.18–0.21 for thread and nut face lubricated, machine oil, f = 0.15
K = 0.14–0.17 for thread and nut face lubricated, API SA2, f = 0.117
K = 0.08–0.10 for thread and nut face lubricated, Select-a-torque 503, f = 0.06

ANCHORS AND THRUST BLOCKS

Anchors or thrust blocks are necessary at angle points, side outlets, valves, and on steep slopes. The type of pipe joint used determines the necessary anchoring at these points.

All-welded pipelines laid in trenches will ordinarily not need anchors or thrust blocks except on extremely steep slopes and at discontinuities where the pipe has been cut for valves and appurtenances. An all-welded pipeline laid aboveground on piers may be stable when filled and under pressure but may require heavy anchorage at angle points and particularly on steep slopes to resist stresses resulting from temperature changes when the pipe is empty.

When other types of joints are used that have little or no ability to resist tension, all of the previously mentioned critical points must be adequately blocked or anchored. In order to provide resistance to thrust at angles in large-diameter pipelines, whether buried or exposed, welded joints should be provided on each side of the angle point for a distance sufficient to resist the thrust. Under high-pressure conditions, lap-welded field joints should be analyzed for proper strength near valves and at large deflection angles.

Where pipe is laid on piers, antifriction material should separate the pipe from the supporting structure; 90–120 degrees of the pipe surface should be made to bear on the pier. For pipe on piers, the thrust resulting from an elbow or bend tends to overturn the anchor pier.

Pipelines laid on slopes, particularly aboveground, always have a tendency to creep downhill. It is necessary to provide anchor blocks placed against undisturbed earth at sufficiently frequent intervals on a long, steep slope to reduce the weight of pipe supported at each anchorage to a safe figure. When disturbance of the trench is unlikely, concrete thrust blocks may be used to resist the lateral thrust. Vertical angles causing downward thrust require no special treatment if the pipe is laid on a firm and carefully trimmed trench bottom, but vertical angles causing upward thrust should be properly anchored.

Soil resistance to thrust. A force caused by thrust against soil, whether applied horizontally or vertically downward, may cause consolidation and shear strains in the soil, allowing a thrust block to move. The safe load that a thrust block can transfer to a given soil depends on the consolidation characteristics and the passive resistance (shear strength) of that soil, the amount of block movement permissible, the area of the block, and the distance of force application below ground line. Methods of calculating passive resistance are available (Terzaghi and Peck 1948). For all lines, detailed calculations are necessary. Data on permissible soil grip for anchoring ordinary lines are given in chapter 13. Some data for the calculation of thrust at angle points are also included in chapter 13.

Table 12-2 Torque requirements for AWWA C207 Class D ring flange bolts

Nominal Pipe Size	Bolts Number	Bolts Diameter *in.*	Bolt Load* 750 psi *lbf*	Bolt Load* 1,600 psi *lbf*	Bolt Load* 3,000 psi *lbf*	Required Torque, f = 0.06 750 psi *lbf-ft*	f = 0.06 1,600 psi *lbf-ft*	f = 0.06 3,000 psi *lbf-ft*	Required Torque, f = 0.117 750 psi *lbf-ft*	f = 0.117 1,600 psi *lbf-ft*	f = 0.117 3,000 psi *lbf-ft*	Required Torque, f = 0.15 750 psi *lbf-ft*	f = 0.15 1,600 psi *lbf-ft*	f = 0.15 3,000 psi *lbf-ft*	Required Torque, f = 0.4 750 psi *lbf-ft*	f = 0.4 1,600 psi *lbf-ft*	f = 0.4 3,000 psi *lbf-ft*
4	8	5/8	1,021	2,124	3,982	5	9	17	8	16	29	10	20	38	24	50	94
6	8	3/4	1,577	2,566	4,811	11	13	24	14	23	42	18	19	55	45	72	136
8	8	3/4	2,454	3,661	6,865	13	19	35	22	32	60	28	42	78	69	103	193
10	12	7/8	2,391	3,314	6,214	14	20	37	26	37	68	32	43	82	80	112	209
12	12	7/8	3,505	5,105	9,572	21	30	56	39	56	105	46	67	126	118	172	321
14	12	1	3,646	6,239	11,699	25	42	78	46	78	147	58	99	186	146	250	468
16	16	1	3,459	6,034	11,314	24	41	76	44	76	142	57	96	180	142	242	453
18	16	1 1/8	3,947	5,639	10,572	34	48	90	60	85	159	71	101	189	181	259	486
20	20	1 1/8	3,837	5,343	10,018	29	46	85	58	81	151	69	96	179	177	245	461
22	20	1 1/8	4,524	6,032	11,310	38	51	96	67	91	170	81	108	202	208	277	520
24	20	1 1/4	5,323	6,959	13,047	50	63	123	89	116	218	111	145	272	272	356	666
30	28	1 1/4	5,708	7,005	12,970	54	66	122	96	117	217	119	146	271	292	358	662
36	32	1 1/2	6,981	8,471	14,901	79	96	168	140	170	298	175	212	373	445	540	950
42	36	1 1/2	8,394	10,161	17,671	95	115	199	165	204	354	210	254	442	535	648	1,127
48	44	1 1/2	8,822	10,611	17,894	100	120	201	177	213	358	221	266	448	563	677	1,141
54	44	1 3/4	12,547	14,703	21,554	165	193	253	293	343	503	385	451	660	993	1,094	1,603
60	52	1 3/4	11,380	13,551	21,722	150	178	286	266	316	507	349	415	666	847	1,008	1,616
66	52	1 3/4	13,757	16,375	26,180	181	215	344	321	382	611	422	502	802	1,023	1,218	1,947
72	60	1 3/4	14,079	16,706	26,272	185	220	345	329	390	613	432	512	805	1,047	1,243	1,954
78	64	2	15,262	17,996	27,336	255	300	456	433	510	775	535	630	967	1,323	1,560	2,369
84	64	2	17,610	20,720	31,105	294	346	519	499	587	882	617	726	1,089	1,527	1,796	2,696
90	68	2 1/4	18,805	22,014	32,087	318	372	542	565	660	963	741	867	1,263	1,799	2,105	3,068
96	68	2 1/4	21,303	24,891	35,888	360	420	606	639	747	1,077	839	980	1,413	2,037	2,381	3,432
102	72	2 1/2	22,235	25,736	35,016	464	537	730	788	912	1,240	973	1,126	1,532	2,409	2,788	3,794
108	72	2 1/2	24,728	28,519	37,907	516	595	790	876	1,010	1,343	1,082	1,248	1,659	2,679	3,090	4,107
114	76	2 3/4	25,641	29,330	36,893	588	673	846	999	1,143	1,438	1,234	1,412	1,775	3,056	3,495	4,397
120	76	2 3/4	28,096	31,971	38,753	644	733	888	1,093	1,246	1,510	1,353	1,539	1,865	3,348	3,810	4,618
126	80	3	29,272	33,227	39,545	586*	586*	791*	1,098*	1,246*	1,483*	1,390*	1,579*	1,879*	3,659*	4,154*	4,943*
132	80	3	32,147	36,501	43,547	643*	730*	871*	1,206*	1,369*	1,633*	1,527*	1,734*	2,069*	4,019*	4,563*	5,444*
138	84	3 1/2	33,160	37,487	43,275	663*	813*	938*	1,347*	1,523	1,769*	1,707*	1,928*	2,227*	4,491*	5,077*	5,860*
144	84	3 1/2	35,959	40,571	46,128	780*	879*	1,000*	1,461*	1,648*	1,874*	1,851*	2,088*	2,374*	4,870*	5,494*	6,247*

*NOTE: Bolt load is based on 175 psi (1,207 kPa) for 4 in. (100 mm) through 12 in. (300 mm) and 150 psi (1,034 kPa) for nominal pipe greater than 12 in. (300 mm) with gasket yield pressures of 750 psi (5.17 MPa) for 80 durometer rubber, 1,600 psi (11.03 MPa) for 1/8 in. (3.18 mm) mineral gaskets, and 3,000 psi (20.68 MPa) for Garlock "Blue-Gard" compressed nonasbestos gaskets. Torque is based on standard UNC thread through 2 3/4 in. (70 mm) and 8 thread/in. on 3 in. (76 mm) and 3 1/4 in. (82.6 mm) 8 thread studs 1 in. to 2 3/4 in. (70 mm) require slightly less torque.

To convert nominal in. to mm, multiply by 25; to convert in. to mm, multiply by 25.4; to convert lbf to N, multiply by 4.4482; to convert psi to MPa, multiply by 0.0068948; to convert lbf-ft to N-m, multiply by 1.3558.

Table 12-3 Torque requirements for steel pipe flange bolts and studs

Flange Bolts

Bolt Diameter		Maximum Load for A307, Gr. B Bolts*		Required Torque							
				f = 0.40		f = 0.15		f = 0.117		f = 0.06	
in.	*(mm)*	*lb*	*(N)*	*ft-lb*	*(N-m)*	*ft-lb*	*(N-m)*	*ft-lb*	*(N-m)*	*ft-lb*	*(N-m)*
5/8	15.88	7,119	31,667	167	226	67	90	52	70	30	40
3/4	19.05	10,521	46,800	296	401	119	161	92	125	53	71
7/8	22.23	14,553	64,735	489	663	191	259	160	217	85	115
1	25.40	19,089	84,912	764	1,035	303	411	239	324	128	174
1 1/8	28.58	24,034	106,909	1,105	1,498	428	580	361	489	203	275
1 1/4	31.75	30,523	135,773	1,558	2,112	636	862	509	690	287	389
1 1/2	38.10	44,257	196,865	2,821	3,825	1,107	1,501	886	1,201	498	675
1 3/4	44.45	51,300	228,194	3,815	5,173	1,571	2,130	1,197	1,623	673	913
2	50.80	67,500	300,255	5,850	7,932	2,363	3,203	1,913	2,593	1,125	1,525
2 1/4	57.15	87,750	390,331	8,391	11,377	3,455	4,685	2,633	3,569	1,481	2,008
2 1/2	63.50	108,000	480,408	11,700	15,863	4,725	6,406	3,825	5,186	2,250	3,051
2 3/4	69.85	133,110	592,103	15,862	21,506	6,406	8,685	5,186	7,031	3,050	4,136

Stud Bolts

Bolt Diameter		Maximum Load for A307, Gr. B Bolts*		Required Torque							
				f = 0.40		f = 0.15		f = 0.117		f = 0.06	
in.	*(mm)*	*lb*	*(N)*	*ft-lb*	*(N-m)*	*ft-lb*	*(N-m)*	*ft-lb*	*(N-m)*	*ft-lb*	*(N-m)*
5/8	15.88										
3/4	19.05										
7/8	22.23										
1	25.40	39,541	175,887	1,582	2,144	626	851	495	670	264	360
1 1/8	28.58	51,547	229,292	2,368	3,214	918	1,245	725	1,048	387	590
1 1/4	31.75	65,250	290,246	3,465	4,516	1,360	1,843	1,088	1,474	612	831
1 1/2	38.10	97,353	433,048	6,206	8,415	2,434	3,301	1,947	2,642	1,095	1,485
1 3/4	44.45	135,720	603,712	10,094	13,686	3,959	5,635	3167	4,294	1,584	2,415
2	50.80	180,742	803,980	15,062	21,238	5,724	8,577	4,519	6,943	2,410	4,084
2 1/4	57.15	232,290	1,033,277	21,777	30,116	8,275	12,401	6,533	9,448	3,484	5,315
2 1/2	63.50	289,710	1,288,694	30,178	42,553	11,468	17,185	9,053	13,911	4,828	8,183
2 3/4	69.85	329,800	1,467,023	37,790	53,285	14,360	21,519	11,337	17,420	6,046	10,247
3	76.20	390,600	1,737,475	48,825	68,846	18,553	27,803	14,648	22,507	7,812	13,240
3 1/4	82.55	461,400	2,052,409	62,481	88,102	23,743	35,580	18,744	28,802	9,997	16,943

* Maximum bolt load is figured on 75 percent of bolt proof load.

FIELD COATING OF JOINTS

Acceptable procedures for coating of field joints are described in applicable AWWA standards.

PIPE-ZONE BEDDING AND BACKFILL

The following discussion on pipe bedding and backfill is somewhat general in nature. A foundation study should be performed to provide more precise design criteria for large projects or those with unusual problems.

Pipe-zone bedding and backfill may be classified as Class C1, C2, or C3 (Figure 12-1), or as otherwise defined. Bedding and backfill should be densified around the pipe to the specified height over the top of the pipe. In the absence of a specific height, the backfill should be densified to not less than that described in Note 4 of Figures 12-1 and 12-2.

The dry density of compacted cohesive soil for each class of bedding and backfill, as shown in Figure 12-1, should not be less than previously discussed

Soil densities should be expressed as a percent of the laboratory standard maximum dry-soil density, as determined according to AASHTO T99, The Moisture-Density Relations of Soils Using a 5.5-lb (2.5-kg) Rammer and a 12-in. (305-mm) Drop (AASHTO 1981), or ASTM D698, Tests for Moisture-Density Relations of Soils and Soil-Aggregate Mixtures, Using 5.5-lb (2.5-kg) Rammer and 12-in. (304.8-mm) Drop (DOD Adopted)(ASTM 1978a). In-place tests of soil density are usually made in accordance with ASTM D1556, Test for Density of Soil in Place by the Sand-Cone Method (ASTM 1964), or ASTM D2167, Test for Density of Soil in Place by the Rubber-Balloon Method (ASTM 1966).

Densification

Regardless of the densification method used, materials must be brought up at relatively the same rate on both sides of the pipe. Care also should be taken so that the pipe is not floated or displaced before backfilling is complete.

Mechanical Compaction

Cohesive soils should be densified by compaction using mechanical or hand tamping. Care must be taken not to damage coatings during compaction. Equipment with suitably shaped tamping feet for compacting the material will generally provide that the specified soils density is obtained under the lower quadrant of the pipe. At the time of placement, the backfill material should contain the optimum moisture content required for compaction. The moisture content should be uniform throughout each layer. Backfill should be placed in layers of not more than 6 in. (150 mm) in thickness after compaction.

Hydraulic Consolidation

Soils identified as free draining may be densified by tamping or by consolidation with water using any or all of the following devices or methods: water jets, immersion-type vibrators, bulkheading, and flooding or sluicing. Material should be placed in a minimum of two layers, the first layer being placed loose to the spring line of the pipe. Consolidation of earth backfill by hydraulic methods should be used only if both the backfill and the native soil are free draining. Materials used in hydraulic consolidation should be able to pass through a $1\frac{1}{2}$-in. (38-mm) screen, with not more than 10 percent passing through a 200-mesh sieve. The thickness of layers should not exceed the penetrating depth of the vibrators if consolidation is performed by jetting and internal vibration.

Trench Backfill Above Pipe Zone

Native backfill material above the pipe zone up to the required backfill surface should be placed to the density required in the contract specifications. Trench backfill should not be placed until confirmation that compaction of pipe-zone bedding and backfill complies with the specified compaction. Cohesive materials should always be compacted with tamping or rolling equipment. To prevent excessive line loads on the pipe, sufficient densified backfill should be placed over the pipe before power-operated hauling or rolling equipment is allowed over the pipe.

Interior Bracing of Pipe

When required, the design, installation, and performance of pipe bracing during transportation and installation is generally the responsibility of the contractor. Such bracing limits the maximum vertical deflection of the pipe during installation and backfilling. CAUTION: Internal bracing designed for shipments will not necessarily protect the pipe during backfill operations.

HYDROSTATIC FIELD TEST

The purpose of the hydrostatic field test is primarily to determine if the field joints are watertight. The hydrostatic test is usually conducted after backfilling is complete. It is performed at a fixed pressure above the design working pressure of the line. If thrust resistance is provided by concrete thrust blocks, the blocking must be allowed to cure before the test is conducted.

Field Testing Cement–Mortar-Lined Pipe

Cement–mortar-lined pipe to be tested should be filled with water of approved quality and allowed to stand for at least 24 hours to permit maximum absorption of water by the lining. Additional water should be added to replace water absorbed by the cement–mortar lining. (Pipe with other types of lining may be tested without this waiting period.) Pipe to be cement–mortar lined in place may be hydrostatically tested before or after the lining has been placed.

Bulkheads

If the pipeline is to be tested in segments and valves are not provided to isolate the ends, the ends must be provided with bulkheads for testing. A conventional bulkhead usually consists of a section of pipe 2–3 ft (0.6–0.9 m) long, with a flat plate or dished plate bulkhead welded to the end containing the necessary outlets for accommodating incoming water and outgoing air.

Air Venting

The pipeline should be filled slowly to prevent possible water hammer, and care should be exercised to allow all of the air to escape during the filling operation. After filling the line, a pump may be necessary to raise and maintain the desired test pressure.

Allowable Leakage

The hydrostatic test pressure is usually applied for 24 hours before the test begins, principally to allow for the lining material to absorb as much water as possible. After that, the pipeline should be carefully inspected for evidence of leakage. The amount of leakage that should be permitted depends on the kind of joints used in the pipeline.

When conducting the test, the water pressure should be raised (based on the elevation at the lowest point in the section of the line under test) to a level such that the test section is subjected to not more than 125 percent of the actual (or design) operating pressure or pipe class, whichever is the greater. The test pressure should be maintained for at least 2 hours. There should be no significant leakage in an all-welded pipeline or one that has been joined with properly installed mechanical couplings. On pipe joined with O-ring rubber gaskets, a small amount of leakage should be allowed. A leakage of 10 gal (37.85 L) per in. (25.4 mm) of diameter per mile per 24 hours is usually permitted. Pinhole leaks that develop in welded joints should not be stopped by peening; instead, they should be marked for proper repair by welding. Such welding frequently can be accomplished without emptying the pipeline, providing pressure can be relieved.

If a section fails to pass the hydrostatic field test, it will be necessary to locate, uncover, and repair or replace any defective pipe, valve, joint, or fitting. The pipeline must then be retested.

REFERENCES

American Association of State Highway and Transportation Officials. 1974. AASHTO Standard T180-74, *Moisture-Density Relations of Soils Using a 10-lb (4.54 kg) Rammer and an 18-in. (457 mm) Drop.* 1974. Washington, D.C.: American Association of State Highway and Transportation Officials.

———. 1981. AASHTO Standard T99-81, *The Moisture-Density Relations of Soils Using a 5.5-lb (2.5 kg) Rammer and a 12-in. (305 mm) Drop.* Washington, D.C.: American Association of State Highway and Transportation Officials.

American Society of Civil Engineers. 1969. *Design and Construction of Sanitary and Storm Sewers.* ASCE Manual No. 37. New York: American Society of Civil Engineers.

American Society for Testing and Materials. 1964. ASTM Standard D1556-64, *Test for Density of Soil in Place by the Sand-Cone Method.* Philadelphia, Pa.: American Society for Testing and Materials.

———. 1966. ASTM Standard D2167-66, *Test for Density of Soil in Place by the Rubber-Balloon Method.* Philadelphia, Pa.: American Society for Testing and Materials.

———. 1969. ASTM Standard D2049-69, *Relative Density of Cohesionless Soils.* Philadelphia, Pa.: American Society for Testing and Materials. (Withdrawn.)

———. 1978a. ASTM Standard D698-78, *Tests for Moisture-Density Relations of Soils and Soil-Aggregate Mixtures, Using 5.5-lb (2.5-kg) Rammer and 12-in. (304.8-mm) Drop.* Philadelphia, Pa.: American Society for Testing and Materials.

———. 1978b. ASTM Standard D1557-78, *Test methods for Moisture-Density Relations of Soils and Soil-Aggregate Mixtures Using 10-lb (4.54-kg) Rammer and 18-in. (457-mm) Drop.* Philadelphia, Pa.: American Society for Testing and Materials.

———. 1983a. ASTM Standard D4253-83, *Test Methods for Maximum Index Density of Soils Using Vibratory Table.* Philadelphia, Pa.: American Society for Testing and Materials.

———. 1983b. ASTM Standard D4254-83, *Test Methods for Minimum Index Density of Soils and Calculation of Relative Density.* Philadelphia, Pa.: American Society for Testing and Materials.

American Water Works Association. ANSI/AWWA Standard C206, *Field Welding of Steel Water Pipe.* Latest edition. Denver, Colo.: American Water Works Association.

Garrett, G.H. 1948. Design of Long-Span Self-Supporting Steel Pipe. *Jour. AWWA,* 40(11):1197.

Price, H.A., and G.H. Garrett. 1943. Field Welding of Steel Water Pipe. *Jour. AWWA,* 35(10):1295.

Reitz, H.M. 1956. Soil Mechanics and Backfilling Practices. *Jour. AWWA,* 48(12):1497.

Sowers, G.F. 1956. Trench Excavation and Backfilling. *Jour. AWWA,* 48(7):854.

Terzaghi, K., and R.B. Peck. 1948. *Soil Mechanics in Engineering Practice.* New York: John Wiley & Sons.

Chapter 13

Supplementary Design Data and Details

The illustrations, tables, and descriptions in this chapter are intended as practical aids to engineers and draftsmen engaged in actual design work. References have been made to these data at various points in preceding chapters where the subjects were discussed in detail. Captions and explanatory matter in this chapter have been kept to a minimum on the assumption that the user is familiar with basic design methods.

LAYOUT OF PIPELINES

The problems involved in surveying and laying out a pipeline are affected by both the size of the line and its location. More detail and care are necessary as the size increases and as a line passes from rural to urban areas.

In general, a plan and profile, as well as certain other details, are necessary for any water pipeline. These should show

1. Horizontal and vertical distances, either directly or by survey station and elevation (if slope distances are given, this fact should be stated).
2. Location of angles or bends, both horizontal and vertical (point of intersection preferred).
3. Degree of bends, degree or radius of curves, tangent distances for curves, or external distances if clearance is required.
4. Points of intersection with pipe centerline for tees, wyes, crosses, or other branches, including direction—right- or left-hand, up or down—or angle of flow, as viewed from the inlet end.
5. Location and lengths of all valves, pumps, or other inserted fittings not supplied by the pipe manufacturer.
6. Location of adjacent or interfering installations or structures.
7. Tie-ins with property lines, curb lines, road or street centerlines, and other pertinent features necessary to define right-of-way and locate pipe centerline clearly.

8. Details or descriptions of all specials, including other data required to supplement AWWA standards (Figure 13-1) (see the "Information Regarding Use of This Standard" section of the relevant standard).
9. Details, dimensions, and class designation or other description of all flanges and mechanical field joints.
10. Location and length of all sections of casing including size and type of casing and position of the carrier within the casing.
11. Any special requirements affecting the manufacture of the pipe or the installation procedures.

Soil conditions should be investigated to determine the protective-coating requirements, excavation procedures, permissible foundation pressures, or design of anchor or thrust blocks. The location of the water table may affect design and installation. Soil borings are desirable for all installations, especially for large water lines.

Pipe may be identified by a consecutive piece number system, or by using another system in accordance with the common practice of the pipe manufacturer or as established by mutual agreement between the purchaser and the manufacturer. A requirement for consecutive numbering and installation of straight pieces of uniformly cut length is uneconomical if the pieces are interchangeable in the line. Special sections may best be marked to show their survey station number. (NOTE: General marking requirements are provided in the relevant AWWA standards.)

A pipe-laying schedule is a valuable tool for the manufacture and installation of a pipeline system. Such a schedule is shown in Table 13-1. A schedule should show clearly and completely the essential details for each pipe piece. In addition, the schedule should show the necessary data for the proper assembly sequence and for spotting of pipe specials and sections.

CALCULATION OF ANGLE OF FABRICATED PIPE BEND

In many pipeline jobs, combining a plan and profile deflection in one fitting is necessary. The relationship between θ (the angle of the fabricated pipe bend), α and β (deflection angles in the plan and profile, respectively), and γ (the slope angle of one leg of the bend) must be known. Although approximate angles are often used, unless the exact relationship is known, it is impossible to know how accurate the approximations are.

A simple relationship is illustrated in Figure 13-2. If β increases the slope angle relative to γ, it bears a plus value. For the general case:

$$\cos\theta = \sin\gamma\sin(\gamma+\beta) + \cos\gamma\cos(\gamma+\beta)\cos\alpha$$

For the special case when γ equals zero:

$$\cos\theta = \cos\beta\cos\alpha$$

REINFORCEMENT OF FITTINGS

Tees, crosses, laterals, wyes, headers, or other fittings that provide means of dividing or uniting flow in pipelines have less resistance to internal pressure than straight pipe of the same wall thickness and size. This is because a portion of the side wall of the pipe in these fittings is removed to allow for the branching pipe. Also, there are longitudinal stresses in the throat of unrestrained elbows as a result of distortion or unbalanced hydrostatic pressure.

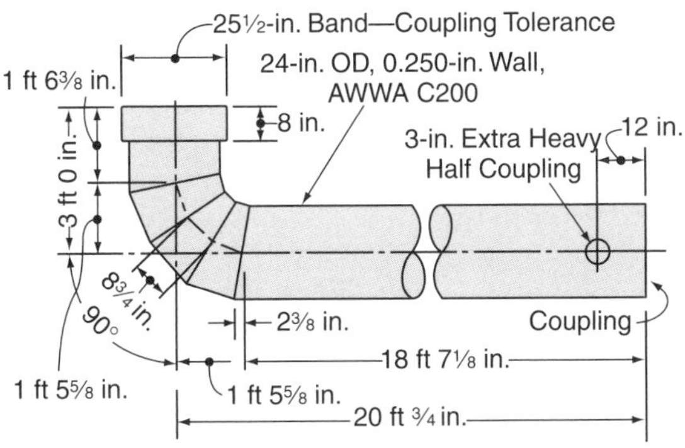

NOTE: Coat inside and outside per relevant AWWA standard.

Figure 13-1 Example of adequately detailed pipe special

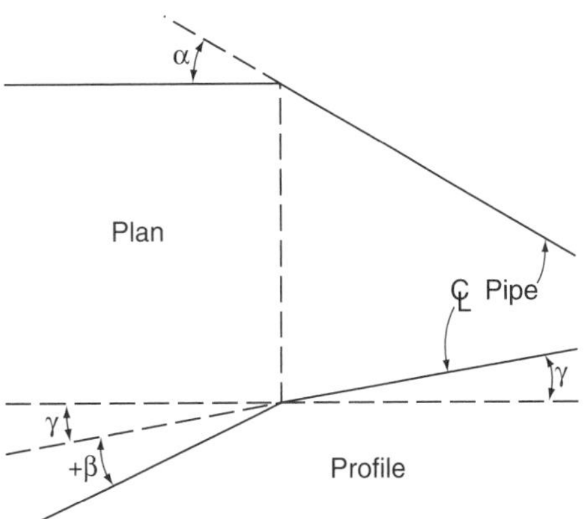

Figure 13-2 Plan and profile of bend in pipe on centerline of pipe

For ordinary waterworks installations, the wall thickness of the pipe commonly used is much greater than pressure conditions require. Consequently, the lowered safety factor of fittings having the same wall thickness as the straight pipe still leaves adequate strength in most cases, and reinforcing may be unnecessary. If the pipe is operating at or near maximum design pressure, however, the strength of the fittings should be investigated, and the proper reinforcement or extra wall thickness provided.

Fittings may be reinforced in various ways for resistance to internal pressure. Typical fitting reinforcements are collars, wrappers, and crotch plates. The design stress in the reinforcement should not be greater than the hoop stress used in the design of the pipe.

The type of reinforcement* can be determined by the magnitude of the pressure-diameter value (PDV) and the ratio of the branch diameter to the main pipe diameter (d/D). The PDV is calculated:

$$PDV = \frac{Pd^2}{D\sin^2\Delta} \tag{13-1}$$

Where:

P = design pressure, in psi
d = branch outside diameter, in in.
D = main pipe outside diameter, in in.
Δ = branch diameter angle of deflection

For PDV values greater than 6,000, the outlet reinforcement should consist of a crotch plate designed in accordance with the method described in the Crotch-Plate (Wye-Branch) Design section. For PDV values less than 6,000, the outlet reinforcement may be either a wrapper or collar, depending on the ratio of the outlet diameter to the main pipe diameter d/D. For a d/D ratio greater than 0.7, a wrapper plate should be used; for a d/D ratio less than 0.7, either a collar or a wrapper plate may be used. The ratio d/D does not include the sin Δ as in the PDV determination because the circumferential dimensions are the controlling factor. Wrappers may be substituted for collars, and crotch plates may be substituted for wrappers or collars.

*Reinforcement for certain crosses, wyes, or double laterals may require additional analyses beyond the criteria discussed in this standard.

Table 13-1 Example of pipe-laying schedule

24-in. (600-mm) OD × 0.250-in. (6.4-mm) wall pipe with mechanical coupling ends unless noted. ¼-in. (6.4-mm) joint allowance for bolted sleeve-type couplings. Stationing is horizontal distance along base or survey line. Bolted sleeve-type joints, for 24-in. (600-mm) OD pipe unless noted, with stops in pipe according to AWWA C200 and coated and wrapped according to AWWA C203.

Station	Length Corr.	No. Sleeve-Type Couplings	PC. MK.*	No. Req.	Description	Fittings (Direction of Stationing) ←
330+53.2					Begin—Settling Basin at Filter Plant	
			1	1	50' Length With Flange	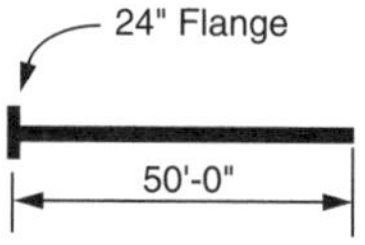 24" Flange; 50'-0"
329+98.9		1	2	1	2 Pc Ell (vertical)	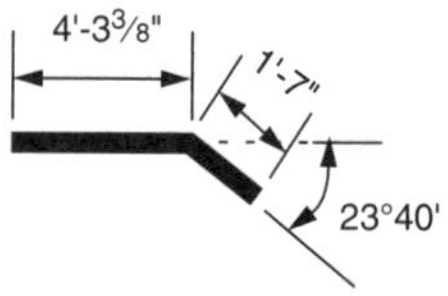 4'-3⅜"; 1'-7"; 23°40'
	Add 1'-9¾" for Slope This Dist.	1	3	1	Flanged Piece—3" Conn.	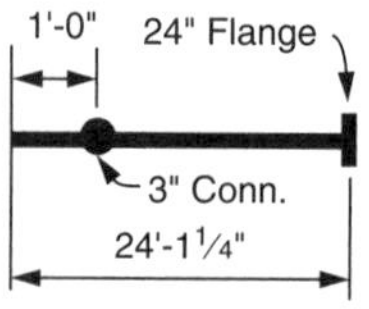 1'-0"; 24" Flange; 3" Conn.; 24'-1¼"
329+75					Begin River Crossing	
323+25					Pipe for Crossing in Place	
					End River Crossing	
	Add 1'-8" for Slope This Dist.		4	1	5 Pc Ell (horizontal) with 3" Conn.	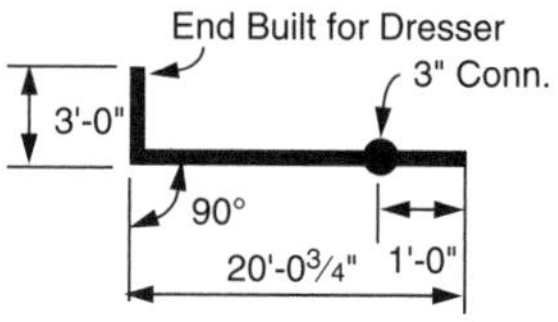 End Built for Dresser; 3" Conn.; 3'-0"; 90°; 20'-0¾"; 1'-0"
		1	5	1	50' Length	
322+55		1	6	1	2 Pc Ell (vertical)	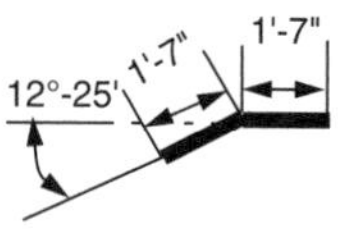 12°-25'; 1'-7"; 1'-7"

* All plain 50' lengths marked 5.

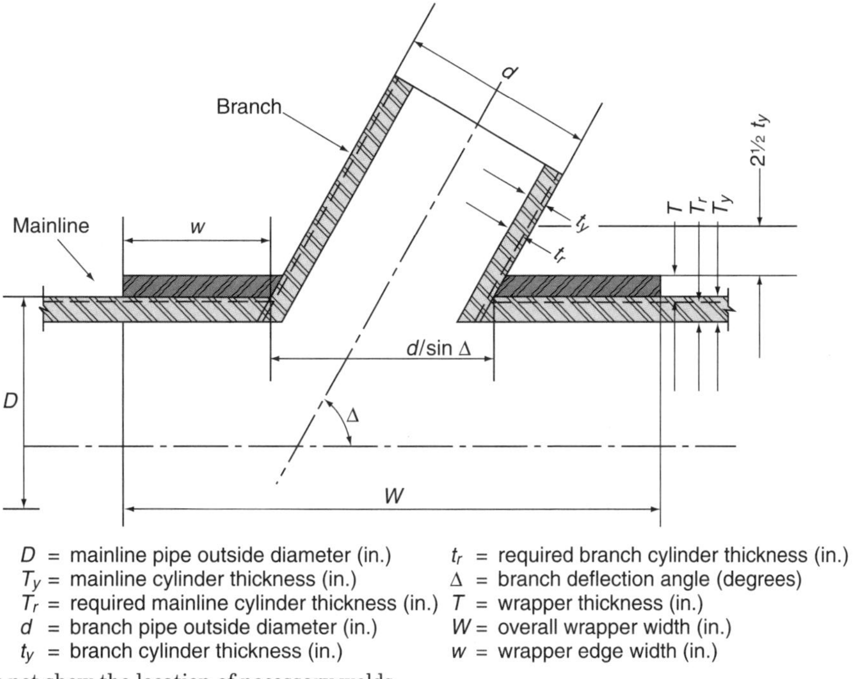

NOTE: Figure does not show the location of necessary welds.

Figure 13-3 Reinforcement of openings in welded steel pipe

Table 13-2 Recommended reinforcement type*

PDV	d/D	*M* Factor	Reinforcement Type
>6,000	all	—	Crotch Plate
4,000–6,000	>0.7	0.00025 PDV	Wrapper
<4,000	>0.7	1.0	Wrapper
4,000–6,000	≤0.7	0.00025 PDV	Collar
<4,000	≤0.7	1.0	Collar

* These reinforcements are for resistance to internal pressure. They should be checked for their ability to resist external loads.

Wrappers and collars are designed similar to the method described in Sec. VIII of the ASME Boiler and Pressure Vessel Code. This code provides that the cross-sectional area of the removed steel at the branch is replaced in the form of a wrapper or collar. In addition to the ASME requirements, when the PDV ranges between 4,000 and 6,000, the cross-sectional area of the replaced steel should be multiplied by an *M* factor of 0.00025 times the PDV. Figure 13-3 shows the reinforcement of wrapper and collar openings for welded steel pipe, and Table 13-2 lists a summary of recommended reinforcement types. The minimum steel thickness for wrappers and collars shall be 12 gauge (0.1046 in. [2.65 mm]).

In determining the required steel replacement, credit should be given to any thickness of material in the main-line pipe in excess of that required for internal pressure, and to the area of the material in the wall of the branch outlet to the allowable distance from the collar or wrapper ($2.5t_y$). Weld areas should not be considered in the design. The overall width of the collar or wrapper should not be less than $1.67d/\sin\Delta$ and should not exceed $2.0d/\sin\Delta$. This width range produces a minimum edge width of $0.33d/\sin\Delta$. Collar edge widths in the circumferential direction should not be less than the longitudinal edge width.

Collars may be oval in shape, or they may be rectangular with rounded corners. The radii at corners should not be less than 4 in. (100 mm) or 20 times the collar thickness, whichever is greater (except for collars with a length or width less than 8 in. [200 mm]). Longitudinal seams should be placed at 90° or more from the center of the removed section.

On the branch outlet centerline, the limit line of the branch reinforcement occurs at a distance 2.5 times the thickness of the branch from the surface of the main pipe run or from the top of the collar or wrapper reinforcement.

In Figure 13-3, the area $T_y(d - 2t_y)/\sin\Delta$ represents the section of the mainline pipe cylinder removed by the opening for the branch. The hoop tension caused by pressure within the pipe that would be taken by the removed section were it present must be carried by the total areas represented by $2wT$ and $5t_y\,(t_y - t_r)$ or $2.5t_y\,(t_y - t_r)$ on each side of outlet.

COLLAR PLATE DESIGN

Criteria-data example—24-in. × 8-in. tee

Main-pipe size (nominal diameter)		24 in.
Main-pipe cylinder OD	D	25¾ in.
Main-pipe cylinder thickness	T_y	0.135 in.
Branch-outlet size (nominal diameter)		8 in.
Branch-outlet cylinder OD	d	8⅝ in.
Branch-outlet thickness	t_y	¼ in.
Deflection angle	Δ	90°
Design Pressure	P	150 psi
Reinforcement steel allowable stress	f_s	16,500 psi

(The allowable stress, based on a design stress resulting from working pressure, shall not exceed ½ the minimum yield of the steel used for the pipe cylinder or in the reinforcement, whichever is less.)

Reinforcement Type

$$PDV = \frac{Pd^2}{D\sin^2\Delta} = \frac{(150)(8.625)^2}{25.72\sin^2 90°} = 433$$

$$\frac{d}{D} = \frac{8.625}{25.75} = 0.335$$

Therefore, for PDV ≤ 4,000 and $d/D \le 0.7$, *use collar* unless wrapper is provided.

Multiplier (*M*-factor)

For PDV <4,000, $M = 1.0$

Collar Design

Theoretical cylinder thicknesses.

Main pipe (T_r)

$$T_r = \frac{PD}{2f_s} = \frac{(50)(25.75)}{2(16{,}500)} = 0.117 \text{ in.}$$

Branch outlet (t_r)

$$t_r = \frac{PD}{2f_s} = \frac{(150)(8.625)}{2(16{,}500)} = 0.039 \text{ in.}$$

Theoretical reinforcement area.
Theoretical reinforcement area = A_r

$$A_r = M\left[T_r\left(\frac{d - 2t_y}{\sin\Delta}\right)\right]$$
$$= 1.0\left[0.117\left(\frac{8.625 - 2(0.25)}{\sin 90°}\right)\right]$$
$$= 0.951 \text{ in.}^2$$

Area available as excess T_y and allowable outlet area.
Area available = A_a

$$A_a = \frac{(d - 2t_y)}{\sin\Delta}(T_y - T_r) + 5t_y(t_y - t_r)$$
$$= \frac{8.625 - 2(0.25)}{\sin 90°}(0.135 - 0.117) + (5 \times 0.25)(0.25 - 0.039)$$
$$= 0.410 \text{ in.}^2$$

Reinforcement area.
Reinforcement area = A_w

$$A_w = A_r - A_a$$
$$= 0.951 - 0.410$$
$$= 0.541 \text{ in.}^2$$

Minimum reinforcement thickness.
Minimum reinforcement thickness = T

$$w = \frac{d}{2\sin\Delta} = \frac{8.625}{2\sin 90°} = 4.313 \text{ in.}$$

$$T = \frac{A_w}{2w} = \frac{0.541}{2(4.313)} = 0.063 \text{ in.}$$

Therefore, use not less than 12-gauge (0.105-in.) steel.

$T = 0.105$ in.

Trial reinforcement width.

$$w = \frac{A_w}{2T_{min}} = \frac{0.541}{2(0.105)} = 2.576 \text{ in.}$$

Minimum allowable width.

$$w_{min} = \frac{d}{3\sin\Delta} = \frac{8.625}{3\sin 90°} = 2.875 \text{ in.}$$

2.875 in. > 2.576 in.

Use: $w = 2.875$ in.

Overall reinforcement width.

$$W = 2w + \frac{d}{\sin\Delta} = 2(2.875) + \frac{8.625}{\sin 90°} = 14.375 \text{ in.}$$

Use: $T = 0.105$ in.

$W = 14^3/_8$ in.

WRAPPER-PLATE DESIGN

Criteria-data example—60-in. × 48-in. lateral

Main-pipe size (nominal diameter)	60 in.	
Main-pipe cylinder OD	D	61⅞ in.
Main-pipe cylinder thickness	T_y	3⁄16 in.
Branch-outlet size (nominal diameter)	48 in.	
Branch-outlet cylinder OD	d	49⅞ in.
Branch-outlet thickness	t_y	3⁄16 in.
Deflection angle	Δ	75°
Design pressure	P	100 psi
Reinforcement steel allowable stress	f_s	16,500 psi

(The allowable stress psi shall not exceed ½ the minimum yield of the steel used for the pipe cylinder or in the reinforcement, whichever is less.)

Reinforcement Type

$$PDV = \frac{Pd^2}{D\sin^2\Delta} = \frac{(100)(49.875)^2}{(61.875)\sin^2 75°} = 4{,}309$$

$$\frac{d}{D} = \frac{49.875}{61.875} = 0.81$$

Therefore, for PDV 6,000 and d/D >0.7, use wrapper.

Multiplier (M-factor)

For 4,000 <PDV <6,000

$M = 0.00025\,PDV = (0.00025)(4{,}309) = 1.077$

Therefore, use $M = 1.08$.

Wrapper Design

Theoretical cylinder thicknesses.

Main pipe (T_r)

$$T_r = \frac{PD}{2f_s} = \frac{(100)(61.875)}{2(16{,}500)} = 0.188 \text{ in.}$$

Branch outlet (t_r)

$$t_r = \frac{Pd}{2f_s} = \frac{(100)(49.875)}{2(16{,}500)} = 0.151 \text{ in.}$$

Theoretical reinforcement area.

Theoretical reinforcement area $= A_r$

$$A_r = M\left[T_r\left(\frac{d-2t_y}{\sin\Delta}\right)\right]$$

$$= (1.08)\left[0.188\left(\frac{49.875-2(0.188)}{\sin 75°}\right)\right]$$

$$= 10.405 \text{ in.}^2$$

Area available as excess T_y and allowable outlet area.
Area available = A_a

$$A_a = \frac{(d-2t_y)}{\sin\Delta}(T_y - T_r) + 5t_y(t_y - t_r)$$

$$= \frac{49.875 - 2(0.188)}{\sin 75°}(0.188 - 0.188) + (5 \times 0.188)(0.188 - 1.151)$$

$$= 0.035 \text{ in.}^2$$

Reinforcement area.
Reinforcement area = A_w

$$A_w = A_r - A_a$$

$$= 10.405 - 0.035 = 10.370 \text{ in.}^2$$

Minimum reinforcement thickness.
Minimum reinforcement thickness = T

$$w = \frac{d}{2\sin\Delta} = \frac{49.875}{2\sin 75°} = 25.817 \text{ in.}$$

$$T = \frac{A_w}{2w} = \frac{10.370}{2(25.817)} = 0.201 \text{ in.}$$

Round up to the nearest standard thickness but not less than 12 gauge (0.105 in.).

$T = 1/4$ in. (0.25 in.)

Minimum reinforcement width.

$$w = \frac{A_w}{2T} = \frac{10.370}{2(0.25)} = 20.740 \text{ in.}$$

Minimum allowable width.

$$w_{min} = \frac{d}{3\sin\Delta} = \frac{49.875}{3\sin 75°} = 17.211 \text{ in.}$$

17.211 in. < 20.740 in.

Use: w = 20.740 in.

Overall reinforcement width.

$$W = 2w + \frac{d}{\sin\Delta} = 2(20.740) + \frac{49.875}{\sin 75°} = 93.114 \text{ in.}$$

Use: $T = 1/4$ in.

$W = 93\ 1/8$ in.

CROTCH-PLATE (WYE-BRANCH) DESIGN

When the PDV exceeds 6,000, crotch-plate reinforcement should be used. Several types of plate reinforcement are illustrated in Figures 13-4 through 13-6. The following section on nomograph use was taken from a published study on crotch-plate (wye-branch) design at Los Angeles (Swanson et al. 1955).

A single curved plate serves as reinforcement for each branch of this 96-in. × 66-in., 90° included angle wye.

Figure 13-4 One-plate wye

This 15-ft × 15-ft × 15-ft, 90° wye has two crotch plates and one back plate.

Figure 13-5 Three-plate wye

This 126-in. × 126-in., 45° wye section has two plates.

Figure 13-6 Two-plate wye

NOMOGRAPH USE IN WYE-BRANCH DESIGN

The nomograph design, based on design working pressure plus surge allowance, includes a safety factor that will keep stresses well below the yield point of steel. The minimum yield strength of the steel used here is 30,000 psi. The design pressure used in the nomograph was based on maximum pressure in order to approximate an allowable stress of 20,000 psi. Stress values are generated from the familiar Mc/I and T/A relations. Both I and A vary linearly with thickness for the rib shapes involved. Therefore, if a yield point other than 30,000 psi is used, the thickness can be linearly corrected for the new value.

Step 1. Lay a straightedge across the nomograph (Figure 13-7) through the appropriate points on the pipe diameter (see step 2b) and internal-pressure scales; read off the depth of plate from its scale. This reading is the crotch depth for 1-in. (25.4-mm) thick plate for a two-plate, 90°, wye-branch pipe. (Thinner plates may be used, provided localized buckling is addressed).

Step 2a. If the wye branch deflection angle is other than 90°, use the N-factor curve (Figure 13-8) to get the factors that, when multiplied by the depth of plate found in step 1, will give the wye depth d_w and the base depth d_b for the new wye branch.

Step 2b. If the wye branch has unequal-diameter pipe, the larger diameter pipe will have been used in steps 1 and 2a, and these results should be multiplied by the Q factors found on the single-plate stiffener curves (Figure 13-9) to give d'_w and d'_b. These factors vary with the ratio of the radius of the small pipe to the radius of the large pipe.

Step 3. If the wye depth d_w found so far is greater than 30 times the thickness of the plate (1 in.), then d_w and d_b should be converted to conform to a greater thickness t using the general equation:

$$d = d_1\left(\frac{t_1}{t}\right)^{\left(0.917 - \frac{\Delta}{360}\right)} \tag{13-2}$$

Where:

d_1 = existing depth of plate
t_1 = existing thickness of plate
d = new depth of plate
t = new thickness of plate selected
Δ = deflection angle of the wye branch

Step 4. To find the top depth d_t or d'_t, use Figure 13-10, in which d_t or d'_t is plotted against d_b or d'_b. This dimension gives the top and bottom depths of the plate at 90° from the crotch depths.

Step 5. The interior curves follow the cut of the pipe, but the outside crotch radius in both crotches should equal d_t plus the radius of the pipe, or in the single-plate design, d'_t plus the radius of the smaller pipe. Tangents connected between these curves complete the outer shape.

The important depths of the reinforcement plates, d_w, d_b, and d_t (Figure 13-11), can be found from the nomograph. If a curved exterior is desired, a radius equal to the inside pipe radius plus d_t can be used, both for the outside curve of the wye section and for the outside curve of the base section.

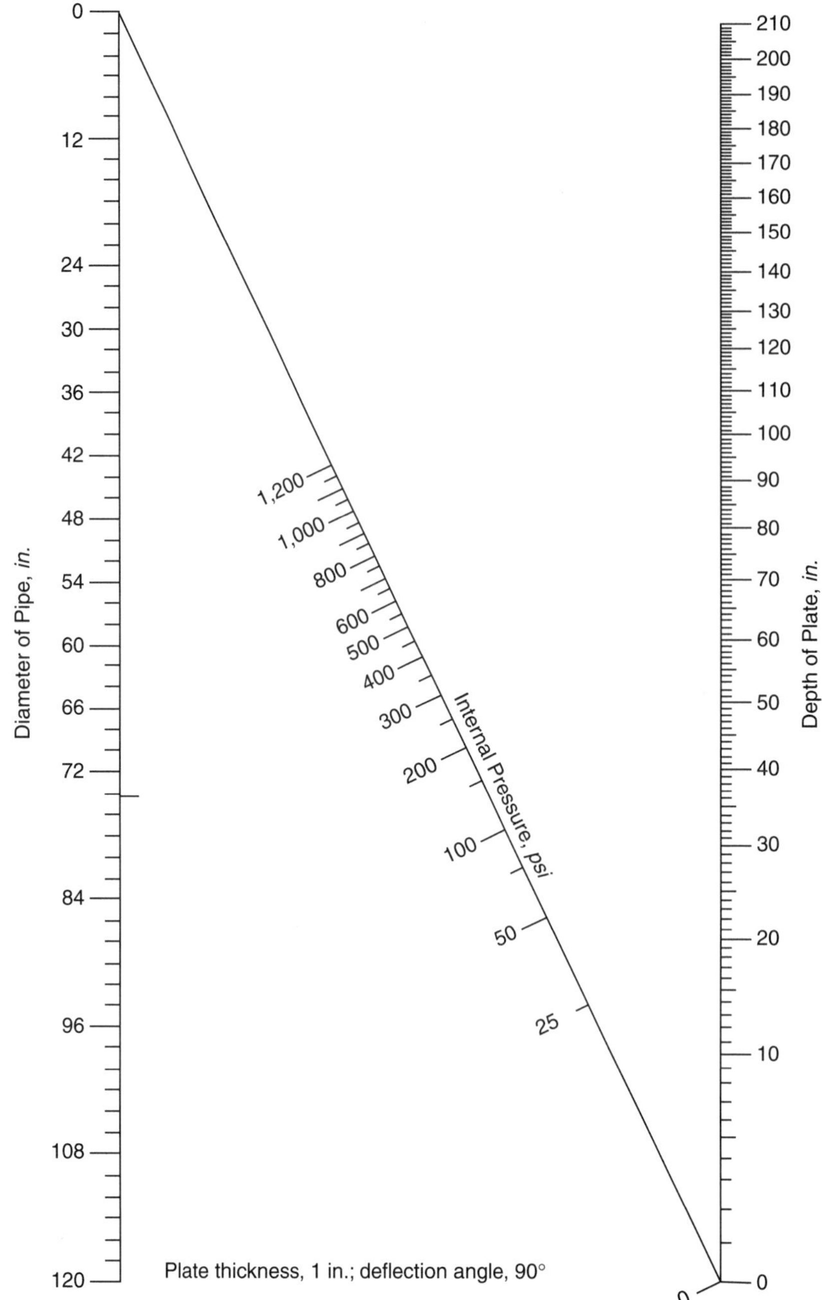

Source: Swanson, H.S., et al., Design of Wye Branches for Steel Pipe. Jour. AWWA, *47:6:581 (June 1955).*

Figure 13-7 Nomograph for selecting reinforcement plate depths of equal-diameter pipes

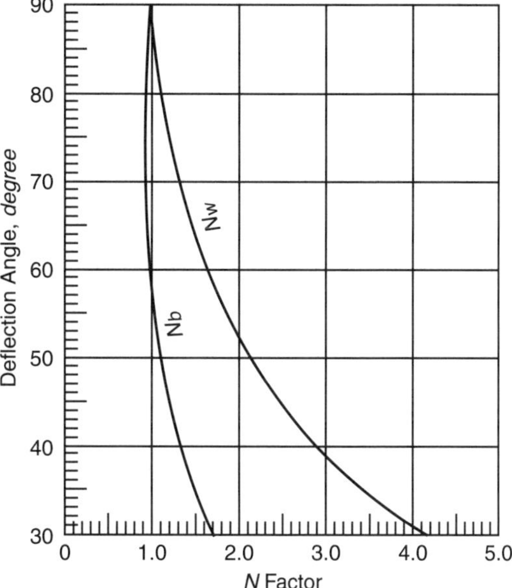

Source: Swanson, H.S., et al., Design of Wye Branches for Steel Pipe. Jour. AWWA, *47:6:581 (June 1955).*

NOTE: For wyes with deflection angles from 30° to 90°, the N factors obtained from the above curves are applied to the plate depth d, found from the nomograph (Figure 13-7), in accordance with the equations.

$$d_w = N_w d \text{ and } d_b = N_b d$$

Figure 13-8 *N* factor curves

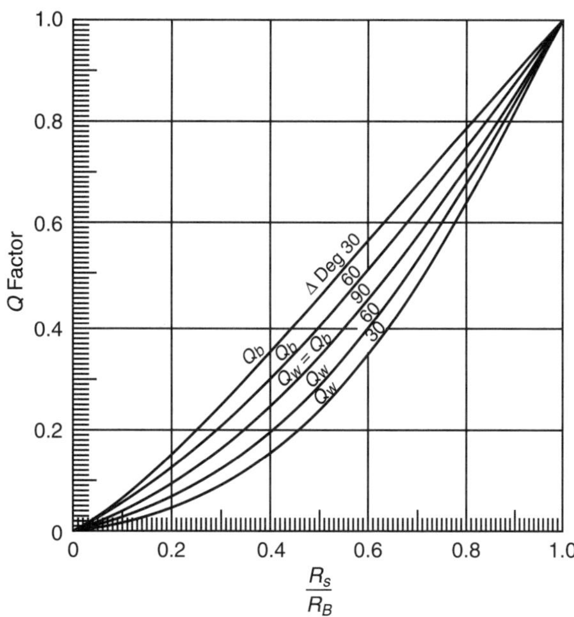

Source: Swanson, H.S., et al. Design of Wye Branches for Steel Pipe. Jour. AWWA, *47:6:581 (June 1955).*

NOTE: For pipes of unequal diameter, find d_w and d_b for the larger-diameter pipe (from Figures 13-7 and 13-8); then $Q_w d_w = d'_w$ crotch depth of single plate stiffener; and $Q_b d_b = d'_b$, base depth of single-plate stiffener.

Figure 13-9 *Q* factor curves

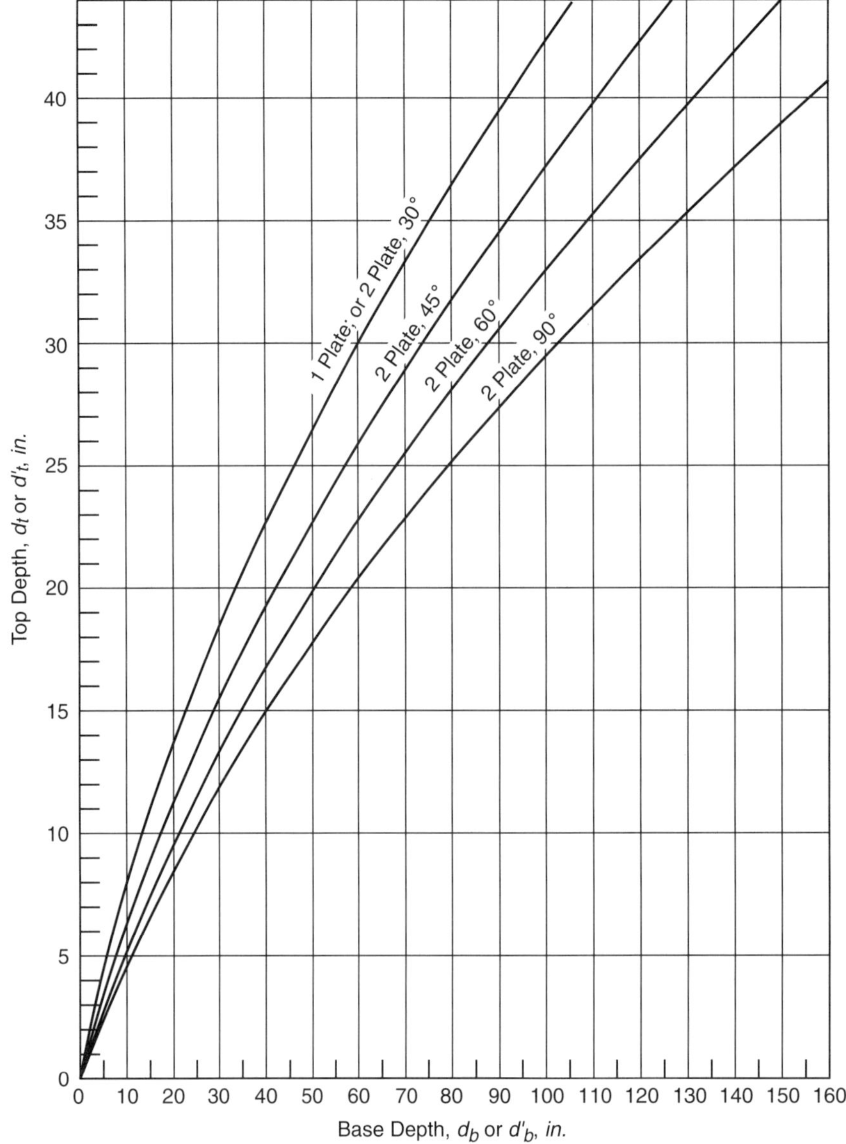

Source: Swanson, H.S., et al., Design of Wye Branches for Steel Pipe. Jour. AWWA, *47:6:581 (June 1955).*

NOTE: d'_t and d'_b are one-plate design dimensions; d_t and d_b are two-plate design dimensions.

Figure 13-10 Selection of top depth

Example 1—One-plate design.

R_B = 30 in.
R_s = 21 in.
Δ = 45°

Working pressure, 230 psi
Design pressure, 230 (1.5) = 350 psi

Step 1. With the larger pipe diameter 60 in. and the design pressure 350 psi, read the critical plate depth d from the nomograph (t = 1 in., Δ= 90°):

d = 50 in.

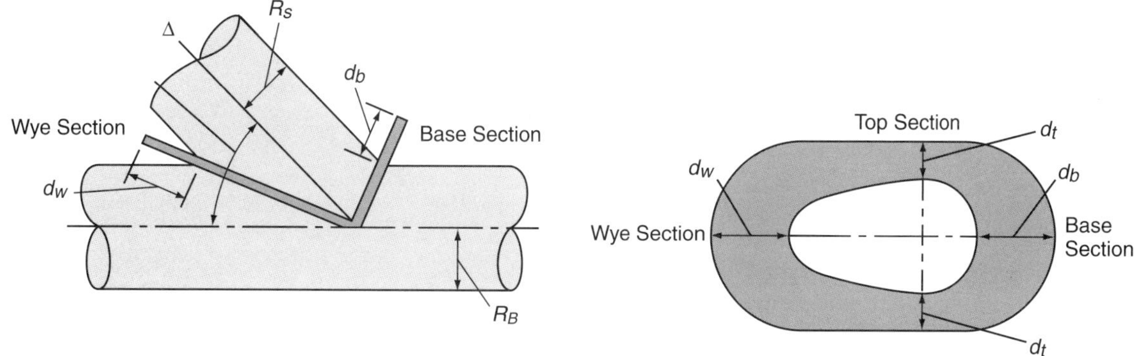

Source: Swanson, H.S., et al. Design of Wye Branches for Steel Pipe. Jour. AWWA, *47:6:581 (June 1955).*

Figure 13-11 Wye branch plan and layout

Step 2. Using the deflection angle 45°, find the factors on the N factor curve that will convert the depth found in step 1 to apply to a 45° wye branch (t = 1 in.):

$d_w = N_w d = 2.45(50) = 122$ in.
$d_b = N_b d = 1.23(50) = 61.5$ in.

Step 3. With the ratio of the smaller pipe radius divided by the larger pipe radius $(R_s/R_B) = (21/30) = 0.70$ and the deflection angle ($\Delta = 45°$), use Figure 13-9 to find the Q factors that give the crotch depths for a single-plate pipe wye stiffener (t = 1 in.):

$Q_w = 0.52$
$Q_b = 0.66$
$d'_w = 0.52(122) = 63.4$ in.
$d'_b = 0.66(61.5) = 40.5$ in.

Step 4. Because the depth d'_w is greater than 30 times the thickness t, the conversion equation should be used:

Try a thickness of 1½ in.:

$$d = d_1\left(\frac{t_1}{t}\right)^{\left(0.917 - \frac{\Delta}{360}\right)}$$

$$d = d_1\left(\frac{1}{1.5}\right)^{\left(0.917 - \frac{45}{360}\right)} = d_1\left(\frac{2}{3}\right)^{0.792}$$

$$= d_1(0.725)$$

$$d'_w = 63.4(0.725) = 46 \text{ in.}$$

$$d'_b = 40.5(0.725) = 29 \text{ in.}$$

Step 5. Find the top depth d'_t from the curve for one-plate design in Figure 13-10:
For $d'_b = 29$ in., $d'_t = 18$ in.

Final results:

Thickness of reinforcing plate, $t = 1\frac{1}{2}$ in.
Depth of plate at acute crotch, $d'_w = 46$ in.
Depth of plate at obtuse crotch, $d'_b = 29$ in.
Depth of plate at top and bottom, $d'_t = 18$ in.

Outside radius of plate at both crotches equals the top depth plus the inside radius of the small pipe $d'_t + R_s = 18 + 21 = 39$ in.

Example 2—Two-plate design.

$R_B = R_s = 36$ in.
$\Delta = 53°$

Working pressure, 150 psi
Design pressure, 150 (1.5) = 225 psi

Step 1. With a pipe diameter of 72 in. and a pressure of 225 psi, read the critical depth of plate from the nomograph ($t = 1$ in., $\Delta = 90°$):

$d = 49$ in.

Step 2. From the *N*-factor curve, find the two factors at $\Delta = 53°$; then, at $t = 1$ in.:

$d_w = 1.97(49) = 96.5$ in.
$d_b = 1.09(49) = 53.4$ in.

Step 3. Because d_w is greater than 30 times the thickness of the plate, use $t = 2$ in. in the conversion equation:

$$d = d_1\left(\frac{t_1}{t}\right)^{\left(0.917 - \frac{\Delta}{360}\right)} = d_1\left(\frac{1}{2}\right)^{0.770}$$

$$= d_1(0.586)$$

$$d_w = 96.5(0.586) = 57 \text{ in.}$$

$$d_b = 53.4(0.586) = 31 \text{ in.}$$

Step 4. Read the top depth d_1 from the two-plate design curve in Figure 13-10:

$d_t = 15$

Final results:

Thickness of reinforcing plate, $t = 2$ in.
Depth of plate at acute crotch, $d_w = 57$ in.
Depth of plate at obtuse crotch, $d_b = 31$ in.
Depth of plate at top and bottom, $d_t = 15$ in.
Outside radius of plate at both crotches, 51 in.

Three-Plate Design

The preceding nomograph section described the design of one- and two-plate wye branches without addressing a three-plate design because of its similarity to the two-plate design. The function of the third plate is to act like a clamp by holding down the deflection of the two main plates. In doing so, it accepts part of the stresses of the other plates and permits a smaller design. This decrease in the depths of the two main plates is small enough to make it practical simply to add a third plate to a two-plate design. The additional plate should be considered a means of reducing the deflection at the junction of the plates. The two factors that dictate the use of a third plate are diameter of pipe and internal pressure. When the diameter is greater than 60 in. (1,500 mm) ID and the internal pressure is greater than 300 psi (2,068 kPa), a ring plate can be advantageous. If either of these factors is below the limit, the designer should be consider using a third plate.

If a third plate is desired as an addition to the two-plate design, its size should be dictated by the top depth d_t. Because the other two plates are flush with the inside surface of the pipe, however, the shell plate thickness plus clearance should be subtracted from the top depth. This dimension should be constant throughout, and the plate should be placed at right angles to the axis of the pipe, giving it a half-ring shape. Its thickness should equal the smaller of the main plates.

The third plate should be welded to the other reinforcement plates only at the top and bottom, free from the pipe shell, so that none of the shell stresses transfer to the ring plate.

THRUST RESTRAINT

When a water transmission or distribution pipeline is under internal pressure, unbalanced thrust forces develop at changes of size and direction in the pipeline. This applies to bends, tees, reducers, offsets, bulkheads, etc. (Figure 13-12). The magnitude of these thrust forces for tees and bulkheads is equal to the product of the internal pressure and the cross-sectional area of the pipe, or

$$T = PA \tag{13-3}$$

Where:

T = the thrust force (lb)

P = maximum internal pressure including any anticipated surge pressure or static test pressure if greater than operating pressure (psi)

A = cross-sectional area of the pipe (in.^{2})

= 0.7854 D^2, where D is the outside diameter of the steel cylinder (in.)

At elbows or bends, the resultant thrust force T (Figure 13-13) is

$$T = 2PA\sin\frac{\Delta}{2} \tag{13-4}$$

Where:

Δ = the deflection angle of the elbow or bend

There are also small unbalanced forces at bends caused by the velocity of water flow within the pipeline. In general, this velocity is so low in transmission or distribution systems that its effect on thrust is insignificant, and thrust forces caused by velocity can be neglected.

Methods to restrain the thrust forces may be provided by an external reaction from a concrete thrust block or piles, or by the development of friction between the pipe and the soil through restrained or harnessed joints, or by a combination of these two methods.

Thrust Blocks

Thrust blocks increase the ability of fittings to resist movement by increasing the bearing area. Typical thrust of a horizontal bend is shown in Figure 13-14.

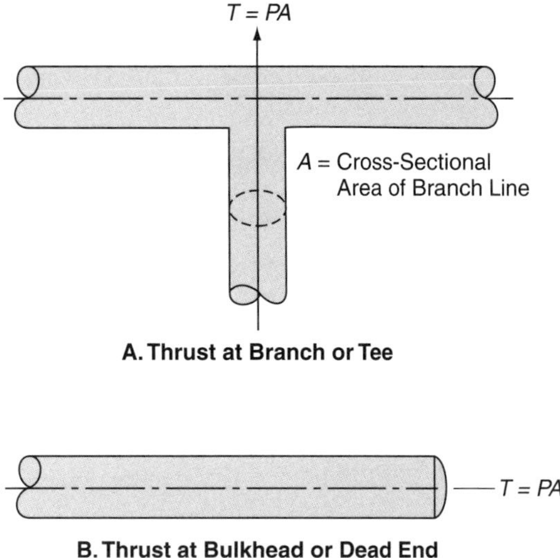

Figure 13-12 Thrust at branch or tee (top), thrust at bulkhead or dead end (bottom)

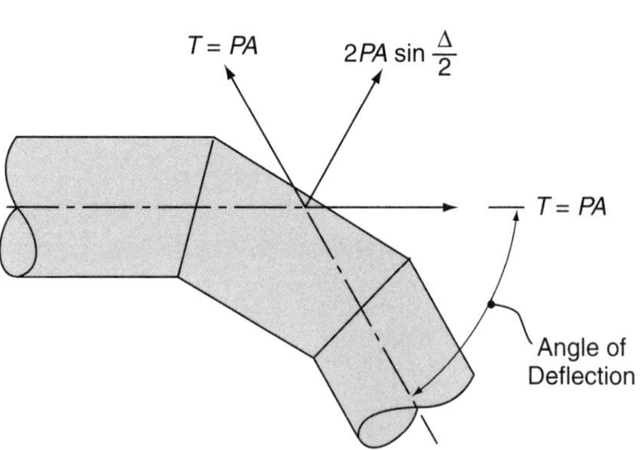

Figure 13-13 Resultant thrust at pipe elbow

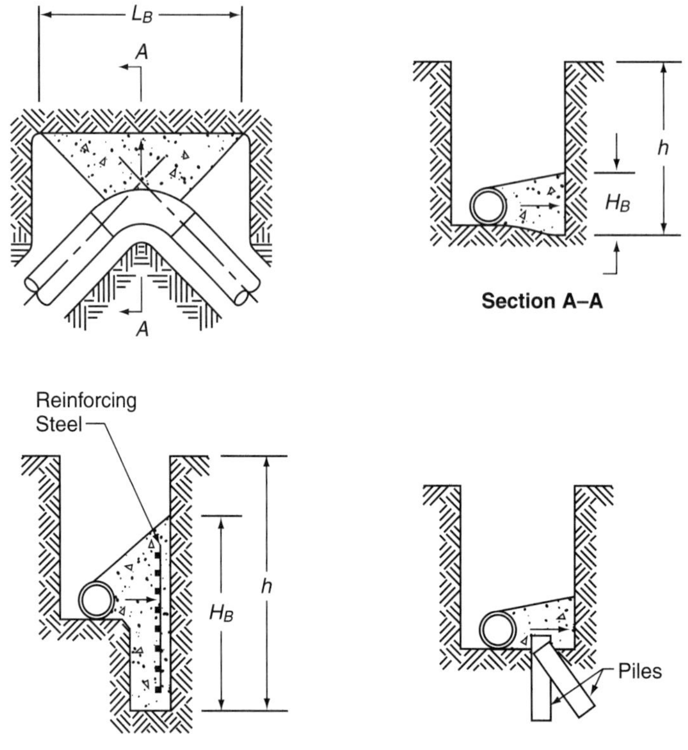

Figure 13-14 Typical thrust blocking of a horizontal bend

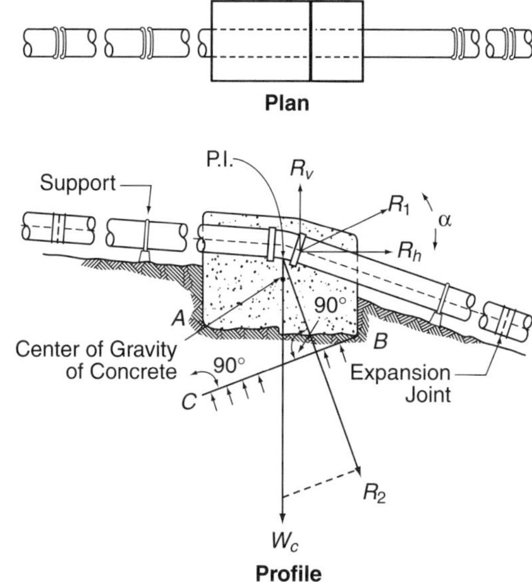

Notation:

R_1 = resultant of anchor forces shown

R_v = vertical component of $R_1 = R_1 \sin \alpha$

R_h = horizontal component of $R_1 = R_1 \cos \alpha$

R_2 = resultant of all forces (including weight of anchor)

W_t = total weight required to balance R_1

W_b = beam reaction at anchor

W_c = weight of anchor

= friction coefficient between anchor and foundation

Assumptions:

It is assumed that the anchor rests on a level foundation and resists horizontal movement by friction on the foundation. Any possible shearing resistance of surface irregularities of the foundation is neglected except where anchor is keyed into rock.

Determination of size:

The design of the anchor must satisfy the following requirements:

1. The resultant R_2 of all forces acting at the anchor must cut the base within the middle third if the base is flat or within the middle third of *BC* if base is stepped. If this requirement is met the entire foundation will be in compression.
2. The anchor must be safe against sliding.
3. The maximum soil pressure must not exceed allowable limits.
4. The base of the anchor must be below the frost line to prevent heaving.

Friction coefficients:

Condition	Friction Coefficient (μ)	Factor of Safety Against Sliding
Concrete on concrete	0.75	1
Masonry on masonry	0.75	1
Concrete on rock	0.75	1
Concrete or masonry on gravel	0.50	1
Concrete or masonry on sand	0.40	1
Concrete or masonry on clay	0.30	1

The factor of safety against sliding is increased by the mechanical resistance of surface irregularities of the foundation, but this does not permit any change in the value of μ.

Depth of frost penetration:

The minimum depth of foundation shall be determined from the minimum average one-day mean temperature as follows:

Minimum Temperature	Minimum Depth of Foundation
Over 32°F	3'-0"
32°F to +10°F	3'-6"
+10°F to –10°F	4'-0"
–10°F to –20°F	4'-6"
–20°F to –30°F	5'-0"
–30°F to –40°F	5'-6"
Under –40°F	6'-0"

Maximum bearing values of foundation: (lb/ft^2)

Soft clay, loam	2,000
Dry clay	4,000
Dry sand	4,000
Coarse sand and gravel	8,000
Shale rock	12,000
Hard rock	30,000

Determine and calculate terms in the following order:

$R_v = R_1 \sin \alpha$

$R_h = R_1 \cos \alpha$

$$W_t = \frac{R_v + R_h}{\mu}$$

W_b = beam reaction

$W_c = W_t - W_b$

The above calculations should be made for expanding and contracting conditions of the pipe when full and empty. Use largest anchor calculated. If the pipe is to be pressure tested after installation, the anchor must be sufficiently large to resist the maximum forces occurring during the test.

Figure 13-15 Thrust blocking of vertical bends

Calculation of Size

Thrust block size can be calculated based on the bearing capacity of the soil:

$$\text{Area of Block} = L_b \times H_b = \frac{T}{\sigma} \tag{13-5}$$

Where:

$L_b \times H_b$ = area of bearing surface of thrust block, ft^2 (m^2)

T = thrust force, lbf (N)

σ = area of bearing surface of thrust block, ft^2 (m^2)

If it is impractical to design the block for the thrust force to pass through the geometric center of the soil-bearing area, then the design should be evaluated for stability.

After calculating the thrust block size on the safe bearing capacity of soil, the shear resistance of the passive soil wedge behind the thrust block should be checked because it may govern the design. For a thrust block having its height, H_b, less than one-half the distance from the ground surface to the base of the block, h, the design of the block is generally governed by shear resistance of the soil wedge behind the thrust block. Determining the value of the safe bearing and shear resistance of the soil is beyond the scope of this manual. Consulting a qualified geotechnical consultant is recommended.

Typical Configurations

Determining the safe bearing value, σ, is the key to sizing a thrust block. Values can vary from less than 1,000 lbf/ft^2 (47.866 N/m^2) for very soft soils to several tons per ft^2 for solid rock. Knowledge of local soil conditions is necessary for proper sizing of thrust blocks. Figure 13-14 shows several details for distributing thrust at a horizontal bend. Section A-A is the more common detail, but the other methods shown in the alternate section A-A may be necessary in weaker soils. Figure 13-15 shows typical thrust blocking of vertical bends. Design of the block for a bottom bend is the same for a horizontal bend, but the block for a top bend must be sized to adequately resist the vertical component of thrust with dead weight of the block, bend, water in the pipe, and overburden.

Restrained or Harnessed Joints

Restrained or harnessed joints may also be used to resist the thrust force through the development of friction between the pipe and the soil surrounding it. The thrust force resisted by restrained or harnessed joints is the force acting along the longitudinal axis of the pipe. If future excavation parallel to the pipe is anticipated, the designer should consider the effect. When this method is used, a sufficient length of pipe must be restrained by welding or harnessing to resist the unbalanced thrust force. The thrust force is equal to PA at bulkheads, tees, valves, and bend angles equal to or greater than 90°. The frictional force developed between the pipeline and the surrounding soil to restrain the thrust force is assumed to be distributed along the restrained length of the pipeline. Figure 13-16 shows a force diagram, wherein axial thrusts are equal to PA (cos Δ) and the unbalanced thrust force to be restrained is PA (1 – cos Δ). The length of pipeline required to be restrained on each side of the bend is

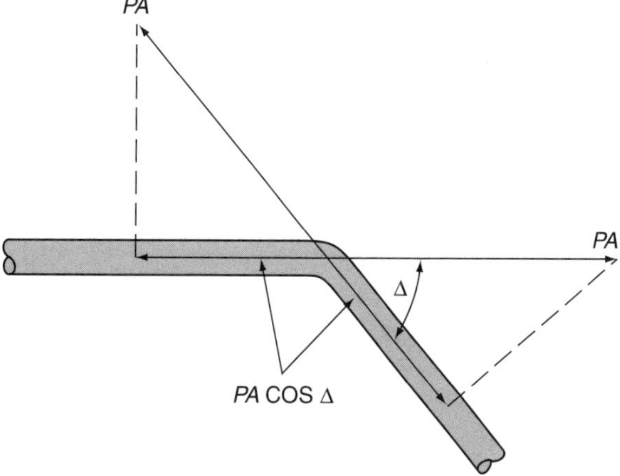

Figure 13-16 Force diagram

$$L = \frac{PA(1\cos\Delta)}{\mu(W_e + W_w + W_p)} \tag{13-6}$$

Where:

L = length of restrained or harnessed joints on each side of the bend or elbow (ft)
P = internal pressure (psi)
A = cross-sectional area of the pipe (in.^{2})
Δ = bend or elbow deflection (degrees), 0° to 90°.
μ = coefficient of friction between the pipe and the soil
W_e (horizontal bends) = two times the weight of the prism of soil over the pipe (lb/ft of pipe length)
W_e (vertical bends) = weight of the prism of soil over the pipe (lb/ft of pipe length)
W_p = weight of the pipe (lb/ft)
W_w = weight of the contained water (lb/ft)

The designer is cautioned that this section does not cover all combinations of design conditions that may be encountered on a particular project. In the preceding equation, all parameters except the value of μ, friction coefficient between the pipe and the soil, can be readily determined. Tests and experience indicate that the value of μ is not only a function of the type of soil, it is also greatly affected by the degree of compaction, moisture content of the backfill, and type of coating. Care must be exercised in the selection of μ. Coefficients of friction are generally in the range of 0.25 to 0.40. When a high water table or submerged conditions are anticipated, the effects of buoyancy of soil weight must be considered.

Eq 13-6, generally in the form presented, has been successfully utilized for computing the required length of restrained joints for steel pipelines. It should be noted that manuals representing other types of pipe, such as concrete, ductile iron, and polyvinyl chloride, may utilize different forms of Eq 13-6, which have also been shown to be successfully utilized for those types of pipe.

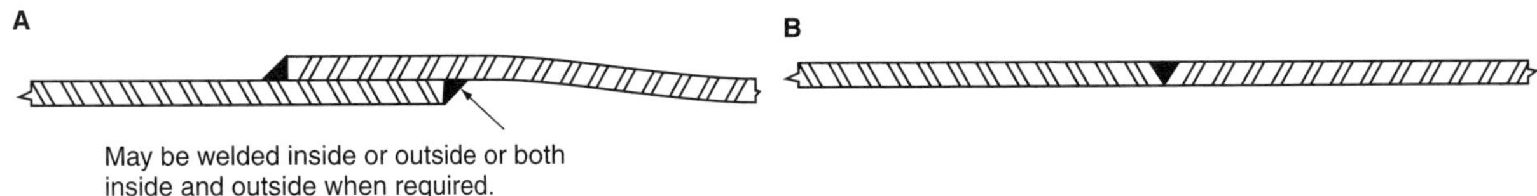

Figure 13-17 (A) Lap welded joint; (B) single-butt weld joint

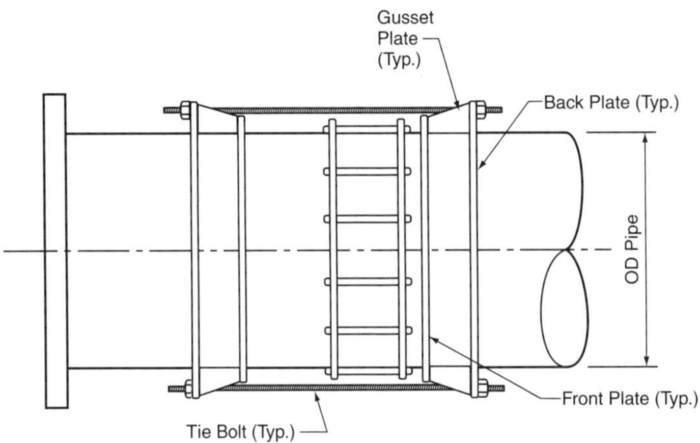

Figure 13-18 Harnessed joint detail

Types of Restrained Joints

Generally, there are two types of restrained joints: (1) welded and (2) harnessed.

Welded joints. Figure 13-17 shows two typical details for a welded joint. Figure 13-17 (A) shows a lap welded joint. Normally, for pipe sizes larger than 30 in. (750 mm), this joint is welded on the inside of the pipe. Figure 13-17(B) shows a butt-welded joint. This joint is usually specified for areas of high internal working pressures (above 400 psi [2,758 kPa]).

Harnessed joints. An alternate approach is to use harnessed joints, which provide a mechanical means of transmitting longitudinal thrust across the joints. Details of typical harnessed joints are shown in Figure 13-18.

Other Uses for Restraints

Tied joints have other uses that are not related to thrust caused by internal pressure. If it is necessary to install a pipeline on a steep slope, it may be desirable to use anchor blocks, harnessed joints, or welded joints to keep the pipe from separating because of downhill sliding. Although the pipe may be capable of resisting downhill movement because of its own frictional resistance with the soil, the backfilling operation can sometimes provide enough additional downhill force to open the joint.

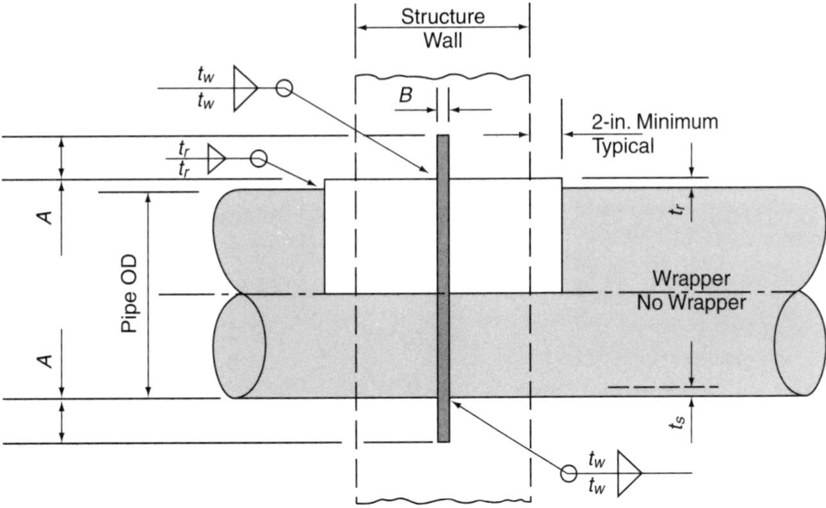

Figure 13-19 Anchor ring

ANCHOR RINGS

Anchor rings for use in concrete anchor blocks or concrete walls are illustrated in Figure 13-19. Corresponding dimensions and thrust or pull loads are given in Table 13-3. Rings are proportional to accept dead-end pull or thrust imposed by 150 psi and 250 psi (1,034 kPa and 1,724 kPa) internal pressures.

JOINT HARNESSES

Information for joint harness tie bolts or studs to be used for given pipe diameters and maximum pressures is shown in Table 13-4. Harness design data applicable to sleeve couplings are shown in Tables 13-5 and 13-5A and Figure 13-20.

Data are based on the following conditions: Stud bolts conforming to ASTM A193, Specifications for Alloy-Steel and Stainless Steel Bolting Materials for High-Temperature Service, Grade B7 or equal (1980a); nuts conforming to ASTM A194, Specifications for Carbon and Alloy Steel Nuts for Bolts for High-Pressure and High-Temperature Service, Grade 2H (1980b); lug material conforming to ASTM A283, Specifications for Low and Intermediate Tensile Strength Carbon Steel Plates, Shapes and Bars, Grade C (1979) or ASTM A36, Specification for Structural Steel (1977), or equal. Stud bolts ⅝-in. (16-mm) through ⅞-in. (22-mm) diameter have UNC threads; stud bolts 1-in. (25-mm) diameter and larger have eight UN threads per inch. Maximum bolt stress allowable is 40,000 psi (275.8 MPa), based on

$$\text{bolt tensile stress area} = 0.7854\left[D - \frac{0.9743}{N}\right]^2$$

Where:

D = nominal bolt diameter (in.)
N = number of threads per in.

Harness lugs are normally spaced equally around the pipe. In assembling the harness, the nuts shall be tightened gradually and equally at diametrically opposite sides until snug, to prevent misalignment and to ensure that all studs carry equal loads. The threads of the studs shall protrude a minimum of ½ in. (12.7 mm) from the nuts.

Table 13-3 Dimensions and bearing loads for anchor rings in concrete—maximum pipe pressure of 150 psi and 250 psi

Pipe OD		Ring Width *A*		Ring Thickness *B*		t_y		Minimum Weld, t_w		Permissible Load on Ring	
in.	*(mm)*	*in.*	*(mm)*	*in.*	*(mm)*	*in.*	*(mm)*	*in.*	*(mm)*	*lb*	*(kg)*
Maximum Pressure of 150 psi (1,034 kPa)											
6⅝	(166)	1	(25.4)	0.375	(9.53)	0.075	(1.91)	0.125	(3.18)	5,170	(146.6)
8⅝	(216)	1	(25.4)	0.375	(9.53)	0.080	(2.03)	0.125	(3.18)	8,760	(248.3)
10¾	(266)	1½	(38.1)	0.375	(9.53)	0.099	(2.51)	0.125	(3.18)	13,620	(386.1)
12¾	(316)	1½	(38.1)	0.375	(9.53)	0.118	(3.00)	0.125	(3.18)	19,150	(542.9)
14	(350)	2	(50.8)	0.375	(9.53)	0.129	(3.28)	0.125	(3.18)	23,090	(654.6)
16	(400)	2	(50.8)	0.375	(9.53)	0.147	(3.73)	0.125	(3.18)	30,160	(855.0)
18	(450)	2	(50.8)	0.375	(9.53)	0.166	(4.22)	0.125	(3.18)	38,170	(1,082.1)
20	(500)	3	(76.2)	0.500	(12.70)	0.184	(4.67)	0.125	(3.18)	47,120	(1,335.8)
24	(600)	3	(76.2)	0.500	(12.70)	0.221	(5.61)	0.125	(3.18)	67,860	(1,923.8)
30	(750)	4	(101.6)	0.625	(15.88)	0.276	(7.01)	0.188	(4.78)	106,030	(3,005.9)
36	(900)	4	(101.6)	0.750	(19.05)	0.331	(8.41)	0.188	(4.78)	152,680	(4,328.4)
Maximum Pipe Pressure of 250 psi (1,724 kPa)											
6⅝	(166)	1	(25.4)	0.375	(9.53)	0.102	(2.59)	0.125	(3.18)	8,620	(244.4)
8⅝	(216)	1	(25.4)	0.375	(9.53)	0.133	(3.38)	0.125	(3.18)	14,610	(414.2)
10¾	(269)	1½	(38.1)	0.375	(9.53)	0.165	(4.19)	0.125	(3.18)	22,690	(643.3)
12¾	(319)	1½	(38.1)	0.375	(9.53)	0.196	(4.98)	0.125	(3.18)	31,920	(904.9)
14	(350)	2	(50.8)	0.375	(9.53)	0.215	(5.46)	0.125	(3.18)	38,490	(1,091.2)
16	(400)	2	(50.8)	0.500	(12.70)	0.245	(6.22)	0.125	(3.18)	50,270	(1,425.1)
18	(450)	2	(50.8)	0.500	(12.70)	0.276	(7.01)	0.188	(4.78)	63,620	(1,803.6)
20	(500)	3	(76.2)	0.625	(15.88)	0.307	(7.80)	0.188	(4.78)	78,540	(2,226.6)
24	(600)	3	(76.2	0.625	(15.88)	0.368	(9.35)	0.188	(4.78)	113,100	(3,206.3)
30	(750)	4	(101.6)	0.750	(19.05)	0.460	(11.68)	0.250	(6.35)	176,710	(5,009.6)
36	(900)	4	(101.6)	0.875	(22.23)	0.552	(14.02)	0.313	(7.95)	254,470	(7,214.1)

NOTE: For minimum cylinder or wrapper thicknesses: If the method of fabrication uses a cylinder plus a wrapper, the minimum thickness for the wrapper shall be $t_r = t_y - t_s$, where t_r = wrapper thickness and t_s = shell thickness. If the method of fabrication does not use a wrapper, the minimum shell thickness shall be t_y. Minimum weld thickness = t_w. See Figure 13-19.

Table 13-4 Tie bolt schedule for harnessed joints

Pipe OD		Design Pressure		Minimum Cylinder Thickness (t)* Under Lug		Lug Type	Tie Bolt Diameter		Number of Bolts	Maximum End Force	
in.	*(mm)*	*psi*	*(kPa)*	*in.*	*(mm)*		*in.*	*(mm)*		*lbf*	*(kN)*
6⅝	(166)	50	(345)	0.188	(4.78)	P	⅝	(15.88)	2	1,725	(7.67)
		100	(689)	0.188	(4.78)	P	⅝	(15.88)	2	3,450	(15.35)
		150	(1,034)	0.193	(4.90)	P	⅝	(15.88)	2	5,170	(23.00)
		200	(1,379)	0.242	(6.15)	P	⅝	(15.88)	2	6,895	(30.67)
		250	(1,724)	0.282	(7.16)	P	⅝	(15.88)	2	8,620	(38.34)
8⅝	(216)	50	(345)	0.188	(4.78)	P	⅝	(15.88)	2	2,925	(13.01)
		100	(689)	0.194	(4.93)	P	⅝	(15.88)	2	5,845	(26.00)
		150	(1,034)	0.239	(6.07)	P	⅝	(15.88)	2	8,765	(38.99)
		200	(1,379)	0.291	(7.39)	P	⅝	(15.88)	2	11,685	(51.98)
		250	(1,724)	0.354	(8.99)	P	⅝	(15.88)	2	14,605	(64.97)
10¾	(269)	50	(345)	0.188	(4.78)	P	⅝	(15.88)	2	4,540	(20.19)
		100	(689)	0.242	(6.15)	P	⅝	(15.88)	2	9,075	(40.37)
		150	(1,034)	0.312	(7.92)	P	⅝	(15.88)	2	13,615	(60.56)
		200	(1,379)	0.386	(9.80)	P	¾	(19.05)	2	18,155	(80.76)
		250	(1,724)	0.466	(11.84)	P	¾	(19.05)	2	22,690	(100.93)
12¾	(319)	50	(345)	0.188	(4.78)	P	⅝	(15.88)	2	6,385	(28.40)
		100	(689)	0.286	(7.26)	P	⅝	(15.88)	2	12,770	(56.80)
		150	(1,034)	0.361	(9.17)	P	¾	(19.05)	2	19,150	(85.18)
		200	(1,379)	0.447	(11.35)	P	⅞	(22.23)	2	25,535	(113.59)
		250	(1,724)	0.540	(13.72)	P	⅞	(22.23)	2	31,920	(141.99)
14	(350)	50	(345)	0.188	(4.78)	RR†	⅝	(15.88)	2	7,695	(34.23)
		100	(689)	0.188	(4.78)	RR	⅝	(15.88)	2	15,395	(68.48)
		150	(1,034)	0.188	(4.78)	RR	¾	(19.05)	2	23,090	(102.71)
		200	(1,379)	0.188	(4.78)	RR	⅞	(22.23)	2	30,790	(136.96)
		250	(1,724)	0.188	(4.78)	RR	⅞	(22.23)	2	38,485	(171.19)
16	(400)	50	(345)	0.188	(4.78)	RR	⅝	(15.88)	2	10,055	(44.73)
		100	(689)	0.188	(4.78)	RR	¾	(19.05)	2	20,105	(89.43)
		150	(1,034)	0.188	(4.78)	RR	⅞	(22.23)	2	30,160	(134.16)
		200	(1,379)	0.188	(4.78)	RR	1	(25.40)	2	40,210	(178.86)
		250	(1,724)	0.188	(4.78)	RR	1⅛	(28.58)	2	50,265	(223.59)

NOTE: It is not recommended that harnessed flexible couplings be located immediately adjacent to pumps as this may cause undo stress on the pumps and pump base.

* The thickness t is the minimum thickness shell allowed under the harness assembly. If wrappers or pads are used, this minimum thickness becomes the thickness of the wrapper or pad.

If wrappers or pads are used, the minimum width or length shall not be less than the A or X dimensions in Figure 13-20, plus $1.56\sqrt{Rt}$, or not less than A or X plus 2 in., whichever is greater.

† The symbol R equals one-half of the pipe OD.

Table continued next page

Table 13-4 Tie bolt schedule for harnessed joints (continued)

Pipe OD		Design Pressure		Minimum Cylinder Thickness (t)* Under Lug		Lug Type	Tie Bolt Diameter		Number of Bolts	Maximum End Force	
in.	*(mm)*	*psi*	*(kPa)*	*in.*	*(mm)*		*in.*	*(mm)*		*lbf*	*(kN)*
18	(450)	50	(345)	0.188	(4.78)	RR	5/8	(15.88)	2	12,725	(56.60)
		100	(689)	0.188	(4.78)	RR	7/8	(22.23)	2	25,445	(113.19)
		150	(1,034)	0.188	(4.78)	RR	1	(25.40)	2	38,170	(169.79)
		200	(1,379)	0.188	(4.78)	RR	1 1/8	(28.58)	2	50,895	(226.39)
		250	(1,724)	0.209	(5.31)	RR	1 1/4	(31.75)	2	63,615	(282.97)
20	(500)	50	(345)	0.188	(4.78)	RR	5/8	(15.88)	2	15,710	(69.88)
		100	(689)	0.188	(4.78)	RR	7/8	(22.23)	2	31,415	(139.74)
		150	(1,034)	0.188	(4.78)	RR	1	(25.40)	2	47,125	(209.62)
		200	(1,379)	0.188	(4.78)	RR	1 1/8	(28.58)	2	62,830	(279.48)
		250	(1,724)	0.227	(5.77)	RR	1 1/4	(31.75)	2	78,540	(349.36)
24	(600)	50	(345)	0.188	(4.78)	RR	3/4	(19.05)	2	22,620	(100.62)
		100	(689)	0.188	(4.78)	RR	3/4	(19.05)	4	45,240	(201.24)
		150	(1,034)	0.188	(4.78)	RR	7/8	(22.23)	4	67,860	(301.86)
		200	(1,379)	0.211	(5.36)	RR	1	(25.40)	4	90,480	(402.48)
		250	(1,724)	0.255	(6.48)	RR	1 7/8	(28.58)	4	113,100	(503.09)
30	(750)	50	(345)	0.188	(4.78)	RR†	3/4	(19.05)	4	35,340	(157.20)
		100	(689)	0.188	(4.78)	RR	1	(25.40)	4	70,685	(314.42)
		150	(1,034)	0.199	(5.05)	RR	1 1/8	(28.58)	4	106,030	(471.64)
		200	(1,379)	0.260	(6.60)	RR	1 1/4	(31.75)	4	141,370	(628.85)
		250	(1,724)	0.314	(7.98)	RR	1 3/8	(34.93)	4	176,715	(786.07)
36	(900)	50	(345)	0.188	(4.78)	RR	7/8	(22.23)	4	50,895	(226.39)
		100	(689)	0.188	(4.78)	RR	1 1/8	(28.58)	4	101,790	(452.78)
		150	(1,034)	0.236	(5.99)	RR	1 1/4	(31.75)	4	152,680	(679.15)
		200	(1,379)	0.308	(7.82)	RR	1 1/2	(38.10)	4	203,575	(905.55)
		250	(1,724)	0.369	(9.37)	RR	1 5/8	(41.28)	4	254,470	(1,131.94)
42	(1,050)	50	(345)	0.188	(4.78)	RR	1	(25.40)	4	69,270	(308.13)
		100	(689)	0.190	(4.83)	RR	1 1/4	(31.75)	4	138,545	(616.28)
		150	(1,034)	0.274	(6.96)	RR	1 1/2	(38.10)	4	207,815	(924.41)
		200	(1,379)	0.346	(8.79)	RR	1 5/8	(41.28)	4	277,090	(1,232.56)
		250	(1,724)	0.409	(10.39)	RR	1 1/2	(38.10)	6	346,360	(1,540.69)
48	(1,200)	50	(345)	0.188	(4.78)	RR	1	(25.40)	4	90,480	(402.48)
		100	(689)	0.213	(5.41)	RR	1 3/8	(34.93)	4	180,955	(804.93)
		150	(1,034)	0.304	(7.72)	RR	1 5/8	(41.28)	4	271,435	(1,207.40)

NOTE: It is not recommended that harnessed flexible couplings be located immediately adjacent to pumps as this may cause undo stress on the pumps and pump base.

* The thickness t is the minimum thickness shell allowed under the harness assembly. If wrappers or pads are used, this minimum thickness becomes the thickness of the wrapper or pad.

If wrappers or pads are used, the minimum width or length shall not be less than the A or X dimensions in Figure 13-20, plus $1.56\sqrt{Rt}$, or not less than A or X plus 2 in., whichever is greater.

† The symbol R equals one-half of the pipe OD.

Table continued next page

Table 13-4 Tie bolt schedule for harnessed joints (continued)

Pipe OD		Design Pressure		Minimum Cylinder Thickness (t)* Under Lug		Lug Type	Tie Bolt Diameter		Number of Bolts	Maximum End Force	
in.	*(mm)*	*psi*	*(kPa)*	*in.*	*(mm)*	Type	*in.*	*(mm)*	Bolts	*lbf*	*(kN)*
		200	(1,379)	0.380	(9.65)	RR	1⅝	(41.28)	6	361,910	(1,609.86)
		250	(1,724)	0.463	(11.76)	RR	1¾	(44.45)	6	452,390	(2,012.33)
54	(1,350)	50	(345)	0.188	(4.78)	RR	1⅛	(28.58)	4	114,510	(509.37)
		100	(689)	0.238	(6.05)	RR	1½	(38.10)	4	229,020	(1,018.73)
		150	(1,034)	0.330	(8.38)	RR	1½	(38.10)	6	343,535	(1,528.12)
		200	(1,379)	0.425	(10.80)	RR	1¾	(44.45)	6	458,045	(2,037.49)
		250	(1,724)	0.501	(12.73)	RR	1¾	(44.45)	8	572,555	(2,546.85)
60	(1,500)	50	(345)	0.200	(5.08)	RR	1¼	(31.75)	4	141,370	(628.85)
		100	(689)	0.252	(6.40)	RR	1⅜	(34.93)	6	282,745	(1,257.71)
		150	(1,034)	0.357	(9.07)	RR	1⅝	(41.28)	6	424,115	(1,886.56)
		200	(1,379)	0.442	(11.23)	RR	1⅝	(41.28)	8	565,485	(2,515.40)
		250	(1,724)	0.519	(13.18)	RR	1⅝	(41.28)	10	706,860	(3,144.27)
66	(1,650)	50	(345)	0.215	(5.46)	RR	1⅜	(34.93)	4	171,060	(760.91)
		100	(689)	0.278	(7.06)	RR	1½	(38.10)	6	342,120	(1,521.83)
		150	(1,034)	0.381	(9.68)	RR	1⅝	(41.28)	8	513,180	(2,282.74)
		200	(1,379)	0.470	(11.94)	RR	1⅝	(41.28)	10	684,240	(3,043.65)
		250	(1,724)	0.570	(14.48)	RR	1¾	(44.45)	12	855,300	(3,804.56)
72	(1,800)	50	(345)	0.230	(5.84)	RR	1½	(38.10)	4	203,575	(905.55)
		100	(689)	0.297	(7.54)	RR	1⅝	(41.28)	6	407,150	(1,811.09)
		150	(1,034)	0.416	(10.57)	RR	1¾	(44.45)	8	610,725	(2,716.64)
		200	(1,379)	0.514	(13.06)	RR	1¾	(44.45)	10	814,300	(3,622.19)
		250	(1,724)	0.605	(15.37)	RR	1⅞	(47.63)	12	1,017,875	(4,527.73)
78	(1,950)	50	(345)	0.245	(6.22)	RR†	1¼	(31.75)	6	238,920	(1,062.77)
		100	(689)	0.313	(7.95)	RR	1½	(38.10)	8	477,835	(2,125.52)
		150	(1,034)	0.439	(11.15)	RR	1¾	(44.45)	10	716,755	(3,188.29)
		200	(1,379)	0.541	(13.74)	RR	1¾	(44.45)	12	955,675	(4,251.05)
		250	(1,724)	0.635	(16.13)	RR	1⅞	(47.63)	14	1,194,590	(5,313.80)
84	(2,100)	50	(345)	0.260	(6.60)	RR	1⅜	(34.93)	6	277,090	(1,232.56)
		100	(689)	0.331	(8.41)	RR	1⅝	(41.28)	8	554,175	(2,465.09)
		150	(1,034)	0.461	(11.71)	RR	1¾	(44.45)	10	831,265	(3,697.65)

NOTE: It is not recommended that harnessed flexible couplings be located immediately adjacent to pumps as this may cause undo stress on the pumps and pump base.

* The thickness t is the minimum thickness shell allowed under the harness assembly. If wrappers or pads are used, this minimum thickness becomes the thickness of the wrapper or pad.

If wrappers or pads are used, the minimum width or length shall not be less than the A or X dimensions in Figure 13-20, plus $1.56\sqrt{Rt}$, or not less than A or X plus 2 in., whichever is greater.

†The symbol R equals one-half of the pipe OD.

Table continued next page

Table 13-4 Tie bolt schedule for harnessed joints (continued)

Pipe OD		Design Pressure		Minimum Cylinder Thickness (t)* Under Lug		Lug Type	Tie Bolt Diameter		Number of Bolts	Maximum End Force	
in.	*(mm)*	*psi*	*(kPa)*	*in.*	*(mm)*		*in.*	*(mm)*		*lbf*	*(kN)*
		200	(1,379)	0.569	(14.45)	RR	1⅞	(47.63)	12	1,108,355	(4,930.21)
		250	(1,724)	0.686	(17.42)	RR	2	(50.80)	14	1,385,445	(6,162.77)
90	(2,250)	50	(345)	0.275	(6.99)	RR	1½	(38.10)	6	318,085	(1,141.91)
		100	(689)	0.357	(9.07)	RR	1¾	(44.45)	8	636,175	(2,829.85)
		150	(1,034)	0.484	(12.29)	RR	1⅞	(47.63)	10	954,260	(4,244.76)
		200	(1,379)	0.612	(15.54)	RR	2	(50.80)	12	1,272,345	(5,696.67)
		250	(1,724)	0.716	(18.19)	RR	2	(50.80)	16	1,590,430	(7,074.59)
96	(2,400)	50	(345)	0.290	(7.37)	RR	1⅝	(41.28)	6	361,910	(1,609.86)
		100	(689)	0.373	(9.47)	RR	1¾	(44.45)	10	723,825	(3,219.73)
		150	(1,034)	0.504	(12.80)	RR	1⅞	(47.63)	12	1,085,735	(4,829.59)
		200	(1,379)	0.637	(16.18)	RR	2	(50.80)	14	1,447,645	(6,439.45)
		250	(1,724)	0.757	(19.23)	RR	2¼	(57.15)	16	1,809,560	(8,049.32)

NOTE: It is not recommended that harnessed flexible couplings be located immediately adjacent to pumps as this may cause undo stress on the pumps and pump base.

* The thickness t is the minimum thickness shell allowed under the harness assembly. If wrappers or pads are used, this minimum thickness becomes the thickness of the wrapper or pad.

If wrappers or pads are used, the minimum width or length shall not be less than the A or X dimensions in Figure 13-20, plus $1.56\sqrt{Rt}$, or not less than A or X plus 2 in., whichever is greater.

† The symbol R equals one-half of the pipe OD.

The end-thrust values shown in Table 13-4 are the maximum end-thrust values that the harness assemblies are designed to withstand. The design pressure must include an anticipated allowance for surge pressure. The field-test pressure must never exceed the design pressure.

SPECIAL AND VALVE CONNECTIONS AND OTHER APPURTENANCES

Special connections are shown in Figure 13-21 (with Tables 13-6 and 13-7), Figures 13-22 through 13-25 (with Table 13-8), and Figure 13-26 (with Table 13-9). Some examples of vault and manhole design are shown in Figures 13-27 through 13-29. Figures 13-30 and 13-31 illustrate blow-off connections. Figure 13-32 shows a relief-valve manifold layout.

Special tapping machines for mains under pressure are available and have been used for many years. Figure 13-33 illustrates the method. The reinforcing pad is eliminated unless pressure requires it. The outlet is ordinarily a piece of extra-heavy, standard-weight pipe with an AWWA standard plate flange attached. The tapping valve is special and allows proper clearance for the cutter on the drilling machine.

FREEZING IN PIPELINES

Design practice should not be determined by the saying that "running water doesn't freeze." Water in a pipeline will freeze, running or not, if its temperature drops a fraction

Table 13-5 Dimensions of joint harness tie bolts and lugs for rubber-gasketed joints

Stud Diameter		T		Lug	A		Y		W		X		HB		E		HF		Hole Diameter	
in.	*(mm)*	*in.*	*(mm)*	Type	*in.*	*(mm)*	*in.*	*(mm)*	in.	*(mm)*	*in.*	*(mm)*	*in.*	*(mm)*	*in.*	*(mm)*	*in.*	*(mm)*	*in.*	*(mm)*
5/8	(15.88)	3/8	(9.53)	P	5	(127.00)	5	(127.00)	1 3/8	(34.93)	5	(127.00)	3 7/8	(98.43)	3	(76.20)	2	(50.80)	3/4	(19.05)
3/4	(19.05)	3/8	(9.53)	P	5	(127.00)	5	(127.00)	1 1/2	(38.10)	5	(127.00)	4 1/8	(104.78)	3 1/8	(79.38)	2	(50.80)	7/8	(22.23)
7/8	(22.23)	1/2	(12.70)	P	5 1/2	(139.70)	5	(127.00)	1 5/8	(41.28)	5	(127.00)	4 1/4	(107.95)	3 1/8	(79.38)	2	(50.80)	1	(25.40)
5/8	(15.88)	3/8	(9.53)	RR	5	(127.00)	Ring		1 3/8	(34.93)	Ring		3 7/8	(98.43)	3	(76.20)	2	(50.80)	3/4	(19.05)
3/4	(19.05)	3/8	(9.53)	RR	5	(127.00)	Ring		1 1/2	(38.10)	Ring		4 1/8	(104.78)	3 1/8	(79.38)	2	(50.80)	7/8	(22.23)
7/8	(22.23)	1/2	(12.70)	RR	5 1/2	(139.70)	Ring		1 5/8	(41.28)	Ring		4 1/4	(107.95)	3 1/8	(79.38)	2	(50.80)	1	(25.40)
1	(25.40)	1/2	(12.70)	RR	5 3/4	(146.04)	Ring		1 3/4	(44.45)	Ring		4 1/2	(114.30)	3 1/4	(82.55)	2	(50.80)	1 1/8	(28.58)
1 1/8	(28.58)	1/2	(12.70)	RR	7	(177.80)	Ring		1 7/8	(47.63)	Ring		4 3/4	(120.65)	3 5/8	(92.08)	2 1/2	(63.50)	1 1/4	(31.75)
1 1/4	(31.75)	5/8	(15.88)	RR	7 1/2	(190.50)	Ring		2	(50.80)	Ring		5	(127.00)	3 3/4	(95.25)	2 1/2	(63.50)	1 3/8	(34.93)
1 3/8	(34.93)	5/8	(15.88)	RR	8 3/4	(222.45)	Ring		2 1/8	(53.98)	Ring		5 3/8	(136.53)	3 3/4	(95.25)	2 1/2	(63.50)	1 1/2	(38.10)
1 1/2	(38.10)	3/4	(19.05)	RR	10	(254.00)	Ring		2 1/4	(57.15)	Ring		5 1/2	(139.70)	3 7/8	(98.43)	2 1/2	(63.50)	1 5/8	(41.28)
1 5/8	(41.28)	3/4	(19.05)	RR	10 3/4	(173.05)	Ring		2 3/8	(60.33)	Ring		5 5/8	(142.88)	3 7/8	(98.43)	2 1/2	(63.50)	1 3/4	(44.45)
1 3/4	(44.45)	7/8	(22.23)	RR	12	(304.80)	Ring		2 1/2	(63.50)	Ring		5 7/8	(149.23)	4	(101.60)	2 1/2	(63.50)	1 7/8	(47.63)
1 7/8	(47.63)	7/8	(22.23)	RR	13	(330.20)	Ring		2 5/8	(66.68)	Ring		6	(152.40)	4	(101.60)	2 1/2	(63.50)	2	(50.80)
2	(50.80)	1	(25.40)	RR	14	(355.60)	Ring		2 3/4	(69.85)	Ring		6 1/4	(158.75)	4 1/4	(107.95)	2 1/2	(63.50)	2 1/8	(53.98)
2 1/4	(57.15)	1	(25.40)	RR	15 3/4	(400.05)	Ring		3	(76.20)	Ring		6 3/4	(171.45)	4 5/16	(109.54)	2 1/2	(63.50)	2 3/8	(60.33)

NOTES:

1. Dimensions shown above are in inches.
2. Use these dimensions with Figure 13-20 and Tables 13-4 and 13-5A.
3. See section on Joint Harnesses, page 199, for design conditions covering maximum allowable pressure and equal placement spacing of the bolts around the circumference of the pipe. The designs represented in Tables 13-4, 13-5, and 13-5A are to resist longitudinal thrust only. Considerations for additional vertical, horizontal, or eccentric loadings are beyond the scope of this application.
4. All fillet welds shall meet the minimum requirements of the American Institute of Steel Construction Specifications, except as follows: (a) fillet welds connecting back plates, front plates, and gusset plates shall be 1/4 in. minimum; (b) fillet welds connecting the harness assembly and rings to the cylinder or wrapper shall meet the minimum requirements of note 4 of Figure 13-20.
5. Dimension E in the above table has been adequate to provide clearance between the harness bolt and the OD of the assembled sleeve-type coupling where the OD of the coupling is 4 in. to 5 in. larger than the OD of the pipe, as normally found in standard couplings thru 72-in. diameter. For sleeve-type couplings designed for higher pressures and for diameters over 72 in., the E dimension should be checked by the designer for adequate clearance of the harness bolt over the OD of the assembled coupling to be provided by the manufacturer.

Table 13-5A Maximum allowable load per tie bolt

Diameter		Number of Threads		Net Area per Tie Bolt		Maximum Load per Tie Bolt	
in.	*(mm)*	*per in.*	*(per mm)*	*in.*^{2}	*(mm^2)*	*lb*	*(kN)*
5/8	(15.88)	11	(0.43)	0.226	(146)	9,040	(40.21)
3/4	(19.05)	10	(0.39)	0.334	(215)	13,360	(59.43)
7/8	(22.23)	9	(0.35)	0.461	(297)	18,440	(82.03)
1	(25.40)	8	(0.31)	0.605	(390)	24,200	(107.65)
1 1/8	(28.58)	8	(0.31)	0.790	(510)	31,600	(140.56)
1 1/4	(31.75)	8	(0.31)	1.000	(645)	40,000	(177.93)
1 3/8	(34.93)	8	(0.31)	1.233	(795)	49,320	(219.39)
1 1/2	(38.10)	8	(0.31)	1.491	(962)	59,640	(265.29)
1 5/8	(41.28)	8	(0.31)	1.774	(1,145)	70,960	(315.65)
1 3/4	(44.45)	8	(0.31)	2.082	(1,343)	83,280	(370.45)
1 7/8	(47.63)	8	(0.31)	2.414	(1,557)	96,560	(429.52)
2	(50.80)	8	(0.31)	2.770	(1,787)	110,800	(492.86)
2 1/4	(57.15)	8	(0.31)	3.557	(2,295)	142,280	(632.89)

NOTES:

1. The net area for tie bolts has been calculated based on bolts with 1-in. diameters and larger having eight UN threads per inch, and on bolts smaller than 1-in. diameter having standard UNC threads.
2. The maximum load per tie bolt has been based on an allowable stress in the bolt of 40,000 psi.

of a degree below 32°F (0°C). If the water is losing its heat to the surrounding medium in a given locality, it will freeze if not moved out of that locality before its temperature drops below 32°F (0°C). In this sense, it is important to run water so that warmer water replaces the water that is near freezing. The heat added to the moving water as a result of frictional resistance to flow is negligible for large pipe with low velocities, but it may be considerable for small pipe with high velocities. Under some circumstances, agitated water may not turn to ice even when the temperature is as low as 28° to 29°F (–2.2° to –1.7°C), but this condition cannot be predicted or depended on. Therefore, it is recommended that the water temperature stays above 32°F (0°C) with a margin, if possible, of 1° or 2°F (0.5° or 1°C) against contingencies.

Calculations relative to the prevention of freezing in pipelines are based on the same general principles of heat transmission and loss that govern similar calculations applied to buildings and other installations. It is well established that complete freezing of water occurs when 144 Btu of heat per pound of water is extracted after the temperature of the mass has been lowered to 32°F (0°C). Also, with certain exceptions, the ratio of the weight of water existing as ice to the weight of liquid water at any time during cooling is directly proportional to the ratio of the British thermal units per pound withdrawn to the 144 Btu per pound required for complete freezing.

Water containing only ice particles (frazil or needle ice) may cause serious trouble because these can quickly block a pipeline by adhering to valves or any minor obstructions. To prevent this from occurring, the water must be maintained at about 32.1° to 32.5°F (0.06° to 0.3°C) to avoid trouble.

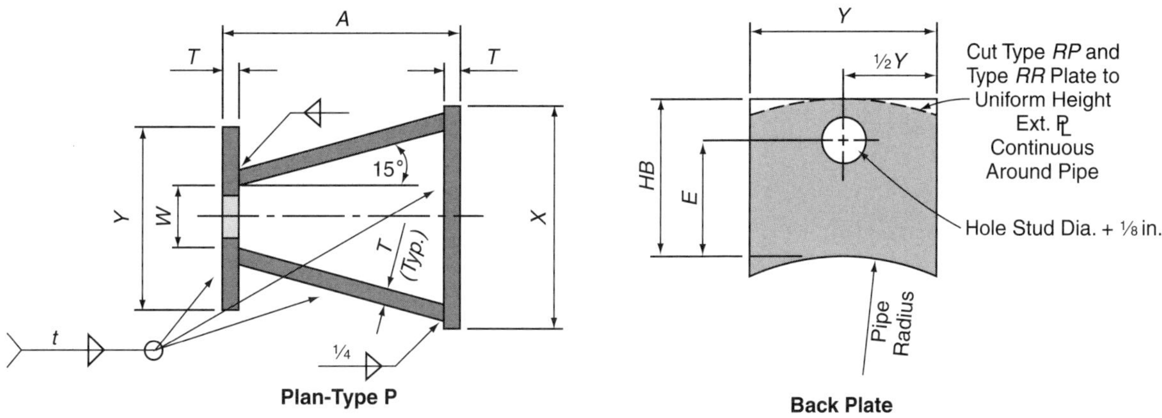

T = Thickness of Steel Harness Plates
t = Pipe Wall Thickness and Size of Fillet Attachment Weld

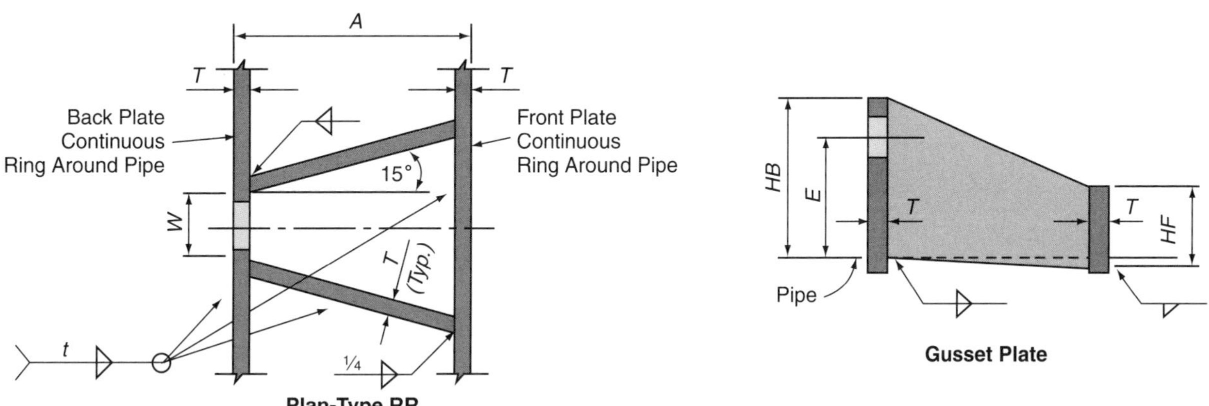

T = Thickness of Steel Harness Plates
t = Pipe Wall Thickness and Size of Fillet Attachment Weld

NOTES:

1. See Tables 13-4 and 13-5 for dimensions.
2. See Joint Harness section for design conditions.
3. For Harness lug-type RR, the gusset plates between the back plate and the front plate may be perpendicular to the front and back plates with a minimum clear distance between each pair of gusset plates dimension W.
4. The minimum wall thickness *t* shall be 3/16 in. for cylinder or wrapper thicknesses through 3/8 in.; and 1/4 in. Minimum thickness for all other cylinder or wrapper thicknesses.

Figure 13-20 Harness lug detail

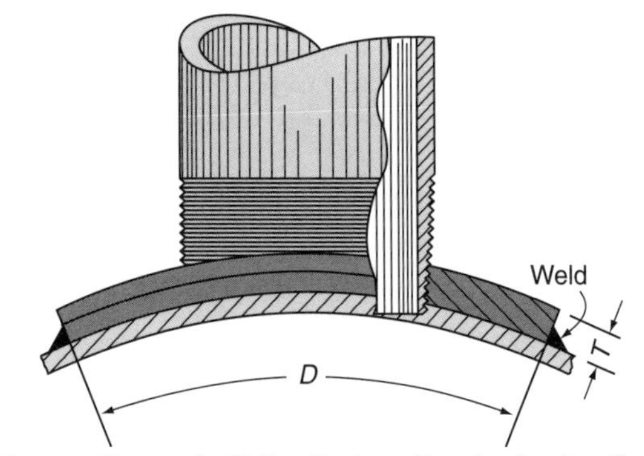

Source: Barnard, R.E., Design Standards for Steel Water Pipe. Jour. AWWA, *40:1:24 (Jan. 1948).*

Figure 13-21 Reinforcing pad for tapped opening

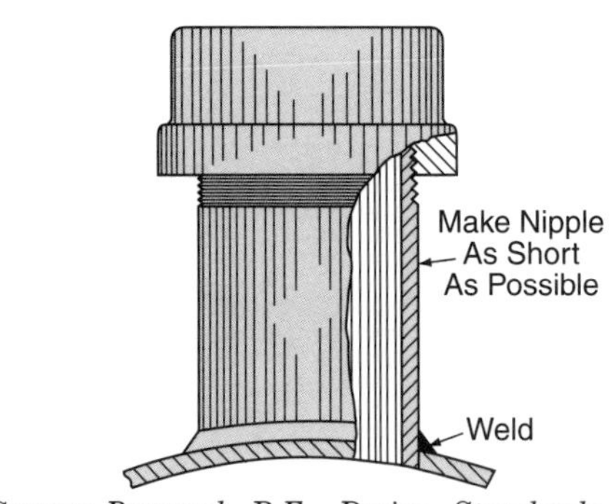

Source: Barnard, R.E., Design Standards for Steel Water Pipe. Jour. AWWA, *40:1:24 (Jan. 1948).*

Figure 13-22 Nipple with cap

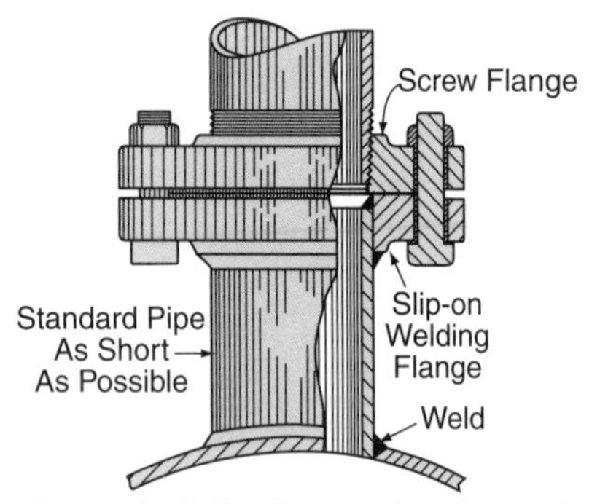

Source: Barnard, R.E., Design Standards for Steel Water Pipe. Jour. AWWA, *40:1:24 (Jan. 1948).*

Figure 13-23 Flanged connection for screw-joint pipe

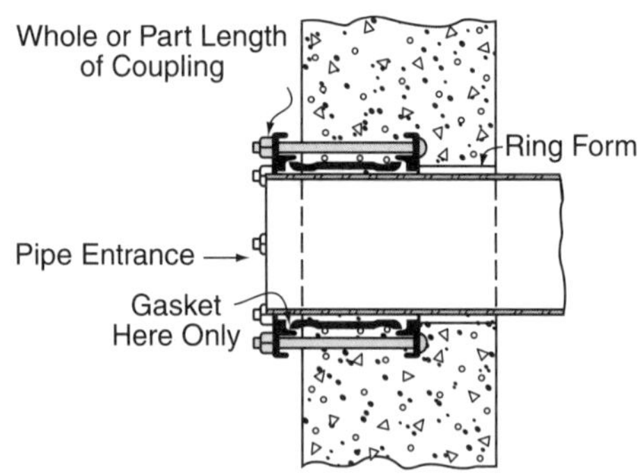

Source: Barnard, R.E., Design Standards for Steel Water Pipe. Jour. AWWA, *40:1:24 (Jan. 1948).*

Figure 13-24 Wall connection using coupling

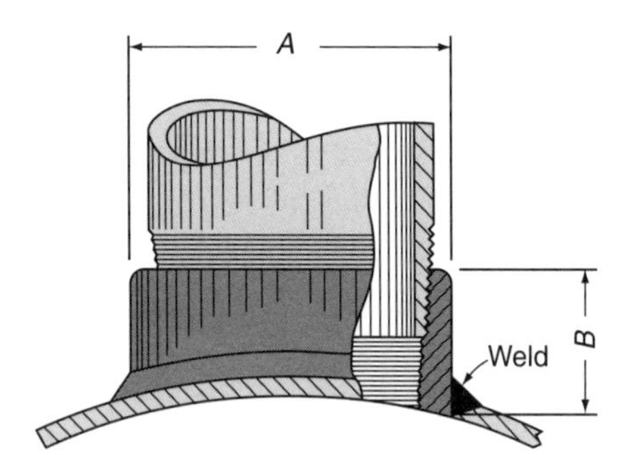

Source: Barnard, R.E., Design Standards for Steel Water Pipe. Jour. AWWA, *40:1:24 (Jan. 1948).*

Figure 13-25 Extra-heavy half coupling welded to pipe as threaded outlet

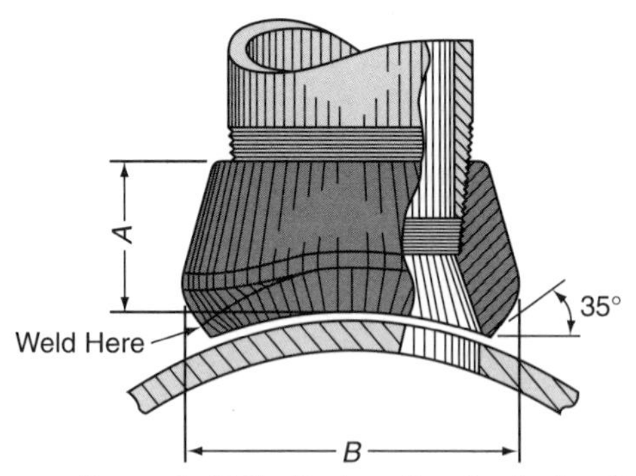

Source: Barnard, R.E., Design Standards for Steel Water Pipe. Jour. AWWA, *40:1:24 (Jan. 1948).*

Figure 13-26 Thredolets

Table 13-6 Plate dimensions and drill sizes for reinforced tapped openings (See Figure 13-21)

Size of Pipe		Tap Size of Drill for Pipe Tap		Dimensions of Plate			
				T		D*	
in.	*(mm)*	*in.*	*(mm)*	*in.*	*(mm)*	*in.*	*(mm)*
3/8	(9.5)	19/32	(15.1)	1/4	(6.4)	1 1/4	(31.8)
1/2	(12.7)	23/32	(18.3)	1/4	(6.4)	1 1/2	(38.1)
3/4	(19.1)	1 15/16	(23.8)	1/2	(12.7)	1 3/4	(44.5)
1	(25.4)	1 5/32	(29.4)	1/2	(12.7)	2 1/8	(54.0)
1 1/4	(31.8)	1 1/2	(38.1)	1/2	(12.7)	2 1/2	(63.5)
1 1/2	(38.1)	1 23/32	(43.7)	1/2	(12.7)	3	(76.2)
2	(50.8)	2 3/16	(55.6)	1/2	(12.7)	3 1/2	(88.9)
2 1/2	(63.5)	2 5/8	(66.7)	3/4	(19.1)	4 1/2	(114.3)
3	(76.2)	3 1/4	(82.6)	3/4	(19.1)	5	(127.0)
3 1/2	(88.9)	3 3/4	(95.3)	3/4	(19.1)	5 1/2	(139.7)
4	(101.6)	4 1/4	(108.0)	3/4	(19.1)	6	(152.6)

Source: Barnard, R.E., Design Standards for Steel Water Pipe. Jour. AWWA, *40:1:24 (Jan. 1948).*

* Diameter of plate pad before curving to fit outside of pipe.

Table 13-7 Maximum size of threaded openings for given size pipe with reinforcing pads (See Figure 13-21)

Pipe Size		Maximum Size Tapped Opening	
in.	*(mm)*	*in.*	*(mm)*
6 5/8	(166)	1 1/4	(32)
8 5/8	(216)	1 1/2	(38)
10 3/4	(269)	2	(51)
12 3/4	(319)	2 1/2	(64)
14	(350)	3	(76)
16	(400)	3 1/2	(89)
18	(450)	3 1/2	(89)
20	(500)	4	(102)

Source: Barnard, R.E., Design Standards for Steel Water Pipe. Jour. AWWA, *40:1:24 (Jan. 1948).*

Table 13-8 Dimensions of extra-heavy half-couplings (See Figures 13-22–13-25)

Coupling Size		Overall Dimensions			
		A		B	
in.	*(mm)*	*in.*	*(mm)*	*in.*	*(mm)*
1/8*	(3.2)				
1/4*	(6.4)				
3/8*	(9.5)				
1/2	(12.7)	1.13	(28.7)	27/32	(21.4)
3/4	(19.1)	1.44	(36.6)	1	(25.4)
1	(25.4)	1.70	(43.2)	1 3/32	(27.8)
1 1/4	(31.8)	2.07	(52.6)	1 3/8	(34.9)
1 1/2	(38.1)	2.31	(58.7)	1 3/8	(34.9)
2	(50.8)	2.81	(71.4)	1 1/2	(38.1)
2 1/2	(63.5)	3.31	(84.1)	1 11/16	(42.9)
3	(76.2)	4.00	(101.6)	1 3/4	(44.5)
3 1/2	(88.9)	4.63	(117.6)	2 1/16	(52.4)
4	(101.6)	5.13	(130.3)	2 1/16	(52.4)

Source: Barnard, R.E., Design Standards for Steel Water Pipe. Jour. AWWA, *40:1:24 (Jan. 1948).*

* Secure these sizes by bushing down 1/2-in. coupling.

Table 13-9 Dimensions figures thredolets (See Figure 13-26)

B In.	1 3/8	1 3/4	2 1/8	2 5/8	2 7/8	3 1/2	4 1/8	4 7/8	5 1/2	6	7 1/8	8 7/8
A In.	7/8	7/8	1 1/8	1 1/4	1 1/4	1 1/2	1 3/4	2	2 1/8	2 1/4	2 3/8	2 1/2
Pipe Size	Outlet Sizes—*in.**											
in.	1/2	3/4	1	1 1/4	1 1/2	2	2 1/2	3	3 1/2	4	5	6
6	6 × 1/2	6 × 3/4	6 × 1	6 × 1 1/4	6 × 1 1/2	6 × 2	6 × 2 1/2	6 × 3	6 × 3 1/2	6 × 4	6 × 5	
8	8 × 1/2	8 × 3/4	8 × 1	8 × 1 1/4	8 × 1 1/2	8 × 2	8 × 2 1/2	8 × 3	8 × 3 1/2	8 × 4	8 × 5	8 × 6
10	10 × 1/2	10 × 3/4	10 × 1	10 × 1 1/4	10 × 1 1/2	10 × 2	10 × 2 1/2	10 × 3	10 × 3 1/2	10 × 4	10 × 5	10 × 6
12	12 × 1/2	12 × 3/4	12 × 1	12 × 1 1/4	12 × 1 1/2	12 × 2	12 × 2 1/2	12 × 3	12 × 3 1/2	12 × 4	12 × 5	12 × 6

Source: Barnard, R.E., Design Standards for Steel Water Pipe. Jour. AWWA, *40:1:24 (Jan. 1948).*

NOTE: To convert in. to mm, multiply by 25.4

* Outlet is tapped to standard iron pipe sizes.

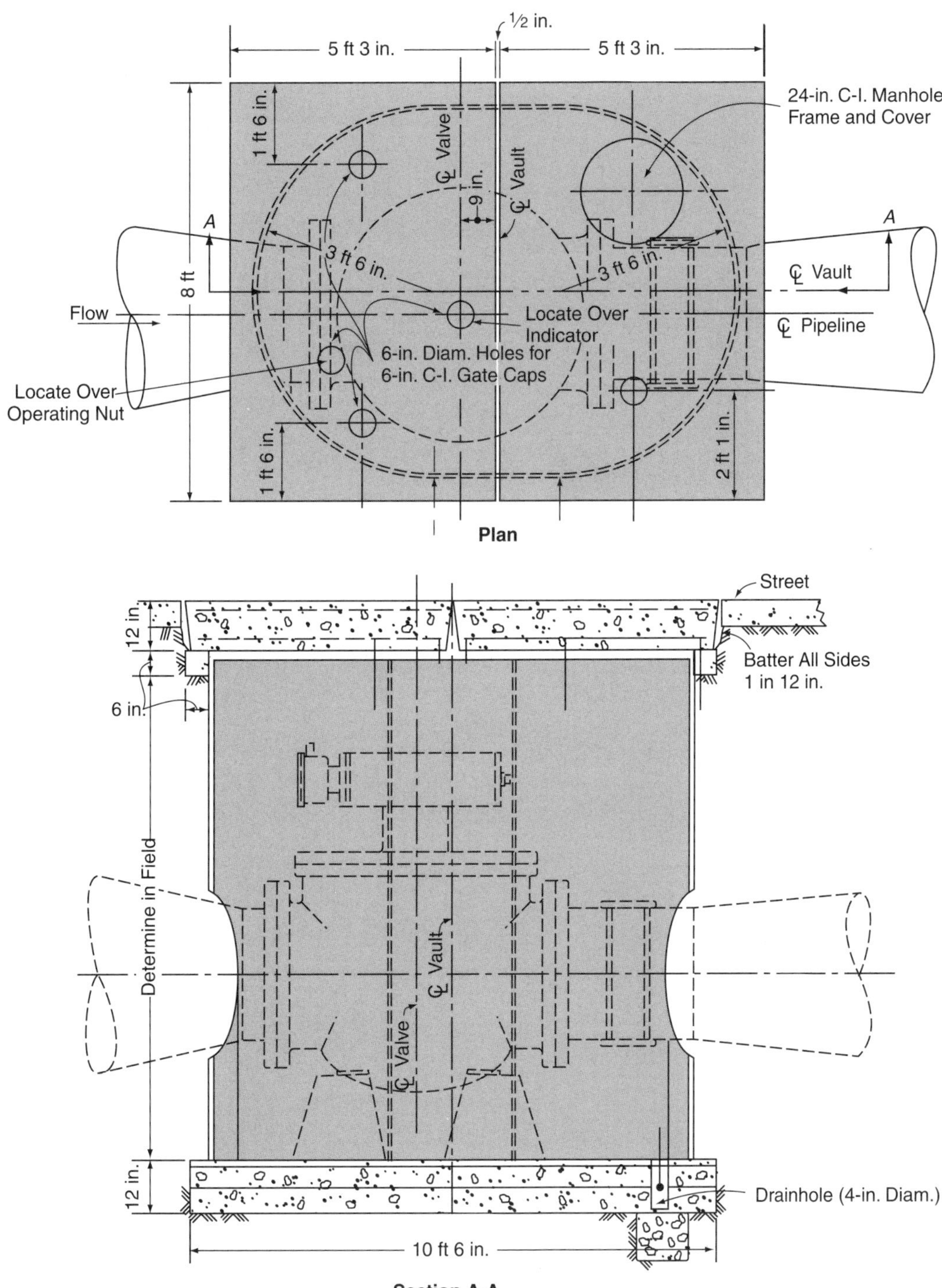

Source: Goit, L.E., Steel Pipeline Appurtenances, Jour. AWWA, *41:1:47 (Jan. 1949)*

Figure 13-27 Casing and removable two-piece roof

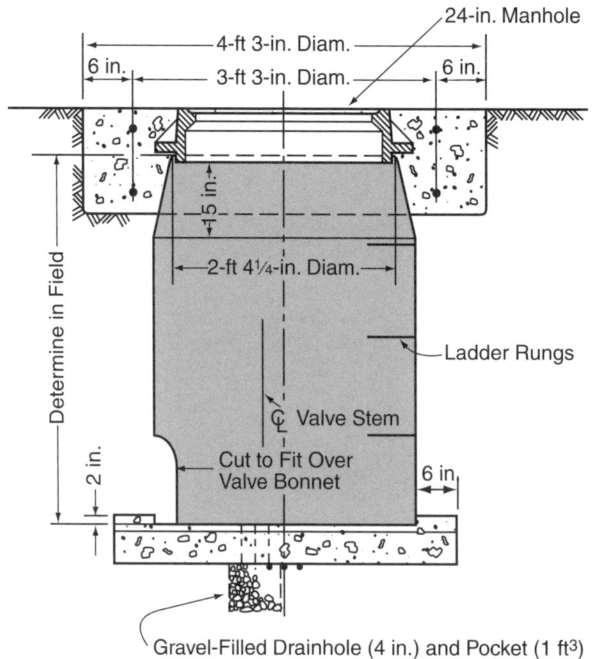

Source: Goit, L.E., Steel Pipeline Appurtenances. Jour. AWWA, *41:1:47 (Jan. 1949).*

Figure 13-28 Section of casing giving access to gate valve gearing

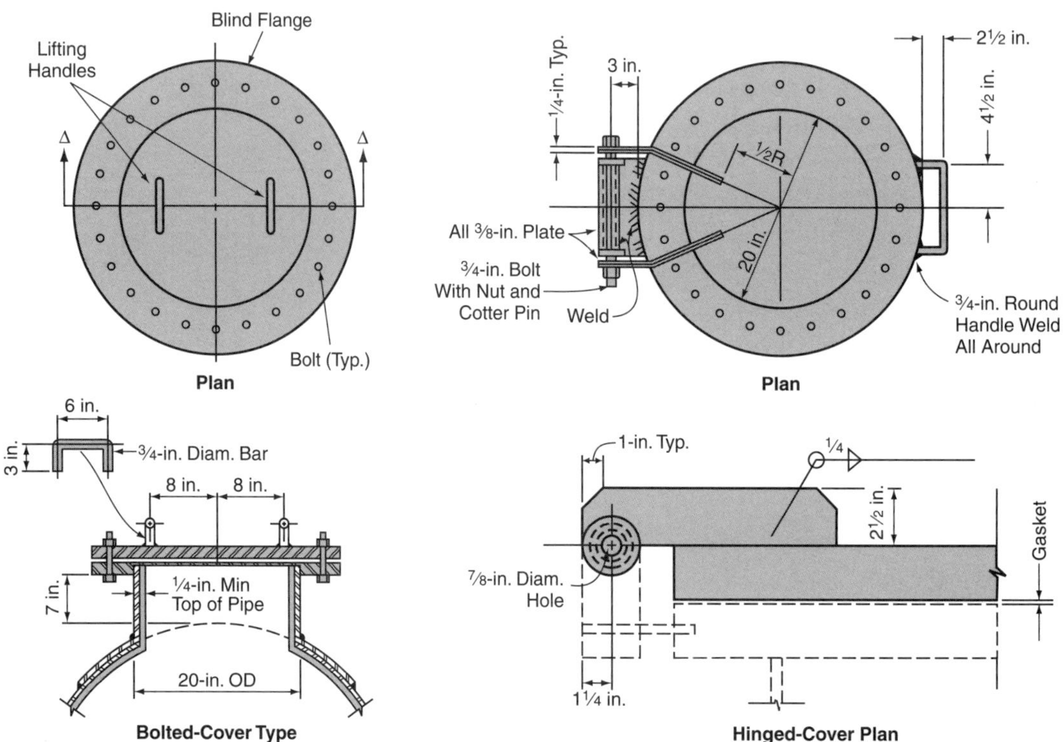

Figure 13-29 Access manhole

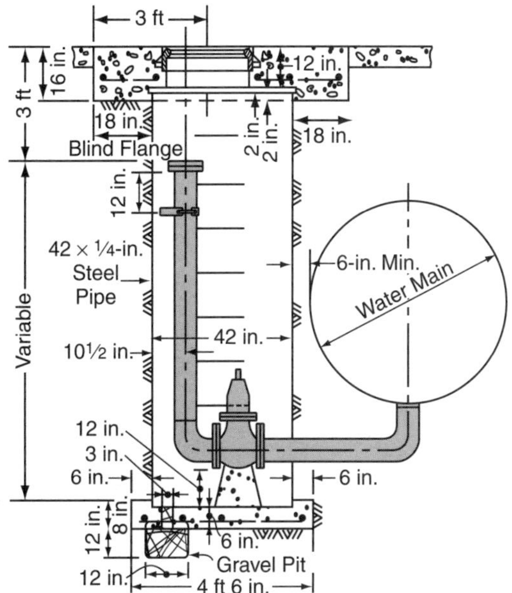

Source: Goit, L.E., Steel Pipeline Appurtenances. Jour. AWWA, *41:1:47 (Jan. 1949).*

Figure 13-30 Blowoff with riser for attaching pump section

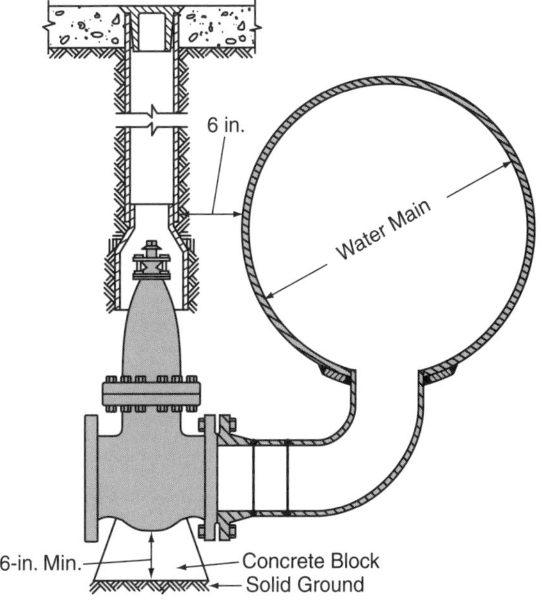

For tangent-type blowoff, See Figure 9-3.

Figure 13-31 Blowoff connection

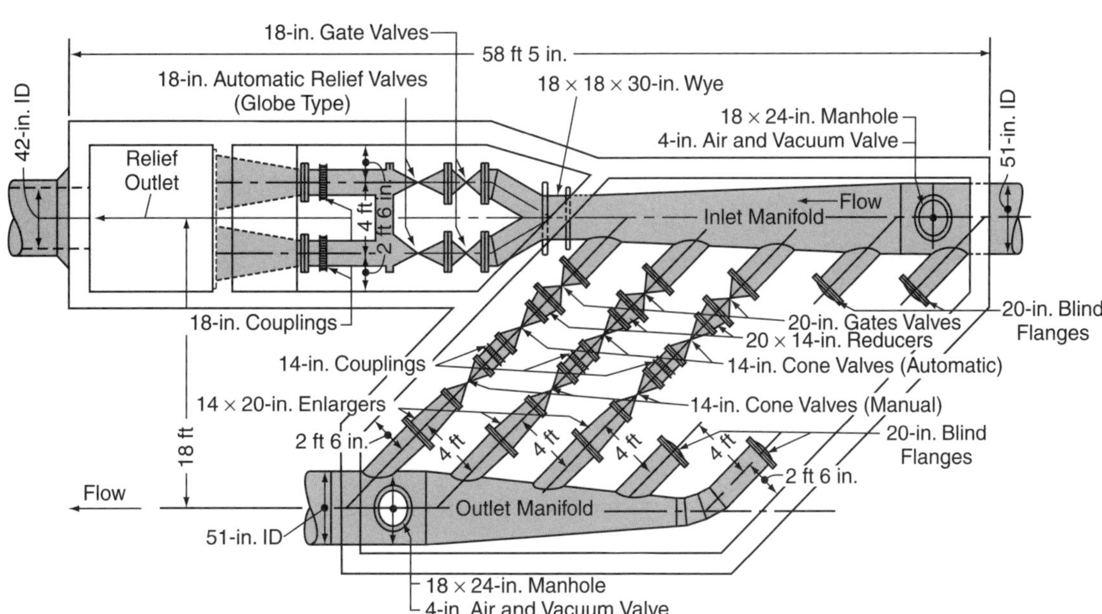

Source: Goit, L.E., Steel Pipeline Appurtenances. Jour. AWWA, *41:1:47 (Jan. 1949).*

Figure 13-32 Manifold layout of relief valves and pressure regulators

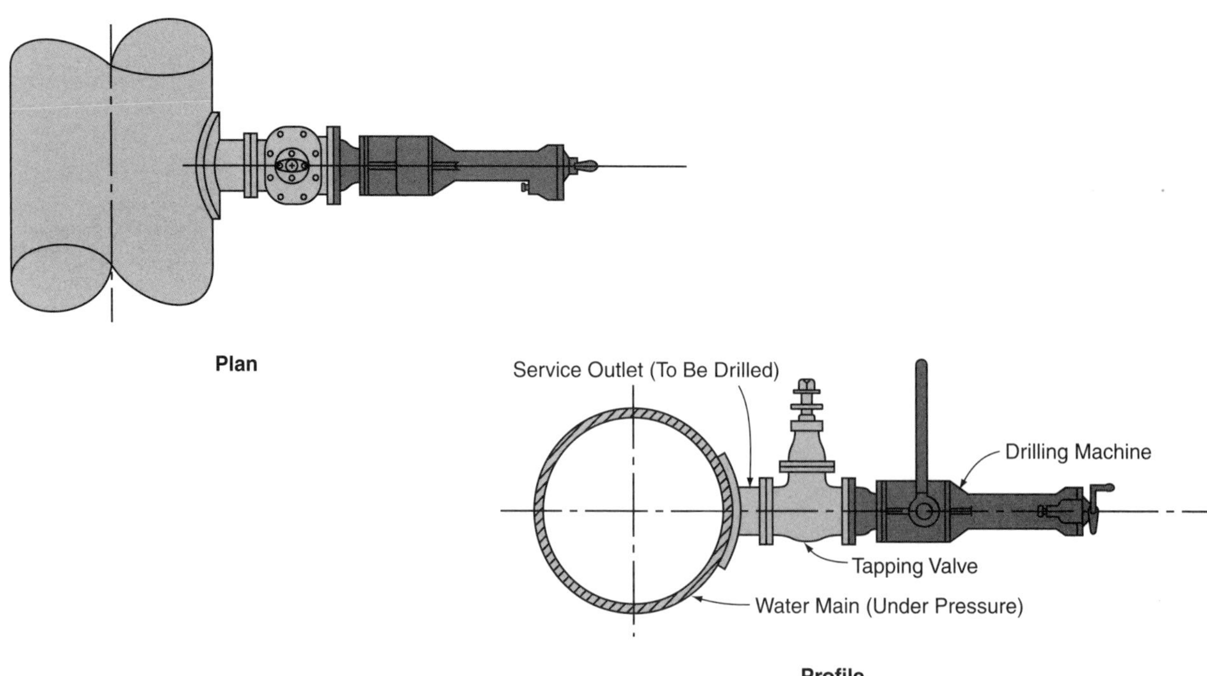

Procedure: (a) weld outlet and saddle to main; (b) bolt on gate valve, adapter (if required), and drilling machine; (c) insert tool and drill hole in main; (d) withdraw tool, close gate, and remove machine.

Figure 13-33 Tapping main under pressure

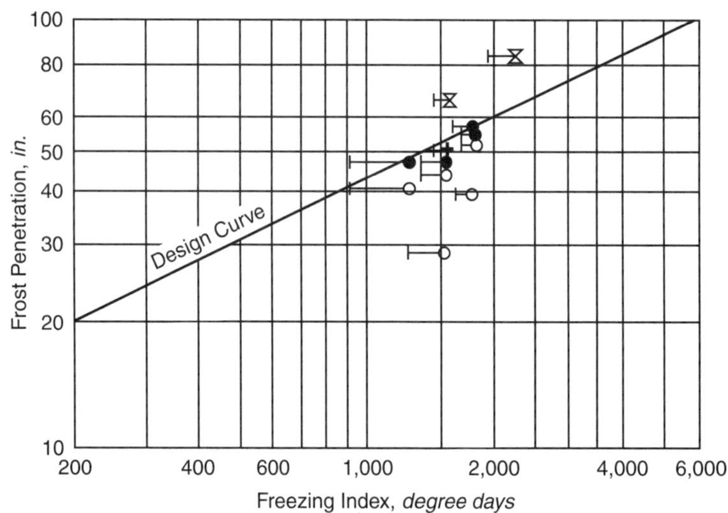

Source: Leggett, R.F. & Crawford, C.B., Soil Temperatures in Water Works Practice. Jour. AWWA, *44:10:923 (Oct. 1952).*

Measurements were made in Ottawa, Ont., 1947–51. Right end of each horizontal line indicates maximum frost depth at maximum freezing index; left end indicates freezing index at time of maximum frost depth; • refers to measurements made in sand (interpolated); ○ in clay (interpolated); × in sand (by excavation); and + in clay (by excavation).

Figure 13-34 Maximum frost penetration and maximum freezing index

Freezing in Underground Pipes

The freezing of water in buried pipes is usually caused by the cooling of the surrounding soil to a point below 32°F (0°C). Soil-temperature variations are related to the flow of heat in soils. Air temperature is the most important factor affecting soil temperature and frost penetration.

The most common method of expressing the seasonal effect of air temperatures on water is the freezing index (Leggett and Crawford 1952). The index is the cumulative total of degree-days below the freezing point in any winter. In this context, a degree-day is a unit representing 1 degree (F) of difference below 32°F in the mean outdoor temperature for one day. Values for midwinter days having temperatures above freezing—that is, negative-degree days—are subtracted from the total.

Temperature data for many localities are available (Heating, Ventilating and Air-Conditioning Guide). A design curve relating frost-depth penetration to the freezing index is shown in Figure 13-34. The curve was developed by the US Army Corps of Engineers (1947, 1949) from an analysis of frost penetration records of the northern United States. The data on the several soils in Figure 13-34 are from observations made at Ottawa, Ont. (Leggett and Crawford 1952).

Experimental work on the subject of frost penetration (Leggett and Crawford 1952) indicates that

- Theoretical equations for computation of frost depth may be erroneous. The US Army Corps of Engineers design curve (Figure 13-34) is the best aid currently available for estimating frost penetration.
- Frost penetration is significantly greater in disturbed soil than in undisturbed soil.
- Water pipes may safely be placed at less depth in clay soils than in sandy soils. Frost penetration has been found about 1½ times as deep in sand as in clay.
- Maximum frost penetration may occur several weeks before or after the freezing index reaches a winter maximum. Water mains have frozen as late as June in Winnipeg, Man.
- Frost penetrates deeper in soils on hillsides with northern exposure than in those with southern exposure.
- Undisturbed continuous snow cover has reduced frost penetration in the Ottawa climate by an amount equal to or greater than the snow-cover thickness.

Freezing in Exposed Pipes

Water in exposed pipelines will freeze when the available heat represented by the degrees of water temperature above the freezing point (32°F, 0°C) has been lost. Heat loss may be caused by radiation, convection, and conduction through the pipe wall and insulation, if any, and through the water film adjacent to the pipe wall. Data have been published supporting these conclusions (Riddick, Lindsay, Tomassi 1950). The heat balance is illustrated by Figure 13-35. The heat input is equal to $H_1 + H_2$, with H_1 being the British thermal units per square foot of exposed pipeline per hour available from the specific heat of water above 32°F at the inlet end, and H_2 being the British thermal units per square foot of pipe per hour generated by frictional energy. (Equations for H_1 and H_2 are given in Table 13-10.) The heat losses are given by

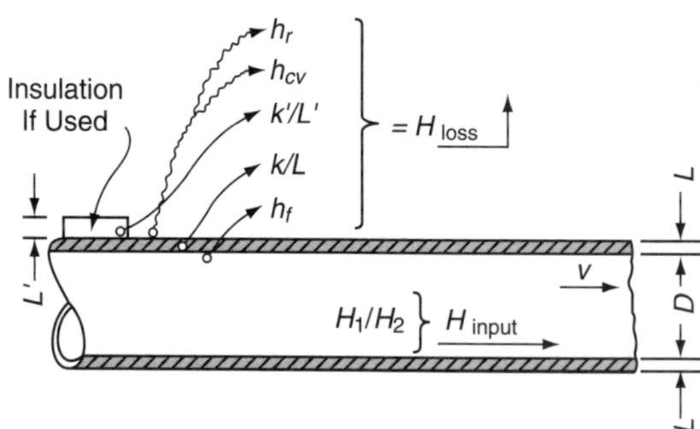

Source: Riddick, T.M., et al., Freezing of Water in Exposed Pipelines. Jour. AWWA, *42:11:1035 (Nov. 1950).*

Figure 13-35 Heat balance in exposed pipelines

$$H_{loss} = \frac{\Delta t}{\frac{1}{h_f} + \frac{L}{k} + \frac{L'}{k'} + \frac{1}{h_r + h_{cv}}} \tag{13-7}$$

Equations for values of factors and explanation of symbols are given in Table 13-10. Values of exponential powers of D and v are given in Table 13-11. Values for conduction heat-transfer, emissivity factors, and wind velocity factors are given in Tables 13-12, 13-13, and 13-14, respectively.

Example of calculation. *Problem*: An exposed, uninsulated steel pipeline is 48 in. in diameter, 0.25-in. thick, 10,000-ft long, and has a C factor of 140. Will this line freeze when carrying 25 mgd of water entering the pipe at 35°F, with an outside air temperature of –5°F, and a 35-mph wind blowing?

For the conditions given:

$$Q = 25, T_w = 35, D = 48, l = 10, f = 0.55, v = 3.1,$$
$$\Delta t = 492 - [460 + (-5)] = 37, V = 35$$

Solution:

Heat input equals $H_1 + H_2$.

$$H_1 = \frac{1{,}325(25)(35-32)}{48(10)} = 207 \text{ Btu/ft}^2\text{/h}$$

$$H_2 = \frac{1.70(25)(0.55)}{48} = 0.5 \text{ Btu/ft}^2\text{/h}$$

$$H_1 + H_2 = 207.5 \text{ Btu/ft}^2\text{/h}$$

Heat losses—calculated using Eq 13-7:

$$h_f = \frac{202(3.1)^{0.8}}{48^{0.2}} + \frac{499.4}{2.17} = 230$$

$$\frac{1}{h_f} = \frac{1}{230} = 0.0043$$

$$\frac{L}{K} = \frac{0.25}{420} = 0.0006$$

Table 13-10 Heat balance factors

Terminology and Required Data

Symbol	Meaning	Unit	Symbol	Meaning	Unit
	Climatological			Pipeline	
T_w	water temp. at pipe inlet	°F	v	velocity	fps
T_1	water freezing temp.	°F abs. = 492	C	friction coefficient*	
T_2	air temp.	°F abs.†	f	friction loss*	ft/1,000 ft
Δt	temp. differential between air and water ($T_1 - T_2$)	°F	l	length	1,000 ft
V	wind velocity	mph	L	wall thickness	in.
			L'	insulation thickness	in.
	Pipeline		k	pipe wall thermal conductivity coefficient	Btu/ft²/hr/°F/in.
Q	flow rate	mgd	k'	insulation thermal conductivity coefficient	Btu/ft²/hr/°F/in.
D	diameter‡	in.	E	emissivity (pipe wall, insulation)	

Heat-Loss Factors

Factor	Heat Transfer§
$h_r = \frac{0.17E}{\Delta t}\left[\left(\frac{T_1}{100}\right)^4 - \left(\frac{T_2}{100}\right)^4\right]$	by radiation
$h_{cv} = \sqrt{\frac{V+0.8}{0.8}}\left[0.55\left(\frac{\Delta t}{D}\right)^{0.25}\right]$	by convection
$\frac{k'}{L'}$	by conduction (through insulation)
$\frac{k}{L}$	by conduction (through pipe wall)
$h_f = \frac{202v^{0.8}}{D^{0.2}}$	through water film

Heat-Input Factors

$H_1 = \frac{1{,}325Q(T_w - 32)}{Dl}$	heat available for water above 32°F
$H_2 = \frac{1.70Qf}{D}$	heat generated by friction

Source: Riddick, T.M.; Lindsay, N.L.; and Tomassi, Antonio. Freezing of Water in Exposed Pipelines. Jour. AWWA, *42:11.1035 (Nov. 1950).*

* Hazen-Williams formula.

† Equals 460 plus air temperature in degrees Fahrenheit.

‡ Inside diameter for thin walls; average diameter for thick walls.

§ Per degree Fahrenheit differential.

Table 13-11 Values of D and v

Pipe Diameter D		$D^{0.2}$		Average Water Velocity v		$v^{0.8}$	
in.	*(mm)*	*in.*	*(mm)*	*fps*	*(m/s)*	*fps*	*(m/s)*
6	(150)	1.4	(2.7)	1	(0.30)	1.0	(0.39)
12	(300)	1.6	(3.1)	2	(0.61)	1.7	(0.67)
18	(450)	1.8	(3.4)	3	(0.91)	2.4	(0.93)
24	(600)	1.9	(3.6)	4	(1.22)	3.0	(1.17)
36	(900)	2.0	(3.9)	5	(1.52)	3.6	(1.40)
48	(1,200)	2.1	(4.1)	6	(1.83)	4.2	(1.62)
60	(1,500)	2.2	(4.3)	8	(2.44)	5.3	(2.04)
72	(1,800)	2.3	(4.5)	10	(3.05)	6.3	(2.44)
84	(2,100)	2.4	(4.6)	12	(3.66)	7.3	(2.82)
96	(2,400)	2.5	(4.7)	14	(4.27)	8.3	(3.19)
				16	(4.88)	9.2	(3.55)
				18	(5.49)	10.1	(3.90)
				20	(6.10)	11.0	(4.25)

Source: Riddick, T.M.; Lindsay, N.L.; & Tomassi, Antonio. Freezing of Water in Exposed Pipelines. Jour. AWWA, *42:11:1035 (Nov. 1950).*

Table 13-12 Conduction heat-transfer values

	Thermal Conductivity k*		Assumed Thickness		Heat Transfer Value	
Substance	*Btu/ft²/hr/ °F/in.*	*(kJ/m²/hr/ °C/mm)*	*in.*	*(mm)*	*Btu/ft²/h/°F*	*(kJ/m²/hr/°C)*
Pipe material						
Steel	420	(104.33)	0.25	(6.35)	1,240	(7,823.4)
Cast iron	385	(95.63)	0.75	(19.05)	515	(3,249.2)
Concrete	5.3	(1.32)	5.0	(127)	1.1	(6.9)
Wood stave	1.0	(0.25)	2.0	(50.8)	0.5	(3.2)
Aluminum	1,410	(350.23)	0.25	(6.35)	5,640	(35,583.8)
Asbestos cement	4.5	(1.12)	1.0	(25.4)	4.5	(28.4)
Insulator						
Dry air	0.17	(0.04)	2	(50.8)	0.08	(0.5)
Water	4.0	(0.99)	2	(50.8)	2.0	(12.6)
Ice	15.6	(3.87)	2	(50.8)	7.8	(49.2)
85% magnesia	0.4	(0.10)	2	(50.8)	0.2	(1.3)
"Foamglas"†	0.4	(0.10)	2	(50.8)	0.2	(1.3)

Source: Riddick, T.M.; Lindsay, N.L.; & Tomassi, Antonio. Freezing of Water in Exposed Pipelines. Jour. AWWA, *42:11:1035 (Nov. 1950).*

* Btu per square foot per hour per degree Fahrenheit differential per inch thickness of material.

† A product of Pittsburgh Corning Corp., Pittsburgh, Pa.

Table 13-13 Emissivity factors

Material	Emissivity Factor E
Asphaltic paint (black)	0.9
White enamel	0.9
Aluminum paint	0.4
Cast iron	0.7
Wood (dressed)	0.9
Asbestos cement	0.9
Aluminum	0.1
Brass or copper (with patina)	0.5

Source: Riddick, T.M.; Lindsay, N.L.; & Tomassi, Antonio. Freezing of Water in Exposed Pipelines. Jour. AWWA, *42:11:1035 (Nov. 1950).*

Table 13-14 Wind velocity factors

Wind Velocity		Factor
mph	*(km/hr)*	$\sqrt{\frac{V+0.8}{0.8}}$
5	(8.0)	2.7
10	(16.1)	3.7
20	(32.2)	5.1
30	(48.3)	6.2
40	(64.4)	7.1

Source: Riddick, T.M.; Lindsay, N.L.; & Tomassi, Antonio. Freezing of Water in Exposed Pipelines. Jour. AWWA, *42:11:1035 (Nov. 1950).*

$$h_r = \frac{0.17(0.7)}{37}\left[\left(\frac{492}{100}\right)^4 - \left(\frac{455}{100}\right)^4\right] = 0.5061$$

$$h_{cv} = \sqrt{\frac{35+0.8}{0.8}}(0.55)\left(\frac{37}{48}\right)^{0.25}$$

$$= 6.69(0.55)(0.937) = 3.4477$$

$$h_r + h_{cv} = 0.5061 + 3.4477 = 3.9538$$

$$\frac{1}{h_r + h_{cv}} = \frac{1}{3.9538} = 0.2529$$

$$H_{\text{loss}} = \frac{37}{0.0043 + 0.0006 + 0.2529} = \frac{37}{0.2569}$$

$$= 143.5 \text{ Btu/ft}^2\text{/h}$$

Because the heat input is 207.5 Btu/ft^2/h and the heat loss is 143.5 Btu/ft^2/h, the pipeline is safe against freezing under the design conditions. Further calculation

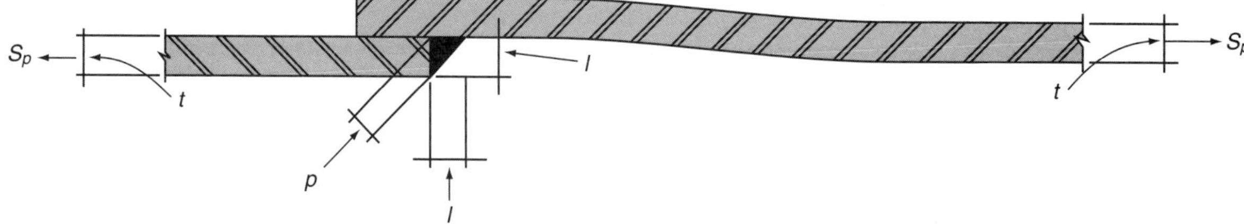

Figure 13-36 Fillet nomenclature

shows that, for the same temperature conditions, heat input and heat loss are equal when the pipeline is about 14,800 ft long and the velocity is 3.1 fps; or, stated conversely, the velocity in the 10,000-ft line could be as low as about 2.1 fps before ice might form near the outlet.

Warming of Water in Exposed Pipelines

In desert areas and in the tropics, the rise in water temperature caused by exposure of pipe to sun and wind should be determined. In this case, the heat input is calculated in accordance with the same basic principles of heat transfer used to determine heat losses in lines undergoing cooling. The values of factors used in the equations in a given instance should be determined locally. Data applicable to calculation of heat loads for air-conditioning and cooling units may be helpful.

DESIGN OF CIRCUMFERENTIAL FILLET WELDS

Any weld that is continuous will contain water, so weld size is insignificant from a seal-weld aspect. Once welded, the weld must withstand any longitudinal forces applied to it; i.e., it does not behave as an unstressed seal weld, because it is the only restraint that prevents motion of the pipe at the joint. In areas of a pipeline not affected by pipeline features that give rise to longitudinal stresses (elbows, valves, reducers, etc.), the only longitudinal stress normally encountered is caused by change in temperature or to beam bending from uneven settlement of the pipeline. To minimize longitudinal stresses, it is customary in specifications to call for one joint every 400–500 ft to be left unwelded until the joints on both sides of it are welded. This joint is later welded at the coolest time during the working day. Determination of weld size then is as follows (see Figure 13-36):

l = fillet weld leg size (in.)
p = throat dimension (in.)
ΔT = temperature change (°F)
T = temperature (°F)
K = constant linear coefficient of thermal expansion for steel = 6.5×10^{-6} in./in./°F
L = length of pipeline (ft)
ΔL = change in length (ft)
E = Young's modulus = 30×10^6 psi
S_p = stress in pipe wall (psi)
S_w = stress in weld (psi)
t = pipe wall (in.)

Assume an anchored straight pipeline is welded at a temperature T_1. The temperature is then reduced to T_2. The pipeline would then tend to reduce in length by an amount that is a function of $T_1 - T_2$, L, and K. Because the ends are anchored, the length cannot be changed.

Therefore, the stress in the line is the same as that which would exist if the line were stretched by the same amount that it would shorten if it were free to do so. This is a conservative assumption. Few pipelines are perfectly straight between anchor points, and temperature changes are usually gradual, so most lines actually can change their length by a small amount, relieving the thermal stress somewhat. To calculate thermal stress:

Shortening caused by temperature change ΔL_T is found as

(1) $$T_1 - T_2 = \Delta T$$

$$\frac{\Delta L_T}{L} = \Delta T(K)$$

(2) $$\Delta L_T = L(\Delta T)(K)$$

Elongation caused by longitudinal tension ΔL_s is found as

(3) $$\frac{\Delta L_s}{L} = \frac{S_p}{E}$$

(4) $$\Delta L_s = \frac{S_{pL}}{E}$$

According to the assumption above

ΔL_s (stress elongation) = ΔL_T (temperature shortening)

Substituting (2) into (4)

(5) $$\Delta L_s = \Delta L_T$$

(6) $$L(\Delta T)(K) = \frac{S_p L}{E}$$

Simplified

(7) $$\Delta T(K) = \frac{S_p}{E}$$

$$E = 30 \times 10^6 \text{psi}$$

$$S_p = \Delta T(6.5 \times 10^{-6})(30 \times 10^6) = \Delta T(195 \text{ psi})$$

For 40°F change in temperature

$$S_p = 40(195) = 7{,}800 \text{ psi (stress in pipe wall)}$$

Calculation for fillet size:

The weld must carry its load through its least dimension (its throat). To be conservative, assume no penetration at the throat. The full force of a unit length of pipe wall in the circumferential direction must be carried by a unit length of fillet weld throat also measured in the circumferential direction. Call this unit length Z.

(8) $$pZS_w = tZS_p$$

$$pS_w = TS_p$$

$$p = \frac{tS_p}{S_w}$$

The weld metal is as strong as the parent metal, so consider the allowable stress to be 15,000 psi (½ yield) in the pipe wall.

(9) then $S_w = 15{,}000$ psi

for $\Delta T = 40°F$

$S_p = 7{,}800$ psi

$$p = \frac{t(7{,}800)}{15{,}000}$$

$$p = 0.52t$$

$$p = 0.707$$

leg size then is

$$0.7071l = 0.5064t$$

$$l = 0.736t$$

NOTE: In areas where a valve anchor block or other pipeline appurtenances can introduce tension into the line, the tension caused by the appurtenances should be checked to determine if it establishes the minimum fillet size. These axial stresses can never exceed half the hoop stress caused by internal pressure. That tension and the thermal tension are not additive because the tension can only exist if the pipe is not restrained, and the thermal stress can only exist if it is restrained. The greater tension applies for design purposes.

SUBMARINE PIPELINES

The type of construction used has a great influence on the design and on total costs of a system. A brief discussion of different available construction techniques will illustrate their effect. There are basically two systems for constructing submarine pipelines: pipe-laying systems and pipe-pulling systems.

Pipe Laying

In a pipe-laying system, the pipe is transported by water to the laying platform, which is a barge equipped primarily with a heavy crane and horse. The horse is a winch capable of moving on skid beams in two directions with cables extending vertically downward into the water. On arrival at the job site, the crane picks up the pipe segment and holds it while the horse is centered above it. The pipe, once attached to the horse, is lowered to the bottom. Divers report the position of the segment in relation to the completed section before it, and the horse is moved up and down, forward and backward, and sideways until the spigot end lines up with the bell end of the completed section.

Pipe Pulling

Pipe pulling has been used for crossing rivers, bays, and in the open ocean. The pipe-pulling method requires pipe capable of withstanding the tensile stresses developed during the pulling operation. The method is usually used with steel pipe because of these high tensile stresses.

A steel-pipe-pulling operation begins on assembly ways established ashore, on which all the pipe is coated and wrapped. To prevent floating, the pipe may be

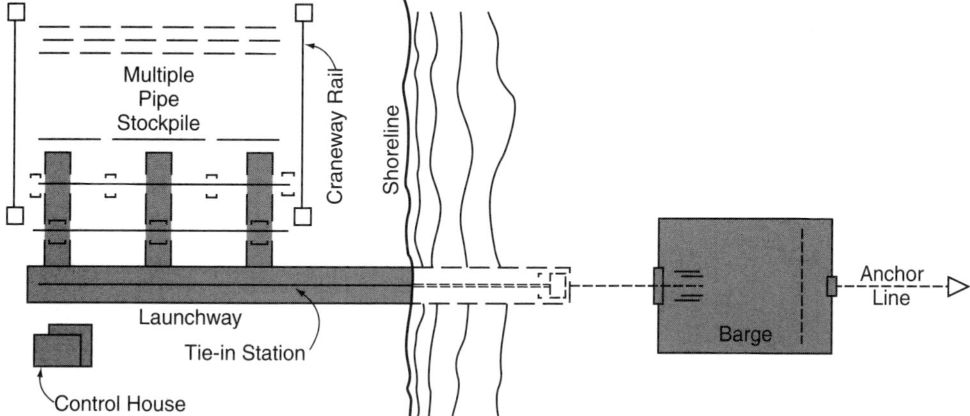

Source: Hayden, W.M. & Piaseckyi, P.J., Economic and Other Design Considerations for a Large Diameter Pipeline. Proc. Sixth Intern. Harbour Congress *(K Vlaam Ingenieursver publisher). Antwerp, Belgium (1974).*

Figure 13-37 Submarine pipeline—assembly and launching

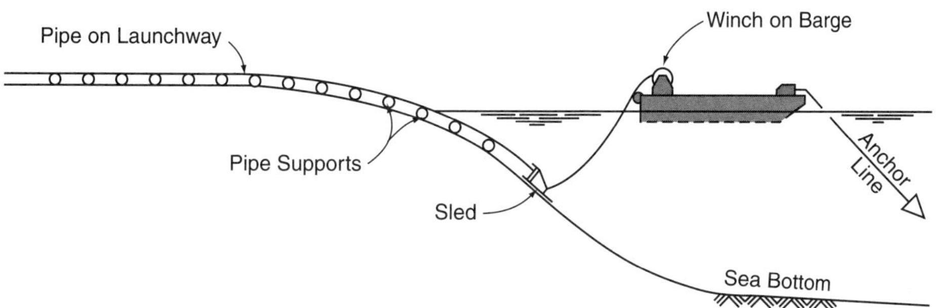

Source: Hayden, W.M. & Piaseckyi, P.J., Economic and Other Design Considerations for a Large Diameter Pipeline. Proc. Sixth Intern. Harbour Congress *(K Vlaam Ingenieursver publisher). Antwerp, Belgium (1974).*

Figure 13-38 Submarine pipeline—positioning by barge

allowed to fill with water as it leaves the assembly way. Alternatively, the pipe may be capped to exclude water, then concrete weighted or coated to overcome its buoyancy. The pipe lengths are welded in continuous strings. The completed pipe string is transferred to launchways (Figure 13-37), which lead to the submerged placement area. Once shore assembly is complete, the reinforced head of the pipe string is attached to a pull barge by wire rope and pulled along the bottom by a winch until it is in position (Figure 13-38).

A variation of the bottom-pull method is the floating-string method of pipe installation. The line is initially assembled in long segments and transferred to the launchways. It is then pulled off the launchway by a tugboat, floated out to location, and sunk (Figure 13-39). Individual strings are connected by divers, as in the pipe-laying method, or strings are joined by picking up the end of the last piece installed and putting it on a deck of a special tie-in platform, where the connection to the beginning of the next string is made.

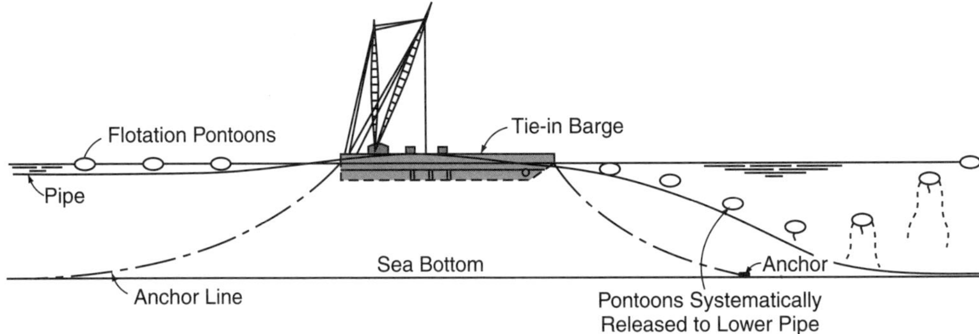

Source: Hayden, W.M. & Piaseckyi, P.J., Economic and Other Design Considerations for a Large Diameter Pipeline. Proc. Sixth Intern. Harbour Congress *(K Vlaam Ingenieursver publisher). Antwerp, Belgium (1974).*

Figure 13-39 Submarine pipeline—floating string positioning

Lay Barge

Smaller diameter pipelines are sometimes laid at sea or across rivers from a lay barge, which has on-board facilities for welding pipe sections together. The pipe string is fed over the end of the barge as the barge moves along the route of the pipeline, adding pipe as it goes. The pipe undergoes bending stresses as it is laid, so the barge should include quality-control facilities for checking the soundness of the circumferential welds.

REFERENCES

American Society for Testing and Materials. 1977. ASTM Standard A36-77, *Specification for Structural Steel.* Philadelphia, Pa.: American Society for Testing and Materials.

———. 1979. ASTM Standard A283-79, *Specifications for Low and Intermediate Tensile Strength Carbon Steel Plates, Shapes and Bars.* Philadelphia, Pa.: American Society for Testing and Materials.

———. 1980a. ASTM Standard A193-80, *Specifications for Alloy-Steel and Stainless Steel Bolting Materials for High-Temperature Service.* Philadelphia, Pa.: American Society for Testing and Materials.

———. 1980b. ASTM Standard A194-80, *Specifications for Carbon and Alloy Steel Nuts for Bolts for High-Pressure and High-Temperature Service.* Philadelphia, Pa.: American Society for Testing and Materials.

ASME Unfired Pressure Vessel Code.

Barnard, R.E. 1948. Design Standards for Steel Water Pipe. *Jour. AWWA*, 40(1):24.

Goit, L.E. 1949. Steel Pipeline Appurtenances. *Jour. AWWA*, 41(1):47.

Heating, Ventilating and Air-Conditioning Guide. New York: American Society of Heating and Air Conditioning Engineers.

Leggett, R.F., and C.B. Crawford. 1952. Soil Temperatures in Water Works Practice. *Jour. AWWA*, 44(10):923.

Riddick, T.M., N.L. Lindsay, and A. Tomassi. 1950. Freezing of Water in Exposed Pipelines. *Jour. AWWA*, 42(11):1035.

Swanson, H.S., et al. 1955. Design of Wye Branches for Steel Pipe. *Jour. AWWA*, 47(6):581.

US Army Corps of Engineers. 1947. *Report on Frost Penetration (1944–45).* Boston: US Army Corps of Engineers, New England Division.

US Army Corps of Engineers. 1949. *Report on Frost Penetration (1944–45). Addendum No. 1, 1945–47.* Boston: US Army Corps of Engineers, New England Division.

The following references are not cited in the text.

Hertzberg, L.B. 1956. Suggested Nontechnical Manual on Corrosion for Water Works Operators. *Jour. AWWA*, 48(6):719.

Shannon, W.L. 1945. Prediction of Frost Penetration. *Jour. NEWWA*, 59:356.

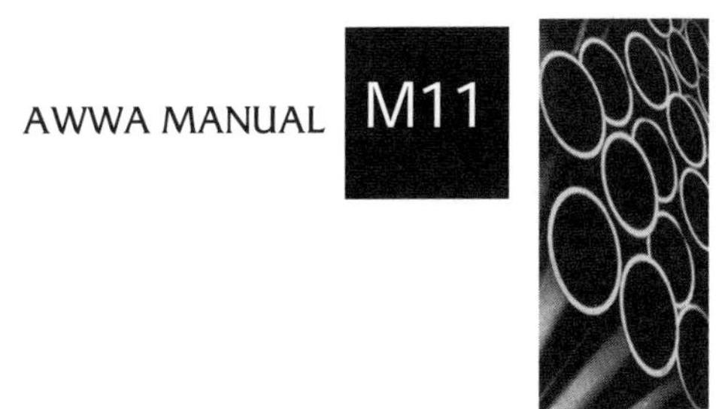

Appendix A

Table of Working Pressures for Allowable Unit Stresses

Table A-1 Working pressures for allowable unit stresses

Pipe OD	Wall Thickness	Pipe Weight	*D/t*	Moment of Inertia	Section Modulus	15,000 psi	16,500 psi	17,500 psi	18,000 psi	21,000 psi	23,000 psi	26,000 psi
in.	*in.*	*lb/LF*	Ratio	*in.*^{4}	*in.*^{3}	*psi*	*psi*	*psi*	*psi*	*psi*	*psi*	*psi*
4	0.0747	3.13	53.55	1.77	0.89	560	616	654	672	784	859	971
	0.1046	4.35	38.24	2.43	1.21	785	863	915	941	1,098	1,203	1,360
	0.1345	5.55	29.74	3.05	1.53	1,009	1,110	1,177	1,211	1,412	1,547	1,749
4.5	0.0747	3.53	60.24	2.54	1.13	498	548	581	598	697	764	863
	0.1046	4.91	43.02	3.49	1.55	697	767	814	837	976	1,069	1,209
	0.1345	6.27	33.46	4.40	1.95	897	986	1,046	1,076	1,255	1,375	1,554
6	0.0747	4.73	80.32	6.10	2.03	374	411	436	448	523	573	647
	0.1046	6.59	57.36	8.42	2.81	523	575	610	628	732	802	907
	0.1345	8.43	44.61	10.66	3.55	673	740	785	807	942	1,031	1,166
6.63	0.0747	5.23	88.69	8.25	2.49	338	372	395	406	474	519	586
	0.1046	7.29	63.34	11.39	3.44	474	521	553	568	663	726	821
	0.1345	9.32	49.26	14.45	4.36	609	670	711	731	853	934	1,056
8	0.0747	6.32	107.10	14.60	3.65	280	308	327	336	392	430	486
	0.1046	8.82	76.48	20.22	5.06	392	431	458	471	549	601	680
	0.1345	11.30	59.48	25.71	6.43	504	555	588	605	706	773	874
8.63	0.0747	6.82	115.46	18.34	4.25	260	286	303	312	364	398	450
	0.1046	9.52	82.46	25.41	5.89	364	400	424	437	509	558	631
	0.1345	12.20	64.13	32.34	7.50	468	515	546	561	655	717	811
10	0.0747	7.92	133.87	28.68	5.74	224	247	261	269	314	344	388
	0.1046	11.06	95.60	39.81	7.96	314	345	366	377	439	481	544
	0.1345	14.17	74.35	50.72	10.14	404	444	471	484	565	619	699
	0.1793	18.81	55.77	66.71	13.34	538	592	628	645	753	825	932
10.8	0.0747	8.52	143.91	35.69	6.64	208	229	243	250	292	320	361
	0.1046	11.89	102.77	49.56	9.22	292	321	341	350	409	448	506
	0.1345	15.25	79.93	63.19	11.76	375	413	438	450	525	576	651
	0.1793	20.24	59.96	83.19	15.48	500	550	584	600	701	767	867
12.0	0.0747	9.52	160.64	49.75	8.29	187	205	218	224	261	286	324
	0.1046	13.29	114.72	69.15	11.52	262	288	305	314	366	401	453
	0.1345	17.05	89.22	88.25	14.71	336	370	392	404	471	516	583
	0.1793	22.64	66.93	116.32	19.39	448	493	523	538	628	687	777
12.8	0.0747	10.11	170.68	59.74	9.37	176	193	205	211	246	270	305
	0.1046	14.13	121.89	83.07	13.03	246	271	287	295	345	377	427
	0.1345	18.12	94.80	106.06	16.64	316	348	369	380	443	485	549
	0.1793	24.08	71.11	139.90	21.94	422	464	492	506	591	647	731
14	0.0747	11.11	187.42	79.21	11.32	160	176	187	192	224	245	277
	0.1046	15.52	133.84	110.21	15.74	224	247	262	269	314	344	389
	0.1345	19.92	104.09	140.81	20.12	288	317	336	346	404	442	500
	0.1563	23.11	89.57	162.87	23.27	335	368	391	402	469	514	581
	0.1793	26.47	78.08	185.91	26.56	384	423	448	461	538	589	666
	0.2188	32.21	63.99	224.95	32.14	469	516	547	563	656	719	813
	0.2500	36.72	56.00	255.30	36.47	536	589	625	643	750	821	929
16	0.0747	12.71	214.19	118.48	14.81	140	154	163	168	196	215	243
	0.1046	17.76	152.96	164.98	20.62	196	216	229	235	275	301	340
	0.1345	22.79	118.96	210.95	26.37	252	277	294	303	353	387	437
	0.1563	26.45	102.37	244.14	30.52	293	322	342	352	410	449	508

NOTE: To convert inches to millimeters, multiply inches by 25.4.
To convert psi to kPa, multiply psi by 6.895.

Table continued next page

Table A-1 Working pressures for allowable unit stresses (continued)

Pipe OD *in.*	Wall Thickness *in.*	Pipe Weight *lb/LF*	*D/t* Ratio	Moment of Inertia *in.*^{4}	Section Modulus *in.*^{3}	15,000 psi *psi*	16,500 psi *psi*	17,500 psi *psi*	18,000 psi *psi*	21,000 psi *psi*	23,000 psi *psi*	26,000 psi *psi*
	0.1793	30.30	89.24	278.85	34.86	336	370	392	403	471	515	583
	0.2188	36.88	73.13	337.76	42.22	410	451	479	492	574	629	711
	0.2500	42.06	64.00	383.66	47.96	469	516	547	563	656	719	813
18	0.0747	14.30	240.96	168.96	18.77	125	137	145	149	174	191	216
	0.1046	19.99	172.08	235.41	26.16	174	192	203	209	244	267	302
	0.1345	25.67	133.83	301.20	33.47	224	247	262	269	314	344	389
	0.1563	29.79	115.16	348.74	38.75	261	287	304	313	365	399	452
	0.1793	34.13	100.39	398.53	44.28	299	329	349	359	418	458	518
	0.2188	41.56	82.27	483.12	53.68	365	401	425	438	511	559	632
	0.2500	47.40	72.00	549.14	61.02	417	458	486	500	583	639	722
20	0.0747	15.90	267.74	232.06	23.21	112	123	131	134	157	172	194
	0.1046	22.23	191.20	323.49	32.35	157	173	183	188	220	241	272
	0.1345	28.54	148.70	414.10	41.41	202	222	235	242	282	309	350
	0.1563	33.13	127.96	479.64	47.96	234	258	274	281	328	359	406
	0.1793	37.96	111.54	548.32	54.83	269	296	314	323	377	412	466
	0.2188	46.23	91.41	665.15	66.51	328	361	383	394	459	503	569
	0.2500	52.74	80.00	756.43	75.64	375	413	438	450	525	575	650
22	0.1046	24.46	210.33	431.18	39.20	143	157	166	171	200	219	247
	0.1345	31.41	163.57	552.17	50.20	183	202	214	220	257	281	318
	0.1563	36.47	140.75	639.76	58.16	213	234	249	256	298	327	369
	0.1793	41.79	122.70	731.60	66.51	245	269	285	293	342	375	424
	0.2188	50.90	100.55	887.97	80.72	298	328	348	358	418	457	517
	0.2500	58.08	88.00	1,010.26	91.84	341	375	398	409	477	523	591
24	0.1046	26.70	229.45	560.46	46.70	131	144	153	157	183	200	227
	0.1345	34.29	178.44	717.97	59.83	168	185	196	202	235	258	291
	0.1563	39.81	153.55	832.07	69.34	195	215	228	234	274	300	339
	0.1793	45.62	133.85	951.76	79.31	224	247	261	269	314	344	388
	0.2188	55.58	109.69	1,155.70	96.31	274	301	319	328	383	419	474
	0.2500	63.42	96.00	1,315.34	109.61	313	344	365	375	438	479	542
	0.3125	79.07	76.80	1,631.33	135.94	391	430	456	469	547	599	677
	0.3750	94.63	64.00	1,942.30	161.86	469	516	547	563	656	719	813
	0.4375	110.11	54.86	2,248.29	187.36	547	602	638	656	766	839	948
	0.5000	125.51	48.00	2,549.35	212.45	625	688	729	750	875	958	1,083
26	0.1046	28.93	248.57	713.29	54.87	121	133	141	145	169	185	209
	0.1345	37.16	193.31	914.02	70.31	155	171	181	186	217	238	269
	0.1563	43.15	166.35	1,059.49	81.50	180	198	210	216	252	277	313
	0.1793	49.45	145.01	1,212.17	93.24	207	228	241	248	290	317	359
	0.2188	60.25	118.83	1,472.47	113.27	252	278	295	303	353	387	438
	0.2500	68.76	104.00	1,676.38	128.95	288	317	337	346	404	442	500
	0.3125	85.74	83.20	2,080.36	160.03	361	397	421	433	505	553	625
	0.3750	102.64	69.33	2,478.42	190.65	433	476	505	519	606	663	750
	0.4375	119.46	59.43	2,870.61	220.82	505	555	589	606	707	774	875
	0.5000	136.19	52.00	3,256.99	250.54	577	635	673	692	808	885	1,000
28	0.1046	31.17	267.69	891.65	63.69	112	123	131	134	157	172	194
	0.1345	40.03	208.18	1,142.86	81.63	144	159	168	173	202	221	250

NOTE: To convert inches to millimeters, multiply inches by 25.4.
To convert psi to kPa, multiply psi by 6.895.

Table continued next page

Table A-1 Working pressures for allowable unit stresses (continued)

Pipe OD *in.*	Wall Thickness *in.*	Pipe Weight *lb/LF*	*D/t* Ratio	Moment of Inertia *in.*^{4}	Section Modulus *in.*^{3}	15,000 psi *psi*	16,500 psi *psi*	17,500 psi *psi*	18,000 psi *psi*	21,000 psi *psi*	23,000 psi *psi*	26,000 psi *psi*
	0.1563	46.49	179.14	1,324.99	94.64	167	184	195	201	234	257	290
	0.1793	53.28	156.16	1,516.22	108.30	192	211	224	231	269	295	333
	0.2188	64.93	127.97	1,842.41	131.60	234	258	274	281	328	359	406
	0.2500	74.10	112.00	2,098.09	149.86	268	295	313	321	375	411	464
	0.3125	92.42	89.60	2,605.05	186.08	335	368	391	402	469	513	580
	0.3750	110.65	74.67	3,105.12	221.79	402	442	469	482	563	616	696
	0.4375	128.80	64.00	3,598.35	257.02	469	516	547	563	656	719	813
	0.5000	146.87	56.00	4,084.80	291.77	536	589	625	643	750	821	929
30	0.1046	33.40	286.81	1,097.51	73.17	105	115	122	126	146	160	181
	0.1345	42.91	223.05	1,407.02	93.80	135	148	157	161	188	206	233
	0.1563	49.82	191.94	1,631.51	108.77	156	172	182	188	219	240	271
	0.1793	57.11	167.32	1,867.28	124.49	179	197	209	215	251	275	311
	0.2188	69.60	137.11	2,269.64	151.31	219	241	255	263	306	335	379
	0.2500	79.44	120.00	2,585.18	172.35	250	275	292	300	350	383	433
	0.3125	99.10	96.00	3,211.28	214.09	313	344	365	375	438	479	542
	0.3750	118.66	80.00	3,829.44	255.30	375	413	438	450	525	575	650
	0.4375	138.15	68.57	4,439.73	295.98	438	481	510	525	613	671	758
	0.5000	157.55	60.00	5,042.20	336.15	500	550	583	600	700	767	867
32	0.1345	45.78	237.92	1,709.04	106.81	126	139	147	151	177	193	219
	0.1563	53.16	204.73	1,981.98	123.87	147	161	171	176	205	225	254
	0.1793	60.94	178.47	2,268.73	141.80	168	185	196	202	235	258	291
	0.2188	74.28	146.25	2,758.28	172.39	205	226	239	246	287	315	356
	0.2500	84.78	128.00	3,142.37	196.40	234	258	273	281	328	359	406
	0.3125	105.77	102.40	3,904.95	244.06	293	322	342	352	410	449	508
	0.3750	126.67	85.33	4,658.47	291.15	352	387	410	422	492	539	609
	0.4375	147.50	73.14	5,403.00	337.69	410	451	479	492	574	629	711
	0.5000	168.23	64.00	6,138.62	383.66	469	516	547	563	656	719	813
34	0.1345	48.65	252.79	2,051.45	120.67	119	131	138	142	166	182	206
	0.1563	56.50	217.53	2,379.37	139.96	138	152	161	165	193	211	239
	0.1793	64.77	189.63	2,723.95	160.23	158	174	185	190	221	243	274
	0.2188	78.95	155.39	3,312.46	194.85	193	212	225	232	270	296	335
	0.2500	90.12	136.00	3,774.37	222.02	221	243	257	265	309	338	382
	0.3125	112.45	108.80	4,691.95	276.00	276	303	322	331	386	423	478
	0.3750	134.69	90.67	5,599.27	329.37	331	364	386	397	463	507	574
	0.4375	156.84	77.71	6,496.42	382.14	386	425	450	463	540	592	669
	0.5000	178.91	68.00	7,383.47	434.32	441	485	515	529	618	676	765
36	0.1345	51.53	267.66	2,436.79	135.38	112	123	131	135	157	172	194
	0.1563	59.84	230.33	2,826.61	157.03	130	143	152	156	182	200	226
	0.1793	68.60	200.78	3,236.33	179.80	149	164	174	179	209	229	259
	0.2188	83.62	164.53	3,936.30	218.68	182	201	213	219	255	280	316
	0.2500	95.47	144.00	4,485.89	249.22	208	229	243	250	292	319	361
	0.3125	119.12	115.20	5,578.16	309.90	260	286	304	313	365	399	451
	0.3750	142.70	96.00	6,658.92	369.94	313	344	365	375	438	479	542
	0.4375	166.19	82.29	7,728.23	429.35	365	401	425	438	510	559	632
	0.5000	189.59	72.00	8,786.19	488.12	417	458	486	500	583	639	722

NOTE: To convert inches to millimeters, multiply inches by 25.4.
To convert psi to kPa, multiply psi by 6.895.

Table continued next page

Table A-1 Working pressures for allowable unit stresses (continued)

Pipe OD *in.*	Wall Thickness *in.*	Pipe Weight *lb/LF*	*D/t* Ratio	Moment of Inertia *in.*^{4}	Section Modulus *in.*^{3}	15,000 psi	16,500 psi	17,500 psi	18,000 psi	21,000 psi	23,000 psi	26,000 psi
						psi	*psi*	*psi*	*psi*	*psi*	*psi*	*psi*
9	0.1563	64.85	249.52	3,597.39	184.48	120	132	140	144	168	184	208
	0.1793	74.35	217.51	4,119.45	211.25	138	152	161	166	193	211	239
	0.2188	90.64	178.24	5,011.69	257.01	168	185	196	202	236	258	292
	0.2500	103.48	156.00	5,712.59	292.95	192	212	224	231	269	295	333
	0.3125	129.14	124.80	7,106.40	364.43	240	264	280	288	337	369	417
	0.3750	154.71	104.00	8,486.67	435.21	288	317	337	346	404	442	500
	0.4375	180.21	89.14	9,853.47	505.31	337	370	393	404	471	516	583
	0.5000	205.62	78.00	11,206.89	574.71	385	423	449	462	538	590	667
40	0.1563	66.52	255.92	3,882.44	194.12	117	129	137	141	164	180	203
	0.1793	76.26	223.09	4,446.06	222.30	134	148	157	161	188	206	233
	0.2188	92.97	182.82	5,409.46	270.47	164	181	191	197	230	252	284
	0.2500	106.15	160.00	6,166.35	308.32	188	206	219	225	263	288	325
	0.3125	132.47	128.00	7,671.81	383.59	234	258	273	281	328	359	406
	0.3750	158.72	106.67	9,163.00	458.15	281	309	328	338	394	431	488
	0.4375	184.88	91.43	10,640.01	532.00	328	361	383	394	459	503	569
	0.5000	210.96	80.00	12,102.93	605.15	375	413	438	450	525	575	650
42	0.1563	69.86	268.71	4,496.92	214.14	112	123	130	134	156	171	194
	0.1793	80.09	234.24	5,150.17	245.25	128	141	149	154	179	196	222
	0.2188	97.65	191.96	6,267.03	298.43	156	172	182	188	219	240	271
	0.2500	111.49	168.00	7,144.71	340.22	179	196	208	214	250	274	310
	0.3125	139.15	134.40	8,891.02	423.38	223	246	260	268	313	342	387
	0.3750	166.73	112.00	10,621.57	505.79	268	295	313	321	375	411	464
	0.4375	194.23	96.00	12,336.47	587.45	313	344	365	375	438	479	542
	0.5000	221.64	84.00	14,035.79	668.37	357	393	417	429	500	548	619
45	0.2500	120.83	182.00	9,096.38	399.84	167	183	194	200	233	253	286
	0.3125	151.25	146.00	11,417.85	500.51	208	229	243	250	292	315	356
	0.3750	181.75	122.00	13,758.48	601.46	250	275	292	300	350	377	426
	0.4375	212.33	104.86	16,118.37	702.71	292	321	340	350	408	439	496
	0.5000	243.00	92.00	18,497.64	804.25	333	367	389	400	467	500	565
	0.5625	273.75	82.00	20,896.37	906.08	375	413	438	450	525	561	634
	0.6250	304.59	74.00	23,314.69	1,008.20	417	458	486	500	583	622	703
	0.6875	335.50	67.45	25,752.70	1,110.63	458	504	535	550	642	682	771
	0.7500	366.51	62.00	28,210.50	1,213.35	500	550	583	600	700	742	839
48	0.2500	128.84	194.00	11,028.16	454.77	156	172	182	188	219	237	268
	0.3125	161.26	155.60	13,839.05	569.22	195	215	228	234	273	296	334
	0.3750	193.77	130.00	16,671.70	683.97	234	258	273	281	328	354	400
	0.4375	226.35	111.71	19,526.22	799.03	273	301	319	328	383	412	465
	0.5000	259.02	98.00	22,402.73	914.40	313	344	365	375	438	469	531
	0.5625	291.78	87.33	25,301.34	1,030.08	352	387	410	422	492	527	595
	0.6250	324.61	78.80	28,222.16	1,146.08	391	430	456	469	547	584	660
	0.6875	357.53	71.82	31,165.30	1,262.39	430	473	501	516	602	641	724
	0.7500	390.54	66.00	34,130.88	1,379.03	469	516	547	563	656	697	788
51	0.2500	136.86	206.00	13,215.74	513.23	147	162	172	176	206	223	252
	0.3125	171.28	165.20	16,580.41	642.34	184	202	214	221	257	278	315
	0.3750	205.78	138.00	19,969.60	771.77	221	243	257	265	309	333	377

NOTE: To convert inches to millimeters, multiply inches by 25.4.
To convert psi to kPa, multiply psi by 6.895.

Table continued next page

Table A-1 Working pressures for allowable unit stresses (continued)

Pipe OD *in.*	Wall Thickness *in.*	Pipe Weight *lb/LF*	*D/t* Ratio	Moment of Inertia *in.*^{4}	Section Modulus *in.*^{3}	15,000 psi *psi*	16,500 psi *psi*	17,500 psi *psi*	18,000 psi *psi*	21,000 psi *psi*	23,000 psi *psi*	26,000 psi *psi*
	0.4375	240.37	118.57	23,383.45	901.53	257	283	300	309	360	388	439
	0.5000	275.05	104.00	26,822.06	1,031.62	294	324	343	353	412	442	500
	0.5625	309.80	92.67	30,285.56	1,162.04	331	364	386	397	463	496	561
	0.6250	344.64	83.60	33,774.07	1,292.79	368	404	429	441	515	550	622
	0.6875	379.57	76.18	37,287.70	1,423.87	404	445	472	485	566	604	683
	0.7500	414.57	70.00	40,826.58	1,555.30	441	485	515	529	618	657	743
54	0.2500	144.87	218.00	15,675.01	575.23	139	153	162	167	194	211	239
	0.3125	181.29	174.80	19,661.80	719.88	174	191	203	208	243	263	297
	0.3750	217.80	146.00	23,676.05	864.88	208	229	243	250	292	315	356
	0.4375	254.39	125.43	27,717.89	1,010.22	243	267	284	292	340	367	415
	0.5000	291.07	110.00	31,787.44	1,155.91	278	306	324	333	389	418	473
	0.5625	327.83	98.00	35,884.83	1,301.94	313	344	365	375	438	469	531
	0.6250	364.67	88.40	40,010.20	1,448.33	347	382	405	417	486	520	588
	0.6875	401.60	80.55	44,163.66	1,595.08	382	420	446	458	535	571	646
	0.7500	438.60	74.00	48,345.34	1,742.17	417	458	486	500	583	622	703
57	0.2500	152.88	230.00	18,421.89	640.76	132	145	154	158	184	200	226
	0.3125	191.31	184.40	23,103.11	801.84	164	181	192	197	230	249	282
	0.3750	229.82	154.00	27,814.89	963.29	197	217	230	237	276	299	338
	0.4375	268.41	132.29	32,557.37	1,125.09	230	253	269	276	322	348	393
	0.5000	307.09	116.00	37,330.68	1,287.26	263	289	307	316	368	397	448
	0.5625	345.85	103.33	42,134.94	1,449.80	296	326	345	355	414	445	503
	0.6250	384.70	93.20	46,970.31	1,612.71	329	362	384	395	461	494	558
	0.6875	423.63	84.91	51,836.90	1,776.00	362	398	422	434	507	542	612
	0.7500	462.64	78.00	56,734.86	1,939.65	395	434	461	474	553	590	667
60	0.2500	160.89	242.00	21,472.28	709.83	125	138	146	150	175	190	215
	0.3125	201.32	194.00	26,924.22	888.22	156	172	182	188	219	237	268
	0.3750	241.83	162.00	32,409.99	1,067.00	188	206	219	225	263	284	321
	0.4375	282.43	139.14	37,929.73	1,246.15	219	241	255	263	306	331	374
	0.5000	323.11	122.00	43,483.58	1,425.69	250	275	292	300	350	377	426
	0.5625	363.88	108.67	49,071.67	1,605.62	281	309	328	338	394	423	479
	0.6250	404.72	98.00	54,694.15	1,785.93	313	344	365	375	438	469	531
	0.6875	445.66	89.27	60,351.17	1,966.64	344	378	401	413	481	515	582
	0.7500	486.67	82.00	66,042.85	2,147.73	375	413	438	450	525	561	634
63	0.3125	211.33	203.60	31,145.01	979.02	149	164	174	179	208	226	255
	0.3750	253.85	170.00	37,485.20	1,176.01	179	196	208	214	250	271	306
	0.4375	296.45	146.00	43,862.80	1,373.39	208	229	243	250	292	315	356
	0.5000	339.13	128.00	50,277.95	1,571.19	238	262	278	286	333	359	406
	0.5625	381.90	114.00	56,730.80	1,769.38	268	295	313	321	375	404	456
	0.6250	424.75	102.80	63,221.50	1,967.98	298	327	347	357	417	447	506
	0.6875	467.69	93.64	69,750.19	2,167.00	327	360	382	393	458	491	555
	0.7500	510.70	86.00	76,317.02	2,366.42	357	393	417	429	500	535	605
	0.8125	553.81	79.54	82,922.15	2,566.26	387	426	451	464	542	578	654
	0.8750	596.99	74.00	89,565.71	2,766.51	417	458	486	500	583	622	703
	0.9375	640.26	69.20	96,247.86	2,967.18	446	491	521	536	625	665	751
	1.0000	683.61	65.00	102,968.75	3,168.27	476	524	556	571	667	708	800

NOTE: To convert inches to millimeters, multiply inches by 25.4.
To convert psi to kPa, multiply psi by 6.895.

Table continued next page

Table A-1 Working pressures for allowable unit stresses (continued)

Pipe OD *in.*	Wall Thickness *in.*	Pipe Weight *lb/LF*	*D/t* Ratio	Moment of Inertia $in.^4$	Section Modulus $in.^3$	15,000 psi	16,500 psi	17,500 psi	18,000 psi	21,000 psi	23,000 psi	26,000 psi
						psi	*psi*	*psi*	*psi*	*psi*	*psi*	*psi*
66	0.3125	221.35	213.20	35,785.36	1,074.23	142	156	166	170	199	216	244
	0.3750	265.87	178.00	43,064.38	1,290.32	170	188	199	205	239	258	292
	0.4375	310.47	152.86	50,384.42	1,506.82	199	219	232	239	278	301	340
	0.5000	355.16	134.00	57,745.61	1,723.75	227	250	265	273	318	343	388
	0.5625	399.93	119.33	65,148.12	1,941.10	256	281	298	307	358	385	436
	0.6250	444.78	107.60	72,592.10	2,158.87	284	313	331	341	398	428	483
	0.6875	489.72	98.00	80,077.71	2,377.07	313	344	365	375	438	469	531
	0.7500	534.74	90.00	87,605.10	2,595.71	341	375	398	409	477	511	578
	0.8125	579.84	83.23	95,174.42	2,814.77	369	406	431	443	517	553	625
	0.8750	625.03	77.43	102,785.83	3,034.27	398	438	464	477	557	594	672
	0.9375	670.30	72.40	110,439.49	3,254.20	426	469	497	511	597	635	718
	1.0000	715.65	68.00	118,135.56	3,474.58	455	500	530	545	636	676	765
72	0.3125	241.37	232.40	46,404.25	1,277.91	130	143	152	156	182	198	224
	0.3750	289.90	194.00	55,830.07	1,534.85	156	172	182	188	219	237	268
	0.4375	338.51	166.57	65,304.60	1,792.24	182	201	213	219	255	276	312
	0.5000	387.20	146.00	74,828.00	2,050.08	208	229	243	250	292	315	356
	0.5625	435.98	130.00	84,400.46	2,308.39	234	258	273	281	328	354	400
	0.6250	484.84	117.20	94,022.13	2,567.16	260	286	304	313	365	392	444
	0.6875	533.78	106.73	103,693.19	2,826.39	286	315	334	344	401	431	487
	0.7500	582.80	98.00	113,413.80	3,086.09	313	344	365	375	438	469	531
	0.8125	631.91	90.62	123,184.13	3,346.26	339	372	395	406	474	508	574
	0.8750	681.11	84.29	133,004.35	3,606.90	365	401	425	438	510	546	617
	0.9375	730.38	78.80	142,874.63	3,868.01	391	430	456	469	547	584	660
	1.0000	779.74	74.00	152,795.14	4,129.60	417	458	486	500	583	622	703
75	0.3125	251.39	242.00	52,422.55	1,386.38	125	138	146	150	175	190	215
	0.3750	301.92	202.00	63,064.29	1,665.06	150	165	175	180	210	228	257
	0.4375	352.53	173.43	73,758.83	1,944.22	175	193	204	210	245	265	300
	0.5000	403.22	152.00	84,506.36	2,223.85	200	220	233	240	280	303	342
	0.5625	454.00	135.33	95,307.05	2,503.96	225	248	263	270	315	340	384
	0.6250	504.86	122.00	106,161.08	2,784.55	250	275	292	300	350	377	426
	0.6875	555.81	111.09	117,068.62	3,065.63	275	303	321	330	385	414	468
	0.7500	606.84	102.00	128,029.85	3,347.19	300	330	350	360	420	451	510
	0.8125	657.95	94.31	139,044.94	3,629.23	325	358	379	390	455	488	551
	0.8750	709.14	87.71	150,114.07	3,911.77	350	385	408	420	490	524	593
	0.9375	760.42	82.00	161,237.42	4,194.79	375	413	438	450	525	561	634
	1.0000	811.79	77.00	172,415.16	4,478.32	400	440	467	480	560	597	675
78	0.3125	261.40	251.60	58,939.94	1,499.27	120	132	140	144	168	183	207
	0.3750	313.93	210.00	70,897.90	1,800.58	144	159	168	173	202	219	248
	0.4375	366.55	180.29	82,912.93	2,102.39	168	185	196	202	236	255	288
	0.5000	419.24	158.00	94,985.23	2,404.69	192	212	224	231	269	291	329
	0.5625	472.03	140.67	107,114.97	2,707.49	216	238	252	260	303	327	370
	0.6250	524.89	126.80	119,302.33	3,010.78	240	264	280	288	337	363	410
	0.6875	577.84	115.45	131,547.49	3,314.58	264	291	308	317	370	398	450
	0.7500	630.87	106.00	143,850.65	3,618.88	288	317	337	346	404	434	491
	0.8125	683.99	98.00	156,211.97	3,923.69	313	344	365	375	438	469	531

NOTE: To convert inches to millimeters, multiply inches by 25.4.
To convert psi to kPa, multiply psi by 6.895.

Table continued next page

Table A-1 Working pressures for allowable unit stresses (continued)

Pipe OD *in.*	Wall Thickness *in.*	Pipe Weight *lb/LF*	*D/t* Ratio	Moment of Inertia *in.*^{4}	Section Modulus *in.*^{3}	15,000 psi *psi*	16,500 psi *psi*	17,500 psi *psi*	18,000 psi *psi*	21,000 psi *psi*	23,000 psi *psi*	26,000 psi *psi*
	0.8750	737.18	91.14	168,631.65	4,229.01	337	370	393	404	471	505	571
	0.9375	790.47	85.20	181,109.87	4,534.83	361	397	421	433	505	540	610
	1.0000	843.83	80.00	193,646.81	4,841.17	385	423	449	462	538	575	650
81	0.3125	271.42	261.20	65,976.29	1,616.57	116	127	135	139	162	176	199
	0.3750	325.95	218.00	79,354.76	1,941.40	139	153	162	167	194	211	239
	0.4375	380.57	187.14	92,794.74	2,266.74	162	178	189	194	227	246	278
	0.5000	435.27	164.00	106,296.43	2,592.60	185	204	216	222	259	280	317
	0.5625	490.05	146.00	119,859.99	2,918.96	208	229	243	250	292	315	356
	0.6250	544.92	131.60	133,485.64	3,245.85	231	255	270	278	324	350	395
	0.6875	599.87	119.82	147,173.55	3,573.26	255	280	297	306	356	384	434
	0.7500	654.90	110.00	160,923.91	3,901.19	278	306	324	333	389	418	473
	0.8125	710.02	101.69	174,736.92	4,229.64	301	331	351	361	421	452	511
	0.8750	765.22	94.57	188,612.76	4,558.62	324	356	378	389	454	486	550
	0.9375	820.51	88.40	202,551.63	4,888.12	347	382	405	417	486	520	588
	1.0000	875.87	83.00	216,553.71	5,218.16	370	407	432	444	519	554	627
84	0.3125	281.43	270.80	73,551.47	1,738.29	112	123	130	134	156	170	192
	0.3750	337.97	226.00	88,458.73	2,087.52	134	147	156	161	188	204	230
	0.4375	394.59	194.00	103,432.09	2,437.28	156	172	182	188	219	237	268
	0.5000	451.29	170.00	118,471.76	2,787.57	179	196	208	214	250	271	306
	0.5625	508.08	151.33	133,577.92	3,138.39	201	221	234	241	281	304	344
	0.6250	564.95	136.40	148,750.77	3,489.75	223	246	260	268	313	337	381
	0.6875	621.90	124.18	163,990.52	3,841.65	246	270	286	295	344	370	419
	0.7500	678.94	114.00	179,297.35	4,194.09	268	295	313	321	375	404	456
	0.8125	736.06	105.38	194,671.47	4,547.07	290	319	339	348	406	436	493
	0.8750	793.26	98.00	210,113.06	4,900.60	313	344	365	375	438	469	531
	0.9375	850.55	91.60	225,622.34	5,254.67	335	368	391	402	469	502	568
	1.0000	907.92	86.00	241,199.48	5,609.29	357	393	417	429	500	535	605
87	0.3125	291.44	280.40	81,685.38	1,864.43	108	119	126	129	151	164	185
	0.3750	349.98	234.00	98,233.66	2,238.94	129	142	151	155	181	197	222
	0.4375	408.61	200.86	114,852.81	2,614.00	151	166	176	181	211	229	259
	0.5000	467.31	176.00	131,543.03	2,989.61	172	190	201	207	241	261	295
	0.5625	526.10	156.67	148,304.52	3,365.78	194	213	226	233	272	294	332
	0.6250	584.97	141.20	165,137.49	3,742.49	216	237	251	259	302	326	368
	0.6875	643.93	128.55	182,042.15	4,119.77	237	261	277	284	332	358	405
	0.7500	702.97	118.00	199,018.68	4,497.60	259	284	302	310	362	390	441
	0.8125	762.09	109.08	216,067.31	4,875.99	280	308	327	336	392	422	477
	0.8750	821.30	101.43	233,188.22	5,254.95	302	332	352	362	422	454	513
	0.9375	880.59	94.80	250,381.63	5,634.47	323	356	377	388	453	485	549
	1.0000	939.96	89.00	267,647.74	6,014.56	345	379	402	414	483	517	584
90	0.3750	362.00	242.00	108,703.41	2,395.67	125	138	146	150	175	190	215
	0.4375	422.62	207.71	127,084.72	2,796.91	146	160	170	175	204	221	250
	0.5000	483.33	182.00	145,542.05	3,198.73	167	183	194	200	233	253	286
	0.5625	544.13	162.00	164,075.59	3,601.11	188	206	219	225	263	284	321
	0.6250	605.00	146.00	182,685.56	4,004.07	208	229	243	250	292	315	356
	0.6875	665.96	132.91	201,372.16	4,407.60	229	252	267	275	321	346	391
	0.7500	727.00	122.00	220,135.62	4,811.71	250	275	292	300	350	377	426

NOTE: To convert inches to millimeters, multiply inches by 25.4.
To convert psi to kPa, multiply psi by 6.895.

Table continued next page

Table A-1 Working pressures for allowable unit stresses (continued)

Pipe OD *in.*	Wall Thickness *in.*	Pipe Weight *lb/LF*	*D/t* Ratio	Moment of Inertia *in.*^{4}	Section Modulus *in.*^{3}	15,000 psi	16,500 psi	17,500 psi	18,000 psi	21,000 psi	23,000 psi	26,000 psi
						psi	*psi*	*psi*	*psi*	*psi*	*psi*	*psi*
	0.8125	788.13	112.77	238,976.12	5,216.40	271	298	316	325	379	408	461
	0.8750	849.34	104.86	257,893.90	5,621.67	292	321	340	350	408	439	496
	0.9375	910.63	98.00	276,889.16	6,027.52	313	344	365	375	438	469	531
	1.0000	972.01	92.00	295,962.10	6,433.96	333	367	389	400	467	500	565
96	0.3750	386.03	258.00	131,822.79	2,725.02	117	129	137	141	164	178	202
	0.4375	450.66	221.43	154,093.50	3,181.29	137	150	160	164	191	208	235
	0.5000	515.38	194.00	176,450.58	3,638.16	156	172	182	188	219	237	268
	0.5625	580.18	172.67	198,894.26	4,095.63	176	193	205	211	246	266	301
	0.6250	645.06	155.60	221,424.77	4,553.72	195	215	228	234	273	296	334
	0.6875	710.02	141.64	244,042.32	5,012.42	215	236	251	258	301	325	367
	0.7500	775.07	130.00	266,747.14	5,471.74	234	258	273	281	328	354	400
	0.8125	840.20	120.15	289,539.45	5,931.67	254	279	296	305	355	383	433
	0.8750	905.42	111.71	312,419.49	6,392.21	273	301	319	328	383	412	465
	0.9375	970.71	104.40	335,387.46	6,853.38	293	322	342	352	410	441	498
	1.0000	1,036.10	98.00	358,443.61	7,315.18	313	344	365	375	438	469	531
102	0.3750	410.07	274.00	158,007.73	3,075.58	110	121	129	132	154	168	190
	0.4375	478.70	235.14	184,681.07	3,590.40	129	142	150	154	180	196	221
	0.5000	547.42	206.00	211,451.82	4,105.86	147	162	172	176	206	223	252
	0.5625	616.22	183.33	238,320.21	4,621.97	165	182	193	199	232	251	284
	0.6250	685.11	165.20	265,286.48	5,138.72	184	202	214	221	257	278	315
	0.6875	754.08	150.36	292,350.87	5,656.12	202	222	236	243	283	306	346
	0.7500	823.14	138.00	319,513.62	6,174.18	221	243	257	265	309	333	377
	0.8125	892.27	127.54	346,774.96	6,692.88	239	263	279	287	335	361	408
	0.8750	961.49	118.57	374,135.14	7,212.24	257	283	300	309	360	388	439
	0.9375	1,030.80	110.80	401,594.39	7,732.26	276	303	322	331	386	415	469
	1.0000	1,100.18	104.00	429,152.94	8,252.94	294	324	343	353	412	442	500
108	0.4375	506.74	248.86	219,070.10	4,024.25	122	134	142	146	170	185	209
	0.5000	579.47	218.00	250,800.23	4,601.84	139	153	162	167	194	211	239
	0.5625	652.27	194.00	282,639.71	5,180.11	156	172	182	188	219	237	268
	0.6250	725.17	174.80	314,588.79	5,759.06	174	191	203	208	243	263	297
	0.6875	798.14	159.09	346,647.73	6,338.70	191	210	223	229	267	289	327
	0.7500	871.20	146.00	378,816.77	6,919.03	208	229	243	250	292	315	356
	0.8125	944.34	134.92	411,096.17	7,500.04	226	248	263	271	316	341	385
	0.8750	1,017.57	125.43	443,486.18	8,081.75	243	267	284	292	340	367	415
	0.9375	1,090.88	117.20	475,987.05	8,664.16	260	286	304	313	365	392	444
	1.0000	1,164.27	110.00	508,599.03	9,247.26	278	306	324	333	389	418	473
114	0.4375	534.78	262.57	257,483.25	4,482.84	115	127	134	138	161	175	198
	0.5000	611.51	230.00	294,750.28	5,126.09	132	145	154	158	184	200	226
	0.5625	688.32	204.67	332,139.04	5,770.06	148	163	173	178	207	225	254
	0.6250	765.22	184.40	369,649.79	6,414.75	164	181	192	197	230	249	282
	0.6875	842.20	167.82	407,282.78	7,060.16	181	199	211	217	253	274	310
	0.7500	919.27	154.00	445,038.30	7,706.29	197	217	230	237	276	299	338
	0.8125	996.42	142.31	482,916.59	8,353.15	214	235	249	257	299	323	365
	0.8750	1,073.65	132.29	520,917.93	9,000.74	230	253	269	276	322	348	393
	0.9375	1,150.96	123.60	559,042.59	9,649.06	247	271	288	296	345	372	421
	1.0000	1,228.36	116.00	597,290.83	10,298.12	263	289	307	316	368	397	448

NOTE: To convert inches to millimeters, multiply inches by 25.4.
To convert psi to kPa, multiply psi by 6.895.

Table continued next page

Table A-1 Working pressures for allowable unit stresses (continued)

Pipe OD *in.*	Wall Thickness *in.*	Pipe Weight *lb/LF*	*D/t* Ratio	Moment of Inertia *in.*^{4}	Section Modulus *in.*^{3}	15,000 psi *psi*	16,500 psi *psi*	17,500 psi *psi*	18,000 psi *psi*	21,000 psi *psi*	23,000 psi *psi*	26,000 psi *psi*
120	0.4375	562.82	276.29	300,143.17	4,966.17	109	120	128	131	153	166	188
	0.5000	643.55	242.00	343,556.45	5,678.62	125	138	146	150	175	190	215
	0.5625	724.37	215.33	387,104.48	6,391.82	141	155	164	169	197	214	241
	0.6250	805.28	194.00	430,787.55	7,105.77	156	172	182	188	219	237	268
	0.6875	886.26	176.55	474,605.92	7,820.49	172	189	201	206	241	261	295
	0.7500	967.33	162.00	518,559.89	8,535.97	188	206	219	225	263	284	321
	0.8125	1,048.49	149.69	562,649.73	9,252.21	203	223	237	244	284	307	347
	0.8750	1,129.72	139.14	606,875.71	9,969.21	219	241	255	263	306	331	374
	0.9375	1,211.05	130.00	651,238.13	10,686.98	234	258	273	281	328	354	400
	1.0000	1,292.45	122.00	695,737.25	11,405.53	250	275	292	300	350	377	426
126	0.5000	675.60	254.00	397,473.20	6,259.42	119	131	139	143	167	181	205
	0.5625	760.42	226.00	447,822.31	7,045.39	134	147	156	161	188	204	230
	0.6250	845.33	203.60	498,320.16	7,832.14	149	164	174	179	208	226	255
	0.6875	930.32	185.27	548,967.05	8,619.70	164	180	191	196	229	248	281
	0.7500	1,015.40	170.00	599,763.26	9,408.05	179	196	208	214	250	271	306
	0.8125	1,100.56	157.08	650,709.09	10,197.20	193	213	226	232	271	293	331
	0.8750	1,185.80	146.00	701,804.84	10,987.16	208	229	243	250	292	315	356
	0.9375	1,271.13	136.40	753,050.80	11,777.92	223	246	260	268	313	337	381
	1.0000	1,356.54	128.00	804,447.25	12,569.49	238	262	278	286	333	359	406
132	0.5000	707.64	266.00	456,755.00	6,868.50	114	125	133	136	159	173	195
	0.5625	796.47	236.67	514,578.80	7,730.76	128	141	149	153	179	194	220
	0.6250	885.39	213.20	572,565.71	8,593.86	142	156	166	170	199	216	244
	0.6875	974.39	194.00	630,716.05	9,457.79	156	172	182	188	219	237	268
	0.7500	1,063.47	178.00	689,030.11	10,322.55	170	188	199	205	239	258	292
	0.8125	1,152.63	164.46	747,508.20	11,188.15	185	203	215	222	259	280	316
	0.8750	1,241.88	152.86	806,150.64	12,054.59	199	219	232	239	278	301	340
	0.9375	1,331.21	142.80	864,957.73	12,921.87	213	234	249	256	298	322	364
	1.0000	1,420.63	134.00	923,929.77	13,790.00	227	250	265	273	318	343	388
138	0.5000	739.69	278.00	521,656.32	7,505.85	109	120	127	130	152	165	187
	0.5625	832.52	247.33	587,660.23	8,447.95	122	135	143	147	171	186	210
	0.6250	925.44	222.80	653,842.29	9,390.91	136	149	159	163	190	206	233
	0.6875	1,018.45	202.73	720,202.82	10,334.75	149	164	174	179	209	227	257
	0.7500	1,111.53	186.00	786,742.14	11,279.46	163	179	190	196	228	247	280
	0.8125	1,204.70	171.85	853,460.57	12,225.04	177	194	206	212	247	268	303
	0.8750	1,297.96	159.71	920,358.43	13,171.50	190	209	222	228	266	288	326
	0.9375	1,391.29	149.20	987,436.04	14,118.84	204	224	238	245	285	308	349
	1.0000	1,484.71	140.00	1,054,693.73	15,067.05	217	239	254	261	304	329	371
144	0.5625	868.57	258.00	667,352.88	9,196.94	117	129	137	141	164	178	202
	0.6250	965.50	232.40	742,467.98	10,223.31	130	143	152	156	182	198	224
	0.6875	1,062.51	211.45	817,777.26	11,250.59	143	158	167	172	201	218	246
	0.7500	1,159.60	194.00	893,281.06	12,278.78	156	172	182	188	219	237	268
	0.8125	1,256.78	179.23	968,979.70	13,307.88	169	186	197	203	237	257	290
	0.8750	1,354.03	166.57	1,044,873.53	14,337.89	182	201	213	219	255	276	312
	0.9375	1,451.38	155.60	1,120,962.88	15,368.81	195	215	228	234	273	296	334
	1.0000	1,548.80	146.00	1,197,248.08	16,400.66	208	229	243	250	292	315	356

NOTE: To convert inches to millimeters, multiply inches by 25.4.
To convert psi to kPa, multiply psi by 6.895.

Index

Note: *f.* indicates a figure; *t.* indicates a table.

AWWA Manuals

M1, *Principles of Water Rates, Fees, and Charges,* Fifth Edition, 2000, #30001PA

M2, *Instrumentation and Control,* Third Edition, 2001, #30002PA

M3, *Safety Practices for Water Utilities,* Sixth Edition, 2002, #30003PA

M4, *Water Fluoridation Principles and Practices,* Fifth Edition, 2004, #30004PA

M5, *Water Utility Management,* First Edition, 1980, #30005PA

M6, *Water Meters—Selection, Installation, Testing, and Maintenance,* Fourth Edition, 1999, #30006PA

M7, *Problem Organisms in Water: Identification and Treatment,* Third Edition, 2003, #30007PA

M9, *Concrete Pressure Pipe,* Second Edition, 1995, #30009PA

M11, *Steel Pipe—A Guide for Design and Installation,* Fourth Edition, 2004, #30011PA

M12, *Simplified Procedures for Water Examination*, Fifth Edition, 2003, #30012PA

M14, *Recommended Practice for Backflow Prevention and Cross-Connection Control,* Third Edition, 2004, #30014PA

M17, *Installation, Field Testing, and Maintenance of Fire Hydrants,* Third Edition, 1989, #30017PA

M19, *Emergency Planning for Water Utilities,* Fourth Edition, 2001, #30019PA

M21, *Groundwater,* Third Edition, 2003, #30021PA

M22, *Sizing Water Service Lines and Meters,* Second Edition, 2004, #30022PA

M23, *PVC Pipe—Design and Installation,* Second Edition, 2002, #30023PA

M24, *Dual Water Systems,* Second Edition, 1994, #30024PA

M25, *Flexible-Membrane Covers and Linings for Potable-Water Reservoirs,* Third Edition, 2000, #30025PA

M27, *External Corrosion—Introduction to Chemistry and Control,* Second Edition, 2004, #30027PA

M28, *Rehabilitation of Water Mains,* Second Edition, 2001, #30028PA

M29, *Water Utility Capital Financing,* Second Edition, 1998, #30029PA

M30, *Precoat Filtration,* Second Edition, 1995, #30030PA

M31, *Distribution System Requirements for Fire Protection,* Third Edition, 1998, #30031PA

M32, *Distribution Network Analysis for Water Utilities*, First Edition, 1989, #30032PA

M33, *Flowmeters in Water Supply,* First Edition, 1989, #30033PA

M36, *Water Audits and Leak Detection,* Second Edition, 1999, #30036PA

M37, *Operational Control of Coagulation and Filtration Processes,* Second Edition, 2000, #30037PA

M38, *Electrodialysis and Electrodialysis Reversal,* First Edition, 1995, #30038PA

M41, *Ductile-Iron Pipe and Fittings,* Second Edition, 2003, #30041PA

M42, *Steel Water-Storage Tanks,* First Edition, 1998, #30042PA

M44, *Distribution Valves: Selection, Installation, Field Testing, and Maintenance,* First Edition, 1996, #30044PA

M45, *Fiberglass Pipe Design,* First Edition, 1996, #30045PA

M46, *Reverse Osmosis and Nanofiltration,* First Edition, 1999, #30046PA

M47, *Construction Contract Administration,* First Edition, 1996, #30047PA

M48, *Waterborne Pathogens,* First Edition, 1999, #30048PA

M49, *Butterfly Valves: Torque, Head Loss, and Cavitation Analysis,* First Edition, 2001, #30049PA

M50, *Water Resources Planning,* First Edition, 2001, #30050PA

M51, *Air-Release, Air/Vacuum and Combination Air Valves,* First Edition, 2001, #30051PA
